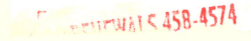

A Concise History of
Solar and Stellar Physics

A Concise History of
Solar and Stellar Physics

Jean-Louis Tassoul
and Monique Tassoul

PRINCETON UNIVERSITY PRESS
PRINCETON AND OXFORD

Published by Princeton University Press,
41 William Street, Princeton, New Jersey 08540

In the United Kingdom: Princeton University Press,
3 Market Place, Woodstock, Oxfordshire OX20 1SY

Library of Congress Cataloging-in-Publication Data

Tassoul, Jean Louis, 1938.
    A concise history of solar and stellar physics / Jean-Louis Tassoul and Monique Tassoul.
        p. cm
    Includes bibliographical references and index.
    ISBN 0-691-11711-X (acid-free paper)
    1. Sun—History. 2. Stars—History. 3. Astrophysics—History. I. Tassoul, Monique,
1942- II. Title
    QB521.T37 2004
    523.01'09—dc22

                                                                          200353681

British Library Cataloging-in-Publication Data is available

This book has been composed in Times Roman with Abadi display

Printed on acid-free paper. ∞

www.pupress.princeton.edu

Printed in the United States of America

10   9   8   7   6   5   4   3   2   1

Contents

List of Figures

Preface

This book is the work of two theoretical astrophysicists who have been active in the field since the early 1960s, and who are convinced that scientific education has become so specialized that scientific literacy is slowly disappearing among scientists. An attempt is made in this book, therefore, to outline the history of ideas about the sun and the stars, from the earliest historical times to the twentieth century's final years, and to present it in a form that is almost entirely devoid of elaborate mathematics. Although intended mainly for the use of students and teachers of astronomy, the present work should also provide a useful reference for practicing astronomers and scientifically curious lay people who are interested in the historical development of solar and stellar physics. To this end we have cited, where appropriate, in footnotes to the text several key references as an aid to independent reading of the original material. Whenever available, we have also quoted secondary references, such as review papers and modern commentaries by science historians. The emphasis on theoretical work is, we think, the main difference between this text and others. Actually, about half of the book includes most of the theoretical research done during the six decades 1940–2000, a large body of work which has so far seldom been explored by historians.

The two opening chapters make no claim to be original contributions to the history of astronomy since they are written largely from second-hand sources. Chapter 1, which covers the period from about 3000 B.C. to A.D. 1700, is intended to show that at every stage in history man has had a particular understanding or view of the sun and the stars, and that his picture of these objects has continually evolved over the centuries. In chapter 2 we deal in a most systematic way with the immense mass of observations that astronomy has accumulated from about 1610 to the late 1910s. (For the sake of clarity, more recent observations are reported in subsequent chapters.) After careful consideration, however, we deliberately abstained from describing observational techniques because the subject of astronomical instrumentation is so well covered in standard astronomy texts and popular books.

Chapters 3 through 6 are concerned with the history of solar and stellar astronomy from the point of view of the physicist. The main emphasis in these four chapters is the theoretical work done from the mid-1840s to the late 1990s. Progress in our understanding of the sun and the stars has at all stages been closely linked to new developments in the physical sciences. Accordingly, the subject matter has experienced several major revolutions in the period covered in these chapters— those of thermodynamics in the 1840s and quantum mechanics in the 1910s, and finally those of nuclear physics and magnetohydrodynamics in the 1940s. The first two of these are discussed in chapters 3 and 4. The other two, which paved the way to remarkable developments in solar and stellar physics from about 1940 to the late 1960s and early 1970s, are recorded in chapter 5. Finally, in chapter 6 we treat the more recent theoretical work, giving precedence to whoever first entered the field or established a new fact. In doing so, we have therefore adopted a method of presentation directly opposite to that of the typical review paper, which covers only the most recent contributions. This and other trends in contemporary scientific research are briefly reviewed in the epilogue.

Admittedly, because research in solar and stellar physics has so many separate branches, most readers will not want to go through all the chapters of this book. This is why we have divided the subject matter into sections, each of which is developed in a logical order. Someone who has an interest in the theory of variable stars, for example, may want to read just section 3.3 and then follow this up with the subsequent developments presented in sections 4.6, 5.6, and 6.5. We therefore suggest that the reader first consult the table of contents if it is desired to read about the history of a specific subject. Another unique feature of this book is the insertion throughout the text of numerous portraits, which remind us that scientific knowledge is the handiwork of individual human beings from all over the world.

We are greatly indebted to Georges Michaud for his careful and constructive evaluation of the entire manuscript and for the numerous suggestions he made for its improvement. Our thanks also go to those who have made available vital information for inclusion in the text: Solveig Berg, Sergei Blinnikov, Hermann A. Brück, Paul Charbonneau, Ene-Kaja Chippendale, John A. Crawford, Bernard R. Durney, Elfriede Gamov, Peter Hingley, Paul Hodge, Robert P. Kraft, Jan Lub, Leon Mestel, Paulo de Tarso Muzy, Hans Olofsson, Yoji Osaki, Donald E. Osterbrock, Jeremiah P. Ostriker, Jean Perdang, Roland Rappmann, Günther Rüdiger, Barbara Schwarzschild, Scott D. Tremaine, Jeremi Wasiutyński, Harold F. Weaver, François Wesemael, Meg Weston Smith, and Galina Zajtseva. The help of these and other individuals is gratefully acknowledged, but of course they are in no way responsible for any errors of fact or judgment that the book may contain.

Jean-Louis and Monique Tassoul
Montréal, Québec
March 2002

ARCHANGEL EDDINGTON

> As well we know, the Sun is fated
>> In polytropic spheres to shine,
> Its journey, long predestinated,
>> Confirms *my theories* down the line.

ARCHANGEL JEANS

> Ideal fluids, hot and spinning,
>> By fission turn to pear-shaped forms.
> *Mine are the theories* that are winning!
>> The atom cannot change the norms.

ARCHANGEL MILNE

> At heat of 10 to 7th power
>> The gas degenerates in flame,
> Permitting us our shining hour
>> Of freest flight in *Fermi*'s name.

THE THREE

> This vision fills us with elation
>> (Though none of us can understand).
> As on the Day of Publication
>> The brilliant Works are strange and grand.[1]

---

[1] From an anonymous parody of Goethe's Faust. Quoted in George Gamow, *Thirty Years that Shook Physics* (New York: Doubleday, 1966), 175–176.

A Concise History of
Solar and Stellar Physics

Chapter One

The Age of Myths and Speculations

And God said, Let there be light, and there was light.
—Genesis 1:3

For thousands of years men have looked up into the star-filled night sky and have wondered about the nature of the "fixed" stars as opposed to that of the five planets wandering among the constellations of the zodiac. The daily course of the sun, its brilliance and heat, and the passing of the seasons are among the central problems that have concerned every human society. Undoubtedly, the appearance of a comet or a shooting star, the passing phenomena of clouds and rain and lightning, the Milky Way, the changing phases of the moon and the eclipses—all of these must have caused quite a sense of wonder and been the source of endless discussions.

Faced with this confusing multiplicity of brute facts, beyond their physical power to control, our ancestors sought to master these unrelated phenomena symbolically by picturing the universe in terms of objects familiar to them so as to make clear the unfamiliar and the unexplained. The cosmologies that these men set up thus inevitably reflect the physical and intellectual environment in which they lived.[1]

For example, according to a Chinese myth going back to the third century A.D., the world was originally shaped like an egg, Chaos, which separated into the earth and the sky. Between these two parts was P'an-ku, a dwarf, who daily for eighteen thousand years grew three meters, his body gradually pushing apart earth and sky. Upon his death, the heavenly bodies, thunder, lightning, rain, and the various constituents of the earth's surface were derived from his body. In particular, his eyes became the sun and the moon. Though this universe was formed from a cosmic egg, its lower half, the earth, was square and lay immobile beneath the inverted round bowl of the sky, which rotated, creating constant stellar motion around the polar star. This structure was often linked to the ancient Chinese chariot—square with an umbrella canopy. The central pole, corresponding to P'an-ku's body, linked heaven and earth.

Of greater interest to the western tradition are the ancient traditions of Egypt and Mesopotamia, from which can be traced the first attempts to liberate the subject of cosmology from the use of myth. Physical cosmology arose in the Greek colonies of the eastern Mediterranean and of southern Italy during the sixth century B.C. It flourished in Athens during the next two centuries, whence it passed

---

[1]There is a wide literature on these earlier cosmologies. Our own preference goes to the following books: V. Ions, *The World's Mythology* (London: Hamlyn, 1974); E. C. Krupp, *Echoes of the Ancient Skies* (New York: Harper & Row, 1983). See also P. Whitfield, *The Mapping of the Heavens* (San Francisco: Pomegranate Artbooks, 1995).

on to Alexandria, which became the center of Hellenistic culture for several centuries.[2]

Unfortunately, although that golden age saw great advances in observational astronomy and mathematics, physics did not keep pace with these sciences. Thus, lacking a firm theoretical base and constrained by metaphysical dogma, physical cosmology stagnated at the dawn of the Christian era until resurrected in the Renaissance. This period, which was essentially the transition of the medieval to the modern order, saw the beginning of a great burst of critical speculation about nature and with it the emergence of two most promising ideas—the demotion of the earth from its position as the center of motion of the universe, and the existence of an infinite universe populated with an infinity of worlds. The times were not ripe for the general acceptance of these ideas, however, since it was not until late into the seventeenth century that the new science of dynamics and the law of gravitation made the age-old geocentric universe obsolete.[3]

1.1 ANCIENT EGYPT AND THE MIDDLE EAST

In the many stories recorded in texts of the third and second millennia B.C. in Egypt, Mesopotamia, and neighboring countries, the tales of creation are presented as a series of births and cosmic battles among the gods. The mother country is always represented as the center, the earth being a flat disk surrounded by a rim of mountains and floating on an ocean.

The Egyptians believed that in the beginning a primordial abysmal ocean, deified as Nun, filled the universe like a cosmic egg. Just as the Nile floodwaters receded, leaving behind fertile lands, so a primeval hill rose out of Nun. In a Heliopolitan text on papyrus from the fifth dynasty (c. 2500–2350 B.C.), Atun—a predynastic sun-god, later identified with the sun-god Ra—arose out of the abyss. He created a son, Shu—god of wind or air—and a daughter, Tefnut—goddess of lifegiving dew. Shu and Tefnut begot the twins Geb and Nut. Nut was the sky-goddess. She was sometimes pictured as an elongated woman bending over the earth-god Geb and separated from him by her father Shu (see figure 1.1). To the Egyptians, the sky was a heavenly Nile, along which the barque of the sun-god Ra sailed from east to west each day. (The sun was occasionally thrown into an eclipse when his barque was attacked by a great serpent.) Below, the earth-god Geb was lying prone in the center of a circular ocean, the Nile flowing through the fertile lands of Egypt and into the subterranean abode of the dead through which the night-barque of the sun-god Ra sailed. It was also believed that the sky-goddess Nut swallowed the sun

---

[2]For a detailed presentation of Greek astronomy, see T. Heath, *Aristarchus of Samos: The Ancient Copernicus* (Oxford: Clarendon Press, 1913; New York: Dover Publications, 1981); O. Neugebauer, *A History of Ancient Mathematical Astronomy*, 3 vols. (New York: Springer-Verlag, 1975). For an overall picture, see G.E.R. Lloyd, *Early Greek Science: Thales to Aristotle / Greek Science after Aristotle*, 2 vols. (London: Chatto & Windus, 1970/1973).

[3]See especially R. K. DeKosky, *Knowledge and Cosmos: Development and Decline of the Medieval Perspective* (Washington, D.C.: University Press of America, 1979); D. C. Lindberg, *The Beginnings of Western Science* (Chicago: University of Chicago Press, 1992); J. C. Pecker, *Understanding the Heavens* (New York: Springer-Verlag, 2001).

Figure 1.1 The Egyptian sun-god Ra traveling in his barque across the arc of the sky formed
by the sky-goddess Nut (figure adorned with stars). She is prevented from falling
upon the earth-god Geb (figure adorned with leaves) by her father Shu, the god
of wind and air. From *Pioneers of Science* by O. Lodge (London: Macmillan,
1893; New York: Dover Publications, 1960).

each evening but that it was reborn from her loins each morning. Then, as the sun
rose in the sky, she would swallow the stars and give birth to them again at sunset.

The theory that the shape of the universe resulted from the forceful separation of
heaven from earth by a third party was also adopted in Mesopotamia.[4] It is recorded
in the great Babylonian epic called, from its opening sentence, *Enuma Elish*, "When
above ...." The composition of its earliest known version goes back to the middle of
the second millennium B.C., but we know that it had its roots in an ancient Sumerian
story of about 3000 B.C. It seems that Enlil, the city-god of Nippur, was the original
hero of the epic and that he was merely replaced by Marduk, the city-god of Babylon,
when the later version was composed. This epic was extremely popular since, later
on, an Assyrian version of the first millennium B.C. substituted the name of Ashur,
the national god of the Assyrians, for that of Marduk. It is recorded on seven clay
tablets.

Enuma Elish presents the earliest stage of the universe as one of watery chaos.
This chaos was closely modeled on the Iraqi seashore, with large pools of sweet
water mingling freely with the salty waters of the Persian Gulf, and with low banks
of clouds hanging over the horizon. It consisted of Apsu (the sweetwater ocean),
his wife Tiamat (the saltwater ocean), and their son Mummu (the cloud banks and
mist rising from the two bodies of water). The myth of creation relates how the
primordial forces of chaos bore the other gods of the Mesopotamian pantheon who,

---

[4] See A. Heidel, *The Babylonian Genesis*, 2nd edition (Chicago: University of Chicago Press, 1951).

in turn, created the earth and the heaven. One of these gods was Ea. To make a long story short, after overcoming Apsu and Mummu, the god Ea fathered Marduk, who possessed outstanding qualities and so was chosen by the gods to fight the remaining evil forces of chaos. After an epic combat, Marduk pierced Tiamat's heart with an arrow and smashed her skull with his mace. Then he completed creation, making half of her dead body into the sky and placing the other half beneath the earth. After his victory, Marduk put the universe in order and organized the calendar. Thus, on the sky he had fashioned he set up the constellations to determine, by their rising and setting, the years, the months, and the days. On both sides of the sky, where the sun comes out in the morning and leaves in the evening, he built gates and secured them with strong locks. Marduk also made the moon shine forth and entrusted the night to her. (It is to be regretted that the only extant clay tablet dealing with the creation and organization of the heavenly bodies breaks off at the point where it reports some detailed orders to the moon.) Then, according to the last two tablets, Marduk decided to create man and to impose upon him the service which the defeated forces of chaos had to render to the gods.

It has long been recognized that *Enuma Elish* contains several points that invite comparison with the biblical account of creation which was compiled by the Priestly school during the fifth century B.C. Indeed, *Enuma Elish* and Genesis (1:1–2:3) both refer to a watery chaos, which was separated into heaven and earth. However, in Genesis it is the Spirit of God which hovers over the waters and divides the light from the darkness; thus, as is fitting to a monotheistic religion, He alone created the heavens and the earth. Both texts also refer to the existence of light and to the alternation of day and night before the creation of the celestial luminaries. And in both texts the succession of creative acts virtually follows in the same order, culminating in the creation of mankind. At this juncture it is worth noting that the Koran, which was written during the first half of the seventh century A.D., also presents an elaborate account of creation that shows quite evident biblical traces. Thus, in sura 21:30 we learn that "the heavens and the earth used to be one entity, and then (God) parted them"; whereas in sura 71:15–16 we are told that "God created seven heavens in layers" and that "He placed the moon therein as a light and the sun as a lamp." The creation of the heavens and the earth in six days, which is mentioned in sura 7:54, also finds its counterpart in Genesis.

1.2 IONIA: THE EASTERN GREEK SCHOOL

Before the sixth century B.C. and the time of rationalism heralded by Ionian philosophers, Greek cosmologies were dependent on traditions that came from more ancient civilizations. The first records containing the early Greek ideas are due to Homer and Hesiod. Homer's two epics, the *Iliad* and *Odyssey*, are mainly devoted to the Trojan war and to the return of Ulysses in his homeland. Hesiod, the father of Greek didactic poetry, probably flourished during the eighth century B.C. His poem *Theogony* tells of the generations of the gods and attempts in mythical terms to provide an account of the events that brought the world into being. His other poem *Works and Days* resembles a modern almanac in that it correlates the agricultural work to be done with the rising or setting of certain stars. Both HOMER and HESIOD agree upon the fact that the earth is a circular disk surrounded by the river Oceanus. Over the flat earth is the vault of heaven; below the earth is Tartarus, the realm of the underworld. The vault of heaven remains forever fixed; the sun, the moon, and the stars move round under it, rising from Oceanus in the east and plunging into it again in the west. We are not told what happens to the heavenly bodies between their setting and rising. They cannot pass under the earth, however, because Tartarus is never lit up by the sun. Possibly they float round Oceanus, past the north, to the points where they rise again in the east.

The great achievement of the Ionian philosophers—Thales, Anaximander, and Anaximenes—was that of dissipating the haze of myth from the origins of the world. The three of them flourished during the sixth century B.C. in Miletus —a rich Greek city on the Aegean, near the mouth of the Maeander River. Tradition tells that Thales (c. 624–547 B.C.) was a man of great versatility: successful businessman, statesman, engineer, mathematician, and astronomer. Anaximander (c. 611–547 B.C.), who was the successor and probably pupil of Thales, was the first to have developed anything like a cosmological system; he was also the first among the Greeks who ventured to draw a map of the inhabited earth. It seems almost certain that Anaximenes (c. 585–528 B.C.), the younger contemporary of Anaximander, was the first to distinguish the planets from the "fixed" stars in respect of their irregular movements; he also extended Anaximander's ideas on cosmology.

THALES thought the earth to be a circular or cylindrical disk that floated on the cosmic ocean, like a log or a cork, and was surrounded by its waters. So far as we can judge of his views of the universe, he would appear to have regarded water as the world's fundamental element. In this conception, therefore, far above the solid earth floating on a mass of water stands the vault of heaven, which is also bounded by the primeval waters. As a matter of fact, the significance of this cosmology does not lie so much in the primacy of water as in the attempt to search for causes within nature itself rather than in supernatural events.

ANAXIMANDER offered a much more detailed picture of the world. He maintained that the earth was in the center of all things, suspended freely and without support, whereas Thales regarded it as resting on water. Anaximander's earth had the shape of a cylinder, round like a stone pillar, its height being one-third of its diameter. Still more original was his conception of the sun, moon, and stars, and of their motions. Indeed, these celestial bodies were not thought to be objects but holes in rotating hoops filled with fire.

To be specific, Anaximander's stars are compressed portions of air, in the shape of rotating hoops filled with fire, that emit visible flames from small openings, thereby producing what we see as the stars. The hoops belonging to the sun, the moon, and the stars were probably assumed to be concentric with the earth. The sun is thus a rotating hoop full of fire, which lets its fire shine out through an opening like the tube of a blowpipe. The moon sometimes appears as waxing, sometimes as waning, to an extent corresponding to the opening or closing of the passages in its hoop. The eclipses of the sun and the moon occur when the openings in their hoops are shut up. For some philosophical reason, however, the radius of the solar hoop is twenty-seven times as large as the earth, whereas that of the lunar hoop is eighteen times as large as the earth. Surprisingly, the hoops from which the stars and planets shine are *nearer* to the earth than that of the moon. Despite some obvious short-comings, this cosmological system represents a definite improvement over previous ones since, instead of moving laterally round the earth, the heavenly bodies describe circles passing *under* the earth, which is freely suspended in space.

For ANAXIMENES the earth is still flat but, instead of resting on nothing, it is supported by air. The sun, the moon, and the planets are all made of fire, and they ride on the air because of their breadth. The sun derives its heat from its rapid motion, whereas the stars, which are also composed of fire, fail to give warmth because they are too far off. It is also assumed that the stars are fastened on a crystal sphere, like nails or studs. Obviously, a definite improvement on Anaximander's system is the relegation of the stars to a more distant region than that in which the sun moves. And, as we know, the rigidity of the sphere of the fixed stars remained the fundamental postulate of all astronomy up to the beginning of modern science.

Highly speculative theories about the nature of the heavenly bodies were also proposed by Xenophanes (c. 570–478 B.C.) of Colophon and Heraclitus (c. 540–480 B.C.) of Ephesus. XENOPHANES was more a poet and satirist than a natural philosopher. According to him, the stars (including the comets and shooting stars) are made of clouds set on fire; they are extinguished each day and are kindled at night like coals, and these happenings constitute their setting and rising, respectively. The sun and the moon are also made of clouds set on fire. On the contrary, HERACLITUS assumes that the stars are bowls turned with their concave sides toward us, which collect bright exhalations arising from the earth and from the sea, thus producing flames. The sun and the moon are also bowl shaped, like the stars, and are thus similarly lit up.

1.3 SOUTHERN ITALY: THE WESTERN GREEK SCHOOL

Much more remarkable developments were to follow in the great colonies of south-ern Italy. The dominant figure is undoubtedly Pythagoras, who gave his name to an order of scientific and religious thinkers. He was an Ionian, born at Samos about 572 B.C. The important part of his career began about the year 530 B.C. with his mi-gration to Crotona, a Dorian colony in southern Italy. Political reaction against his brotherhood brought about his retirement to Metapontum, on the Gulf of Taranto, where he died at the turn of the fifth century B.C.

PARMENIDES of Elea, who was a contemporary of Heraclitus, is said to have been the first to assert that the earth is spherical in shape and lies in the center of the universe. There is, however, an alternative tradition stating that it was PYTHAGORAS who first held that the universe, the earth, and the other heavenly bodies are spherical in shape, that the earth is at rest at the center, and that the sphere of the fixed stars has a daily rotation from east to west about an axis passing through the center of the earth. Pythagoras and his immediate successors left no written exposition of their doctrines. Thus, although it is most likely that Pythagoras arrived at geocentrism and the rotundity of the earth from mathematical considerations, Parmenides—who was closely connected with the Pythagoreans—may have been the first to state the doctrine publicly.

It is improbable, however, that Pythagoras himself was responsible for the astronomical system known as Pythagorean, which reduces the earth to the status of a planet like the others. This cosmology is often attributed to PHILOLAUS of Crotona, who flourished in the latter half of the fifth century B.C.

The Pythagorean system may be described briefly as follows. The universe is spherical in shape and finite in size. At its center is the central fire, in which is located the governing principle—the force that moves the celestial bodies, including the earth, on their circular paths. The outer boundary of the sphere is an envelope of fire, which is called Olympus; below this is the universe. In the universe there revolve in circles round the central fire the following bodies. Nearest to the central fire revolves a body called the counterearth, which always accompanies the earth, the earth's orbit coming next to that of the counterearth; next to the earth, reckoning in order from the central fire outward, come the moon, the sun, the five planets, and last of all, outside the orbits of the planets, the sphere of the fixed stars.

Philolaus explained that the Greeks had never seen the central fire nor the counterearth because the earth turned so as always to place their country away from those objects. Such a situation implies that the earth is rotating around its axis in the same time as it takes the earth to complete a revolution around the central fire. Obviously, this was a giant step forward since it provided a simple explanation for the apparent daily motion of all celestial objects. What evidence motivated Philolaus to include a counterearth is not clear, however, but it could have been invented for the purpose of explaining the frequency of eclipses of the moon. Arguably, it could have been added to bring the number of celestial objects up to ten, the perfect number according to the Pythagoreans.

It is worth noting that the sun is not a body with light of its own in this cosmological model. Instead, the sun is made of a substance comparable to *glass* that concentrates rays of fire from elsewhere and transmits them to us. It is not stated, however, whether these rays of fire come from Olympus or from the central fire. A similar conception for the nature of the sun has been attributed also to EMPEDOCLES (c. 490–430 B.C.) of Acragas in Sicily. More importantly, Empedocles advocated the idea that there are four primary substances—fire, air, water, and earth—out of which all the structures in the world are made by their combination in different proportions.

1.4 THE ATHENIAN PERIOD

During the first half of the fifth century B.C., both the Ionian and the Italian schools of natural philosophy were progressively overshadowed by Athens, which became for two centuries the intellectual center of the Greek world. The Ionian ANAXAGORAS seems to have been the first natural philosopher to move to Athens. He was born in Clazomenae, near modern Izmir, about 500 B.C. He came to Athens about the year 464 B.C., where he enjoyed the friendship of the statesman Pericles (490–429 B.C.). However, when Pericles became unpopular during the late 430s B.C., shortly before the outbreak of the Peloponnesian war, he was attacked through his friends; and so

Anaxagoras was prosecuted on a charge of impiety for holding that the sun is a red-hot mass and the moon like earth. According to one account of the events, Pericles managed to save him, but Anaxagoras was forced to leave Athens. Eventually, he withdrew to Lampsacus in Asia Minor, near modern Gallipoli, where he died about 428 B.C. This is an early case of persecution of scientific ideas opposed to current views.

His greatest achievement was the discovery of the fact that the moon does not shine by its own light but is illuminated by the sun. As a result, he was able to give the correct explanation of eclipses. He thus held that the eclipses of the moon were caused by its falling within the shadow of the earth, which then comes between the sun and the moon, while the eclipses of the sun were due to the interposition of the moon. Whether he reached the true explanation of the phases of the moon is much more doubtful, however.

Anaxagoras also developed a cosmological system according to which the formation of the world began as a huge vortex set up by a *deus ex machina* in the primordial chaos. From the violence of this whirling motion, which is still visible on the sky, he was able to explain the formation of the earth and, thence, of the heavenly bodies. In his view, therefore, the initial vortex caused the heavy earth to collect at the center of the world while the surrounding ether tore stones away from the earth and kindled them into planets and stars revolving around us. Obviously, this is an indefinitely expanding universe since there is no spatial limit set to the expansion of the ripples from the initial vortex. Parenthetically note that another cosmogony initiated by a vortex was suggested by his contemporary LEUCIPPUS of Miletus, with an infinity of *atoms* taking the place of Anaxagoras' initial mixed mass in which "all things were together." There is also the difference that material and space are supposed to be limitless so that there would be an unlimited number of worlds similar to our own world-system in the universe.

Apart from these remarkable innovations, Anaxagoras did not improve much upon the earlier Ionian theories. Following Anaximenes, he thus assumed that the earth was flat and supported by air; and that the sun, the moon, and all the stars were stones on fire. However, he also held the view that the moon was of earthy nature and had in it plains, mountains, and ravines. According to Anaxagoras, the sun was "rather larger than the Peloponnese."

At this juncture it is worth noting that Anaxagoras also put forward a very original hypothesis to explain the Milky Way. As we have seen, he thought the sun to be *smaller* than the earth. Hence, when the sun in its revolution passes under the earth during the night, the shadow cast by the earth extends without limit. Since the stars within this shadow are not illuminated by the sun, it follows that we can see them shining; those stars, on the other hand, that are outside this shadow are overpowered by the rays of the sun, which shines on them even during the night, so that we cannot see them. According to Anaxagoras, the Milky Way is the reflection of the light of those stars that the sun cannot see when it is passing below the earth. Although this conjecture can be easily disproved by observation, it was supported by DEMOCRITUS (c. 460–370 B.C.) of Abdera. As far as we know, Democritus seems to have been the first to appreciate the true character of the Milky Way as a multitude of faint stars so close together that it has the appearance of a continuous body of light.

Important for subsequent astronomical developments were the Athenian Plato (c. 427–347 B.C.) and his pupil Aristotle (384–322 B.C.), who was born at Stagira, a Greek colonial town on the Aegean near modern Salonika. Although Plato is primarily known as a philosophical writer, he himself probably felt that the foundation of the Academy in about 387 B.C. was his chief work. Aristotle entered the Academy in about 367 B.C. and worked there for twenty years by the side of Plato. After acting for some years as tutor to the young Alexander (356–323 B.C.) in Macedonia, he returned to Athens to found his own institute of higher education, the Lyceum, which came to be known as the Peripatetic School from the path in his garden where he walked and talked with his pupils.

The importance of PLATO, so far as the development of astronomy is concerned, is to be sought in his appreciation of the role of mathematics in making intelligible the motions of the heavenly bodies. Tradition tells that the motto "Let none who has not learned geometry enter here" was inscribed over the entrance of the Academy. Following in this the Pythagoreans, the apparently irregular motions of the planets had, in his opinion, to be accounted for in terms of combinations of uniform circular motions only. This was to become the main task of astronomers from his time to the dawn of the seventeenth century—a stretch of two thousand years.

EUDOXUS (c. 408–355 B.C.) of Cnidus, who came to Athens at the age of twenty-three where he heard Plato and the sophists, is one of the key figures in the history of Greek astronomy. His most influential contribution was his cosmological model based on concentric spheres centered on the earth, which is itself a sphere at rest. Thus, each heavenly body is attached to the surface of an imaginary invisible sphere which is rotating uniformly around a fixed direction. Each sphere is itself attached to a larger sphere rotating around another axis. The secondary spheres could be succeeded, in turn, by a set of interlocking spheres according to the observed positions of each heavenly body. In this representation, twenty-seven spheres in all were considered sufficient: three each for the sun and moon, four for each of the five planets, and one for the fixed stars. Further improvements were made by CALLIPPUS (c. 370–300 B.C.) of Cyzicus, who added further imaginary spheres to the model, making use of thirty-four concentric spheres in all.

Of great interest also is the work of ARISTOTLE, since his picture of the universe was to become the orthodox view of cosmology in the universities of western Europe from the thirteenth to the seventeenth century (see figure 1.2). In his astronomy, Aristotle builds upon the mathematical results of Eudoxus and Callippus in their use of concentric spheres centered on the earth. However, what were for these astronomers mere mathematical devices of representation are regarded by him as physical entities. The heavenly bodies must also be thought of as composed of a distinctive ethereal substance, different from the objects found below the sphere of the moon. The latter alone are made of varying combinations of earth, water, air, and fire. Beyond the moon's orbit, both the visible heavenly bodies and their invisible spheres are therefore composed of a fifth element, a special *ether*, the movement of which is always circular. Finally, beyond all these spheres is the outermost heaven, the *primum movens*, that communicates its motion to the whole system. Aristotle's world is of finite extent and has neither beginning nor end.

Figure 1.2 The cosmic system of Aristotle as presented by Petrus Apianus around 1530. This geocentric universe focuses on a round earth articulated into four concentric spheres, one for each of the four elements. Surrounding these spheres are seven planetary spheres, one for each of the seven planets. The "eighth heaven" is the sphere of the fixed stars. The "ninth heaven" is a transparent, crystalline sphere that nonetheless reflects the signs of the zodiac. The "tenth heaven," which is the *primum movens* that communicates its motion to the whole system, is the boundary of the universe, and beyond it lies the habitation of God and all the Saints. The addition of this infinite, God-filled void space was an essentially Christian reaction to Aristotle's claim that there can be "neither place, nor void, nor time beyond the sphere of the fixed stars." From S. K. Heninger, Jr., *The Cosmographical Glass: Renaissance Diagrams of the Universe* (San Marino, Calif.: Huntington Library, 1977). By permission of the Huntington Library.

If so, then, by what mechanism do the sun, stars, and planets shine? This problem is also discussed in Aristotle's treatise *De caelo* (On the heavens), where it is argued that the warmth and light that proceed from the heavenly bodies are caused by *friction* set up by their motion. According to Aristotle, therefore, the heavenly bodies are not themselves fiery, as was suggested by the Ionian philosophers. Rather, they shine because they are carried on moving spheres, so that the substance underneath the sphere of each of them is necessarily heated by its motion, and particularly in that part where the body is attached to its sphere. In his treatise *Meteorologica*, Aristotle further argues that the sun's motion alone is sufficient to produce heat in the place where we live because it moves swiftly and is not so far off as the stars. In his own terms: "For a motion that is to have this effect must be rapid and near, and that of the stars is rapid but distant, while that of the moon is near but slow, whereas the sun's motion combines both conditions in a sufficient degree."

Another brilliant pupil of Plato was HERACLIDES PONTICUS (c. 388–315 B.C.) of Heraclea on the Black Sea (present-day Eregli, Turkey). His great advance in astronomy was his suggestion that the earth is in the center of the universe and rotates around an axis while the sphere of the fixed stars is at rest. This is clearly at variance with the views held by Aristotle. Heraclides has often been credited also with the claim that Venus and Mercury circle round the sun like satellites, but recent studies have revealed this interpretation of ancient texts to be without rational basis. Actually, the first clear evidence for such a hypothesis in relation to the motion of the inner planets is found in the works of THEON of Smyrna, who flourished in the early part of the second century A.D.[5]

Throughout antiquity, natural philosophers were also divided on many other aspects of Aristotelian cosmology, such as the finitude of the universe and the continuity of matter. Indeed, although the Athenian philosopher EPICURUS (341–270 B.C.) of Samos was especially celebrated for his ethical teaching, he contributed to cosmology by holding the views that the universe has no limit and that "nothing can arise out of nothing and nothing can be reduced to nothing." Following the doctrines of Leucippus and Democritus regarding the atoms and the void, he also taught that the number of these irreducible particles of matter and the extent of empty space are infinite. According to Epicurus, the universe therefore consists of numberless worlds strewn throughout an infinite void, with all matter composed of atoms and regulated by natural laws. His most eminent Roman advocate was the poet and philosopher LUCRETIUS (c. 95–55 B.C.), whose poem *De rerum natura* (On the nature of things) is the fullest statement extant of the atomist theory. However, about the same time Epicurus established his cosmology, another system of philosophy was founded by ZENO (c. 320–250 B.C.) of Citium in Cyprus, who came to Athens and set up a school in a roofed colonnade called a *stoa*. The Stoics were also ready to accept that space by its nature is edgeless; however, they argued for a rival cosmological system consisting of a finite starry universe surrounded by a void of infinite extent which is regularly drawn into it and then exhaled. The void is actually needed to provide a region in which the universe, as a living creature, continually breathes in and out in its entirety. Specifically, the Stoics regarded the physical world as a dynamical continuum, made coherent by the all-pervading *pneuma*, a mixture of air and fire, which imparts to matter its structure and all physical qualities. Through its inherent tensions the pneuma was also thought to constitute a kind of elastic medium for the propagation of physical action in the universe as a whole. The Epicurean and Stoic systems—that is, atomism and continuum theory—greatly influenced the Scientific Revolution of the seventeenth century.[6]

---

[5] See G. J. Toomer, *Dictionary of Scientific Biography*, **15**, 202 (1978); and especially B. S. Eastwood, *J. History Astron.*, **23**, 232 (1992).

[6] See E. Harrison, "Newton and the Infinite Universe," *Physics Today*, **39**, No. 2, 24 (1986).

1.5 THE ALEXANDRIAN PERIOD

While Aristotle was teaching in Athens, his former pupil Alexander of Macedonia was carving out a large military empire that brought Greek science into direct contact with the older sources of culture in the East. When Alexander died in 323 B.C., his empire broke down, and Egypt was seized by Ptolemy, one of his generals. Alexandria, which was founded by Alexander in 332 B.C., became the capital of Egypt during the reigns of Ptolemy and his successors. The Ptolemies were patrons of learning; they founded the Museum, temple of the Muses, which contained a large library and an observatory, and the city soon became the center of Greek scientific thought.

Among the earlier members of the Alexandrian school was ARISTARCHUS (c. 310–230 B.C.) of Samos, who was the first to put forward the heliocentric hypothesis. Although his original text is no longer extant, his most important witness is ARCHIMEDES (287–212 B.C.) of Syracuse, who made explicit mention of Aristarchus' contribution in his treatise *Arenarius*, noting that "his hypotheses are that the fixed stars and the sun remain unmoved, that the earth revolves about the sun in the circumference of a circle, the sun lying in the middle of the orbit." On this ground the Stoic philosopher CLEANTHES (c. 301–232 B.C.) declared that it was the duty of Greeks to indict Aristarchus on the charge of impiety for putting in motion the hearth of the universe.

About a century later, the heliocentric hypothesis was also rejected by the greatest of observational astronomers of antiquity, HIPPARCHUS (c. 190–120 B.C.) of Nicaea in Asia Minor. The reasons that weighed with Hipparchus were presumably the facts that a system in which the sun was the exact center did not seem to account for the variations of distance and the irregularities of planetary motion; that the theory of epicycles did apparently suffice to describe the phenomena; and that the latter could be reconciled with the commonly held view according to which the earth was immobile.

When Hipparchus came to examine suitable kinematic patterns for the planets, he had before him the techniques of the eccentric circle and epicycle. The simple device that was found satisfactory in the case of the sun was the use of an *eccentric*, that is, a circular orbit whose center does not coincide with the position of the observer on the earth. However, a more enduring solution was broached by a pure mathematician, APOLLONIUS (c. 240–170 B.C.) of Perga, who lived for some time in Alexandria. In this approach each planet moves uniformly on a small circle, called the *epicycle*, whose center moves on a larger circle, called the deferent. The epicyclic view largely prevailed through the mediation of Claudius Ptolemaeus, commonly known as PTOLEMY, who flourished in Alexandria about the middle of the second century A.D. (He was not related to the ruling dynasty of the Ptolemies.) His great treatise, later known as the *Almagest*,[7] largely refines upon the work of his predecessors, notably Hipparchus, and brings to a systematic culmination all the efforts of Greek astronomy. The *Almagest* also contains a catalog giving the ecliptical coordinates and magnitudes of 1,022 stars, based in part on the lost catalog by

---

[7]The Greeks called it the *megalē syntaxis*, or great composition. The Arabian translators converted *megalē*, great, into *megistē*, greatest, and hence it became known to the Arabs as *Al Magisti*, whence the medieval Latin *Almagestum* and our *Almagest*.

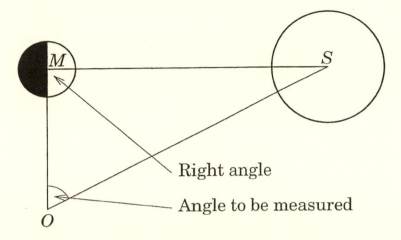

Right angle

Angle to be measured

Figure 1.3 The measurement of the relative distances of sun and moon from earth (not to scale).

Hipparchus. In a later book, *Hypotheses of the Planets*, Ptolemy attempted to produce a physical system that would fulfill the requirements of both the mathematical theory expounded in the *Almagest* and his philosophy of nature.[8] The basic idea was to assume that the planets were carried around by a sequence of nested eccentric shells (of an invisible ethereal substance), within each of which the epicycles would revolve. This geocentric and epicyclic representation of the world remained virtually unchanged until the Renaissance.

Another important advance in astronomy was made when Aristarchus attempted to measure the relative distances of the sun and moon from the earth, and their sizes relative to each other. He knew that the light of the moon was reflected from the sun. Accordingly, if in figure 1.3 the points O, S, and M denote, respectively, the observer and the centers of the sun and moon, the moon appears to the observer half full when the angle OMS is a right angle. When this is the case, the observer can measure the angle MOS. With a knowledge of the angles at the points M and O, the relative lengths of the sides OS and OM can be obtained, thus giving at once the relative distances of the sun and moon from the observer.

Aristarchus made the angle MOS about 87°. Lacking the tools provided by trigonometry, he used purely geometrical reasoning to estimate that the distance to the sun was from 18 to 20 times that of the moon, whereas the correct value is about 390. Aristarchus further estimated that the apparent sizes of the sun and moon were about equal, and correctly inferred that the relative sizes of these bodies were in proportion to their distances. By a method based on eclipse observations, which was afterward developed by Hipparchus, he also found that the diameter of the moon was about one-third that of the earth, a result not far from true. Thus, in spite of large observational errors, Aristarchus was nevertheless able to show convincingly that, while the moon is smaller than the earth, the sun is much larger.

---

[8] See especially A. Murschel, *J. History Astron.*, **26**, 33 (1995).

To ERATOSTHENES (c. 276–194 B.C.) of Cyrene, who was librarian of the Museum at Alexandria, we owe one of the first scientific estimates of the earth's radius. He knew that at noon at the summer solstice the sun threw no shadow at Syene (near modern Aswan), while at the same hour at Alexandria the angular distance of the sun from the zenith was about 7°. Now, if in figure 1.4 the point S denotes the

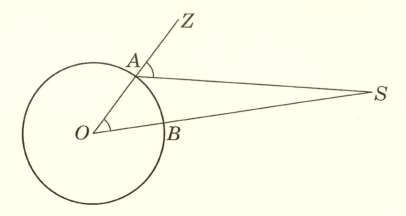

Figure 1.4 The measurement of the earth's radius.

sun, the points A and B Alexandria and Syene, respectively, the point O the center of the earth, and AZ the direction of the zenith, the angle SAZ is almost equal to the angle SOZ owing to the great distance of the sun, so that the arc AB is to the circumference of the earth in the proportion of 7° to 360°. The distance between Alexandria and Syene being about 5,000 stadia, Eratosthenes estimated the earth's circumference to be about 250,000 stadia. If the value of the stadium used was the common Olympic stadium, or 157.5 meters, then his evaluation of the earth's radius is about one percent in error. This good agreement with modern values is perhaps fortuitous, however, since both the angle (7°) and the distance (5,000 stadia) have obviously been rounded.

Hipparchus improved on Aristarchus' method and made a satisfactory estimate of the distance of the moon by observing the angular diameter of the earth's shadow at the distance of the moon during an eclipse. In figure 1.5, which represents

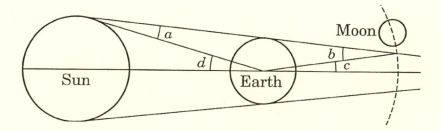

Figure 1.5 The shadow cone of the earth during a lunar eclipse (not to scale).

a plane section of the shadow cone, one can readily see the simple relation that exists between the angles a, b, c, and d. In the triangle moon-earth-sun, one has $a + b = c + d$, that is, the sun's parallax plus the moon's parallax is equal to the sun's angular radius plus the semidiameter of the earth's shadow. Since the sun's parallax is much smaller than the moon's parallax, the latter is therefore almost equal to the sum of the apparent semidiameters of the sun and the shadow. Thus, letting $a = 0'$, Hipparchus found that the distance of the moon was 59 times the earth's radius. However, assuming a solar parallax of $7'$, the smallest perceptible parallax, he found that distance to be 67 ⅓ times the earth's radius.[9]

Although these evaluations were made during the third and second centuries B.C., they look surprisingly modern in the sense that they are neither a mere speculation nor an accumulation of brute facts. Indeed, the three natural philosophers— Aristarchus, Eratosthenes, and Hipparchus—assumed a reasonable physical model and then, from simple measurements only, they were able to deduce some information about the relative sizes and distances of the earth, moon, and sun. The importance of their work lies not so much in the numerical figures they obtained, but in the facts that the relevant questions were asked and that they were formulated correctly. Such an approach, which was also favored by the Sicilian polymath Archimedes, lies at the core of modern science; it is unfortunate that it did not survive the Hellenistic period and had to be rediscovered at the end of the Middle Ages.

Of great historical importance also is the work of the Christian philosopher Johannes PHILOPONUS, who taught in Alexandria in the first half of the sixth century A.D. and was one of the last great commentators of Aristotle. His conception of the laws of motion, opposed to that of Aristotle, is particularly interesting. According to the latter, a body can move only if a mover exercises upon it an action at every instant of the motion; if left alone, it will not continue to move. The fact that missiles—such as stones and arrows—go on moving for some time after having been projected was assumed to be due to the contact with the surrounding air, to which the mover had imparted some of its moving power. Philoponus' innovation was to suggest instead that the continuance of the motion of a projectile was due to the action of a certain immaterial force—later called the "impetus"—which was communicated to the projectile by the mover. This new idea about forced motion had a considerable impact on physical doctrine in the Middle Ages—both in the Muslim world and in the Latin West—and played an important part in the gradual disintegration of Aristotelian orthodoxy. However, of greatest importance are Philoponus' cosmological views, based on the Judeo-Christian doctrine of creation. Thus admitting that the world was created by God from nothing and will be destroyed, he vehemently attacked Aristotle's assumption that the celestial region beyond the moon was immutable. In particular, he advocated the idea that the sun and the stars did not consist of ether but were made of the same stuff as the earth. He also claimed that the difference among stars in magnitude, color, and brightness was to be found in the composition of the matter of which the stars were constituted, being sources of fire of the same kind as terrestrial fires. Moreover, he declared that all matter in the universe is nothing

[9]See A. Van Helden, *Measuring the Universe* (Chicago: University of Chicago Press, 1985), 12–13.

but tri-dimensional extension, in which respect there is no difference between any of the celestial and the terrestrial bodies. Philoponus' cosmological views found no echo in his time, however, and about eleven centuries had to pass until observation showed that creation and decay were by no means confined—as was commonly believed—to the earth and its immediate surroundings.[10]

1.6 FROM THE DARK AGE TO THE RENAISSANCE

With the decline and fall of the Roman Empire, ancient philosophies gave place to a new religious, philosophical, and social movement—Christianity. Most of the early Fathers of the Church had little sympathy for anything that emanated from the heathen Greek and Roman world. For many centuries, therefore, anything that could not be reconciled with the Holy Scriptures was rejected with horror and scorn. Thus the earth became flat again; the heaven was no longer a sphere but a tent or tabernacle; and the sun did not pass under the earth during the night, but traveled laterally round its northern parts as if hidden by a wall. It was also widely believed that the sun, moon, and stars were moved in their orbits by angels, who had to carry on this work until the last day. Although this type of cosmology continued to flourish during the Middle Ages, some writers did study the works of the Greek philosophers, and they were not afraid to accept some of the teaching of antiquity. One of the most eminent figures of the transition period between Antiquity and the Middle Ages, Anicius BOETHIUS (c. 480–524), is often called "The Last Roman." He translated several Greek scientific works into Latin, the only language common to the learned West. Boethius translated Ptolemy's *Almagest* into Latin, but this Latin translation was lost shortly after. As a result, astronomers of the early Middle Ages in the Latin West were ignorant of the epicycle theories of the planets.[11] The scientific heritage of that period was fully displayed by the encyclopedist ISIDORUS HISPALENSIS (c. 560–636), archbishop of Seville, who repeatedly referred to the spherical shape of the earth in his writings, although they otherwise contain passages that may be reconciled only with belief in a flat earth.

Another important step forward was made by an English monk, the Venerable BEDE (673–735), who stoutly maintained that the earth is a sphere and cited as proof the fact that stars visible in one latitude are invisible in another; borrowing heavily from the Latin author PLINY the Elder (23–79), he also gave a detailed account of the nonuniform apparent motions of the planets circling round it. From about the ninth century geocentrism and the rotundity of the earth may be considered to have been acceptable again in the Christian West. In the early Middle Ages, from the eighth to the twelfth century, the geoheliocentric system attributed to Heraclides also enjoyed a widespread popularity propagated by several Latin authors—such as Martianus CAPELLA, who wrote a compendium of science and philosophy in the first

---

[10] See especially S. Sambursky, ed., *Physical Thought from the Presocratics to the Quantum Physicists: An Anthology* (New York: Pica Press, 1975); R. Sorabji, ed., *Philoponus and the Rejection of Aristotelian Science* (Ithaca, N.Y.: Cornell University Press, 1987).

[11] It was not until the 1160s that a poor Latin translation of the *Almagest* was made in Sicily from Greek; soon afterward, in 1175, a better translation was made from Arabic to Latin by GERARD OF CREMONA (1114–1187) in Toledo, Spain.

half of the fifth century. Let us also mention the work of Johannes Scotus ERIGENA (c. 810–877), an Irish-born philosopher active in France in the 850s and 860s, who referred to "Jupiter, Mars, Venus, and Mercury, which ceaselessly pursue their orbits around the sun (*qui circa solem volvuntur*)." Historians have disagreed about whether Erigena meant to claim that these four planets move on heliocentric orbits, or merely that they differ from chilly Saturn in that they are near the sun, are much akin to it, and traverse the same region of the heavens that it does. It is unfortunate that in the original manuscript Erigena's famous passage was not accompanied by a graph.[12]

Meanwhile, carrying the banner of Islam, Arabic tribes were suddenly fused into a powerful nation during the seventh century. War and conquest were rapidly followed by intense intellectual activity in Syria and then in Iraq. The rise of the Abbasid caliphate (750–1258) inaugurated the greatest period of Islamic rule, and Baghdad became the center of a brilliant civilization which spread over the entire Muslim world. Old Indian and Greek books were translated into Arabic during the eighth and ninth centuries. Ptolemy's *Almagest* was translated into Arabic for the caliph al-Ma'mun (c. 786–833) in the late 820s, while large observatories were being built in Baghdad and Damascus. AL-BATTANI (c. 858–929), perhaps the greatest Islamic astronomer, tested many of Ptolemy's results, brought important ameliorations to them, and published improved tables of the sun and the moon. From the same time AL-SUFI (903–986) is chiefly known for his catalog of stellar magnitudes. This work was particularly valuable since its author carefully recorded the magnitudes "as they were seen by his own eyes," whereas other star catalogs of that period had simply borrowed all the magnitudes from the *Almagest*.

The next astronomical center of that time and region was at Cairo, where Ibn YUNUS (c. 950–1009) published a set of astronomical and mathematical tables, the *Hakemite Tables*, which remained the standard ones for about two centuries. Another important observatory was built at Maragha in northwest Iran by the Mongol prince Hulagu Khan (c. 1217–1265), a grandson of Genghis Khan (1162–1227). Its main contribution was a volume of new astronomical tables published under the direction of Nasir al-Din AL-TUSI (1201–1274) in 1271, and known as the *Ilkhanic Tables*. But the most productive observatory in western Asia was that of the Muslim ruler and scholar ULUGHBEK (1394–1449), a grandson of the Mongol conqueror Timur (1336–1405). Working at Samarkand with his assistants, Ulughbek made a star catalog from observations with a 40-meter sextant fixed on the meridian and other precision instruments. This was probably the first substantially independent catalog of stellar positions made since the time of Hipparchus sixteen centuries before. He also published fresh tables of the planets, and did his own amazingly accurate calculations of the length of the year. Unfortunately, Ulughbek was the victim of a cultural and religious backlash: He was murdered at the instigation of his own son, and the Samarkand observatory was reduced to ruins by the beginning of the sixteenth century, although his

---

[12] See especially P. Duhem, *Le système du monde*, vol. 3 (Paris: Hermann, 1915–1958), 58–62; E. von Erhardt-Siebold and R. von Erhardt, *The Astronomy of Johannes Scotus Erigena* (Poughkeepsie, N.Y.: Vassar College Press, 1940). For an overall picture, see S. C. McCluskey, *Astronomies and Cultures in Early Medieval Europe* (Cambridge: Cambridge University Press, 1998).

astronomical work was saved and published to posthumous acclaim in Western Europe.[13]

During the tenth century, astronomy and other branches of knowledge also made some progress in Muslim Spain and the opposite coast of Africa. A great library and an academy were founded at Cordoba about 970, and similar establishments sprang up in Toledo, Seville, and Morocco. Thence, the fame of Arabic astronomy began slowly to spread through Spain into other parts of western Europe. Toledo, despite its reconquest by Christendom in 1085, long remained an important center of Arab and Hebrew culture, where a true spirit of collaboration lingered on among Jewish, Christian, and Moorish scholars. The most important work produced by these astronomers was the volume of astronomical tables known as the *Toledan Tables* (1080), of which translations and adaptations were widely distributed in Europe. Western knowledge of astronomy was very much increased by the activity of Alfonso X (1221–1284), "the Wise," king of Castile and León. He collected at Toledo a body of scholars, mostly Jews, who calculated a set of new astronomical tables for predicting positions of planetary bodies. These *Alfonsine Tables* (1272) spread rapidly through Europe and were in use for about three centuries, up to the middle of the sixteenth century. To Alfonso is also due the publication of a vast encyclopedia of astronomical knowledge compiled by a similar group from Arab sources.

One of the last significant astronomers of Muslim Spain was al-Bitruji, known to the Latins as ALPETRAGIUS (died 1204). Philosophically, he objected to Ptolemy's planetary system on the grounds that it violated Aristotle's physical principles. He therefore advocated an alternative model for planetary motion that was based on concentric spheres rotating about inclined axes. This work, which was translated into Latin in 1217, remained influential for several centuries. In fact, at the beginning of the twelfth century, a great wave of translations from Arabic to Latin had already begun, partly of original Arabic books, partly of Arabic translations of Greek books. The physical works of Aristotle, among which *De caelo* is by far the most important, and Ptolemy's books thus became available to western scholars. The influence of Aristotle over medieval thought soon became almost supreme, and by the end of the thirteenth century most of his cosmological views, or supposed views, were firmly established in the Latin West, with Church approval. With the notable exception of Robert GROSSETESTE (c. 1175–1253), chancellor of Oxford University, and a few others who denied the existence of Aristotle's fifth element, it was therefore widely accepted that the four elements—earth, water, air, and fire—filled the sublunar region of the sky, while the celestial region from the moon outward was composed exclusively of an extraordinary substance, a special ether or fifth element. The earth was of course resting at the center of the universe, which was spherical in shape and of finite extent (see figure 1.6). Not unexpectedly, beyond the general assertion that rotating heavenly spheres move the stars and planets, opinion among medieval thinkers as to the final causes of these rotations varied widely. Of particular interest is the *Guide for the Perplexed*, in which the Jewish philosopher

---

[13] See K. Krisciunas, *Astronomical Centers of the World* (Cambridge: Cambridge University Press, 1988), 23–39.

Figure 1.6 A Christian view of the universe around 1350. In the lowest layer we see the elements earth, water, and air, and their creatures. Just above is the sphere of fire, which is a pure but invisible element. Above these four elements rise the spheres of the seven planets and that of the fixed stars, all of which are apparent to our eyes. Above this level lies the Christian heaven, which is not perceptible to our senses. In this artistic depiction we observe the Son with His Crown, the Holy Ghost in the familiar form of a dove, and the Father. Note that for purposes of symmetry the Father is made two figures, just as there are two symmetrical groups of angels. This type of cosmology, which has the virtue of simplicity and completeness, was commonly accepted until the seventeenth century. From S. K. Heninger, Jr., *The Cosmographical Glass: Renaissance Diagrams of the Universe* (San Marino, Calif.: Huntington Library, 1977). By permission of the Huntington Library.

Moses MAIMONIDES (1135–1204) embraces different patterns of interpretation of heavenly phenomena. His cosmology is basically Aristotelian, and he held that rotating spheres moved the planets by mechanical transmission of motion through contact. Movement of the spheres proceeded from another causation: Elaborating on the thought of his Muslim predecessors, Maimonides identified the separate intelligences of Aristotelianism moving the heavenly spheres with the angels of Scripture. These views influenced the Dominican theologian Thomas AQUINAS (1227–1274) and, through him, the whole thought of Latin Christendom. In fact, such terms as "moving intelligences" or "angels of the spheres" were still familiar to all in the sixteenth and seventeenth centuries.

Thirteenth-century astronomy in western Europe was concerned mainly with a debate as to the relative merits of two rival planetary systems. The first was derived from the works of Aristotle and Alpetragius, where it was assumed that the planets were carried around on concentric spheres. In the second system, derived ultimately from Ptolemy's book *Hypotheses of the Planets*, the planets were assumed to be carried around by a system of material eccentric and epicyclic spheres, whose centers did not coincide with the earth's center. The Ptolemaic system was quickly recognized as being the best geometric device for "saving the astronomical appearances." The practical astronomers, who inevitably demanded quantitative results, thus had no choice but to retain the eccentrics and epicycles of Ptolemaic astronomy, even though they could not be easily reconciled with the principles of the only adequate system of physics known—that of Aristotle. By the end of the thirteenth century the system of concentric spheres had been largely discarded, and scholastic philosophers increasingly came to assume the existence of material eccentrics and epicycles. Special consideration among the medieval astronomers may be given to the Provençal Jew Levi ben Gerson, also known as GERSONIDES (1288–1344), who was the first to criticize the faulty methodology of medieval science. Specifically, he measured the changes in the apparent brightnesses of the planets with the goal of refuting the Ptolemaic system. (He developed specific instruments for this purpose, essentially pinholes and the camera obscura.) Since the results of his observations did not fit Ptolemy's planetary models, Gersonides therefore concluded that the epicyclic hypothesis must be rejected as contrary to the phenomena. He also rejected the model suggested by Alpetragius, who had objected to epicycles for Aristotelian reasons, but was unable to devise a better one. Gersonides' contention was that "no argument can nullify the reality that is perceived by the senses; for true opinion must follow reality, but reality need not conform to opinion"—certainly not the usual position in the scholastic age.[14] During the fourteenth and fifteenth centuries, even more radical developments may have been suggested by ancient Greek speculations, in particular the earth's daily rotation and the semiheliocentric system in which Venus and Mercury revolve around the sun while the sun itself revolves around the earth. The concept of geocentrism remained unchallenged until the 1510s, when the Polish astronomer Nicolaus COPERNICUS (1473–1543) revived Aristarchus' heliocentric hypothesis.

---

[14]See B. R. Goldstein, *Theory and Observation in Ancient and Medieval Astronomy* (London: Variorum Reprints, 1985).

Although few considered the problem during the Middle Ages, prior to the seventeenth century scholars were divided on the nature of the light emanating from the fixed stars and the planets.[15] Of particular interest is the work of the Cairo astronomer and optician Ibn al-Haytham, known in the West as ALHAZEN (965–1038), who pointed out that the stars and planets, unlike the moon, always exhibited the bright shape of a complete circle regardless of their positions with respect to the sun, and regardless of the observer's location. (The phases of Venus had not yet been observed.) He therefore concluded that the stars and planets were selfluminous, and that the moon was unique among heavenly bodies in being the only one that borrowed its light from the sun.[16] The Persian physician and philosopher Ibn Sina, known to the Latins as AVICENNA (980–1037), did also concede that the moon received its light from the sun and that the stars and planets were self-luminous. Ibn Rushd, also known as AVERROËS (1126–1198) and perhaps the most influential Muslim philosopher in Spain, agreed that the moon had its light from the sun, but argued for the sun as the sole source of light for all stars and planets; the Franciscan friar Roger BACON (c. 1220–1292) in Oxford was of the same opinion. Subsequently, the Dominican friar ALBERTUS MAGNUS (c. 1200–1280), also known as Albert of Cologne, and the scholastic philosopher ALBERT OF SAXONY (c. 1316–1390) developed various arguments in defense of the idea that the sun illuminated the planets. Among other problems, they had to explain how the planets could appear visibly different and yet receive their light from the same source. Albert of Saxony coped with this problem by assuming that the solar light can penetrate the diaphanous planetary matter, each planet differing in its ability to absorb light. This was a most popular explanation of planetary light: the planets were visible to us because they were partly transparent and, hence, could be filled more or less completely with the sun's light. The Norman polymath Nicole ORESME (c. 1325–1382), a scholar in Paris and later bishop of Lisieux, thought that the self-luminosity option was more probable.

Interest in the light source of the stars became manifest in the late sixteenth century, but again scholars were hardly of one mind. Some were convinced that the fixed stars received their light from the sun because they were not so far off that the sun's light could not reach them. Others believed that the solar light was not the only source of stellar brightness but that each star might have a small amount of light within itself. Thomas DIGGES (1545/46–1595) and Giordano BRUNO (1548–1600) were the first natural philosophers to break with these traditional views by interpreting the fixed stars as suns, and hence as self-luminous bodies. In 1576 Digges published an English version of Copernicus' *De revolutionibus orbium coelestium* (1543), adding to his translation the assertion that the heliocentric universe should be conceived as infinite, with the fixed stars located at varying distances throughout

---

[15] See E. Grant, *Planets, Stars, & Orbs: The Medieval Cosmos, 1200–1687* (Cambridge: Cambridge University Press, 1994), 390–421.

[16] Alhazen was actually the first to experiment with different media in the hope of finding a working theory of reflection and refraction. His disciple WITELO, a Polish physicist who spent most of his life in Italy, also tried to establish the laws of refraction. In his *Perspectiva*, a huge treatise on optics of the 1270s, Witelo explained the twinkling of stars as due to moving air currents in the earth's atmosphere and showed that the effect was intensified when stars were viewed through running water. See H. C. King, *The History of the Telescope* (Cambridge, Mass.: Sky Publishing Corporation, 1955), 26.

infinite space. Thus Digges clearly perceived that, the moment the displacement of the earth was conceded, there was no longer any necessity for picturing the stars as attached to a celestial sphere at a finite distance from the earth. Contrary to Digges, Bruno was a mystic, and his own speculations had their roots in some earlier suggestion by the Rhinelander theologian Nicolaus CUSANUS (1401–1464), who had been led to a belief that the universe has neither center nor circumference; that it is not infinite, yet it cannot be conceived as finite, since there are no limits within which it is enclosed; and that the earth, which cannot be the center of the universe, must in some way be in motion, as are all other bodies in space, with a velocity that is not absolute but relative to the observer. Actually, it was this doctrine of relativity, in space and in motion, that served as the background for Bruno's conception of an infinite universe and innumerable worlds.[17] The essential conclusions of Bruno's work are that not only does the earth move round the sun but the sun itself moves; that our planetary system is not the center of the universe, and that the latter is of infinite size and the worlds therein without number; and that the stars are at large but varying distances from the sun and are themselves centers of planetary systems. These assertions, which were published in England in 1584, marked the change from medieval to modern scientific thought. Betrayed and arrested in Venice, Bruno was transferred to Rome in 1593 and remained in prison for seven years. Then, after a long trial, he was burned at the stake in Rome for his writings which, in addition to containing religious and political heresies, supported the new Copernican system and opposed the traditional Aristotelian cosmology.

1.7 THE EMERGENCE OF MODERN ASTRONOMY

About the end of the sixteenth century and the beginning of the seventeenth century were made the first decided advances since antiquity in observational astronomy. In this field the leading names are those of Tycho BRAHE (1546–1601) and Galileo GALILEI (1564–1642), the former relying upon the naked eye for his measurements, the latter on his recently built refracting telescope.

In 1572 Tycho observed a new star and, failing to detect parallax in his observations, he correctly deduced that it was located well beyond the planets. The star's eighteen-month life span left a considerable impact since it cast serious doubt on the existence of an eternally unalterable celestial region. His observations of a brilliant comet in 1577 also convinced Tycho that its parallax placed it well beyond the sublunar region where comets were supposed to be confined. This was another severe blow to the medieval belief that the celestial region beyond the moon's orbit was immutable. However, because his observations did not disclose a stellar parallax, Tycho rejected the Copernican theory of the earth's revolution. He proposed as a compromise that the sun revolved around a stationary earth, while all the planets—Mercury, Venus, Mars, Jupiter, and Saturn—went around the sun.

---

[17] For English translations of this and other pioneering works, see M. K. Munitz, ed., *Theories of the Universe: From Babylonian Myth to Modern Science* (New York: Free Press, 1957).

Galileo's first contribution to astronomical discovery was made in 1604, when a new star appeared in Ophiuchus, and was shown by him to be without parallax, a result confirming Tycho's conclusions that changes take place in the celestial region beyond the planets. With his telescopic observations made in the 1610s, Galileo added to this a picture of celestial bodies that exhibited surface irregularities: the moon, whose mountains and valleys made it akin to the earth; and the sun, whose spots continually changed in shape and then disappeared while others again succeeded them. The discovery of sunspots is of particular importance because it caused irremediable damage to the Aristotelian principle of stability and incorruptibility of the heavens beyond the moon. Of the new discoveries, Galileo's telescopic observations of the phases of Venus and of the satellites of Jupiter were perhaps the most dramatic events in transforming heliocentrism from a convenient hypothesis for mathematicians into a most plausible physical description of our planetary system. This was a deadly blow to the traditional geocentric and anthropocentric universe.

In 1616 the Copernican doctrines were thus declared "false and absurd, formally heretical and contrary to Scripture" and Galileo was admonished to abandon these opinions. In 1632 Galileo published his epoch-making treatise *Dialogo sopra i due massimi sistemi del mondo* (Dialogue on the two chief world systems), that is, the Ptolemaic and the Copernican (see figure 1.7). Although the book came out bearing both the Roman and the Florentine imprimaturs, it was soon suspended and its sale prohibited on the ground that it was a compelling and unabashed plea for the heliocentric system.[18] In 1633 Galileo was condemned by the Roman Inquisition for believing and holding that the Copernican doctrines were true and, in punishment, was required to "abjure, curse, and detest the aforesaid errors" and forced to spend the last eight years of his life under house arrest. While confined to his villa at Arcetri, near Florence, he published in 1638 his *Discorsi e dimostrazioni matematiche interno a due nuove scienze* (Dialogue on two new sciences), in which he summarized the results of his early work with falling bodies and projectiles and presented his penetrating views on the principles of mechanics. This important work, which was not published in Italy but in Holland, paved the way for Newtonian mechanics.

Copernicus' heliocentric theory made continued support for the medieval geocentric cosmology untenable for the majority of astronomers, if not for the majority of scholastic philosophers. Yet Copernicus was a conservative because he did not venture to abandon the use of patterns of circular motion and indeed continued to make use of epicycles in describing planetary trajectories. The great contribution of Johannes KEPLER (1571–1630) was to improve the theory of planetary motion by replacing circles and epicycles by *ellipses* (see figure 1.8). He himself was not an observer, but he had great faith in the accuracy of the observations of his teacher, Tycho, which he had inherited at the time of Tycho's death. Although Kepler had strong mystical leanings, so that a large portion of his work is now unreadable, his

---

[18]After many unsuccessful attempts, the ban on Galileo's *Dialogue* and other books advocating heliocentrism was finally lifted in 1822; and after 1835 the names of Copernicus, Galileo, and Kepler no longer appeared on the *Index of Prohibited Books*.

claim to immortal fame rests upon his discovery of the following empirical laws of planetary motion:

1. Each planet describes an ellipse, with the sun being at one focus.
2. The speed of each moving planet changes with distance from the sun, so that the straight line joining each planet to the sun sweeps out equal areas in any two equal intervals of time.
3. The squares of the periods of revolution of any two planets (including the earth) about the sun are in the same ratio as the cubes of their mean distances from the sun.

Figure 1.7 Frontispiece of Galileo's *Dialogue on the Two Chief World Systems*, printed in Florence in 1632. The three figures are inscribed *Aristotle* (*left*), *Ptolemy* (*middle*), who carries a model of nested geocentric spheres, and *Copernicus* (*right*), who bears an emblem of his own heliocentric theory. Note that the last figure bears more resemblance to Galileo's portraits than to those of Copernicus. Courtesy of Owen Gingerich.

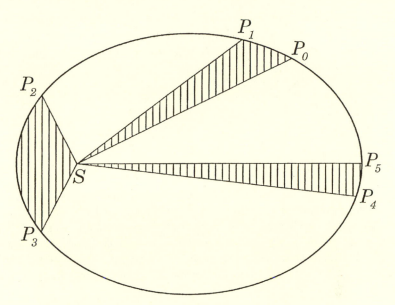

Figure 1.8 Kepler's law of equal areas. Planets sweep out equal areas in equal times. P_0P_1,
P_2P_3, and P_4P_5 are distances along its elliptical orbit around the sun S traversed
by the planet in equal times. The areas SP_0P_1, SP_2P_3, and SP_4P_5 are equal. The
path presented here is far more elliptical than those pursued by the planets, which
are very much more nearly circular.

(The "mean distance" in the third law is half the major axis of the ellipse.) Kepler
announced his third law in 1619, ten years after he had given the first two. Not
unexpectedly, his books were promptly banned and placed on the Roman *Index of
Prohibited Books*.

Although the geocentric system continued to command widespread support
throughout the seventeenth century, heliocentrism with its Keplerian modifications
was cautiously gaining acceptance. Attempts were thus made to find out what causes
the planets to move on their elliptic orbits with the sun occupying one of the focal
points. The solution proposed by the French philosopher René DESCARTES (1596–
1650) in 1644 was eagerly accepted.[19] It assumes that all space is filled with a fluid,
or ether, the parts of which act on each other and generate a whole spectrum of
vortices of different size, velocity, and density. There is an immense vortex around
each star and, in particular, around the sun, which carries in its circular motion the
earth and the other planets. Each of these planets is located, in turn, in a smaller
vortex by which gravitational attraction is produced. Descartes made no attempt

---

[19]From 1629 to 1649 Descartes was to remain almost permanently in Holland, making only three
short visits to France during all this time. Most of his more important works were written and published
in Holland, where he could profit from greater intellectual freedom. But even in this land of toleration
he was to meet with enmity, the bitterest coming from the president of Utrecht University. Late in 1649
Descartes moved to Sweden at the invitation of Queen Christina (1626–1689), but he caught a severe
chill there and died after a short illness.

to reconcile his theory with Kepler's laws. In fact, it was unsupported by any experimental evidence and did not explain a single phenomenon satisfactorily. Yet "Cartesianism" became extremely fashionable, especially in France, and its vogue undoubtedly contributed to the overthrow of the Aristotelian system.

Descartes' vortex theory required one motion to be the cause of another and explained the mutual attraction of the sun and the planets in terms of actual bodily motions. As early as the 1660s, however, Isaac NEWTON (1642–1727) rejected Descartes' hypothesis of material vortices as the cause of gravitation. According to him, gravity is the power of attraction that one body exercises upon another without the first being in motion or coming in contact with the second. Although this approach raises the problem of action at a distance, which Newton properly acknowledged, it has the merit of fitting the observed phenomena.[20] What Newton brought to the subject is to be found in the simplicity that his *laws of motion* combined with the principle of *gravitational attraction* were able to introduce into the description of planetary orbits and ordinary terrestrial phenomena. Newton's unique achievement was to prove that the force that causes a stone to fall on earth is also that which keeps the planets in their orbits. This is expressed in Pope's famous epigram:

> Nature and Nature's laws lay hid in Night,
> God said, Let Newton be, and all was light,

which celebrates the illumination of nature by Newton's mechanical approach to the physical world.

When, in 1687, Newton published his epoch-making treatise *Philosophiae naturalis principia mathematica*, or "Mathematical Principles of Natural Philosophy" (figure 1.9), the medieval geocentric world had become utterly irrelevant because it no longer represented a viable alternative to a world in which gravitational attraction forces each planet to move along an ellipse. In other words, Newton's theory of gravitation made physical sense of heliocentrism while providing a simple, natural derivation of Kepler's empirical laws from first principles alone. The law of gravity is basically simple: it merely states that *two bodies attract each other with a force that is proportional to the product of their masses and inversely proportional to the square of the distance between them.* The success of the theory was all the more evident that the general solution of the two-body problem was found to be a conic (i.e., a circle, an ellipse, a parabola, or a hyperbola), in perfect agreement with the

---

[20]It was not until 1915 that Albert EINSTEIN (1879–1955) resolved this dilemma by transforming Newton's theory of action at a distance, in which gravity acts instantaneously across the entire universe, into a local-action theory in which gravity is equivalent to a curvature of four-dimensional spacetime. Specifically, whereas Newton assumed that space and time form the stage upon which mechanical phenomena are displayed, Einstein took gravity as an intrinsic feature of spacetime itself. In his *general theory of relativity*, as it became known, the geometrical properties of this four-dimensional continuum are determined by the masses present in space and time, and these geometrical properties in turn have a causal influence on the motions of these masses. An important inference we can draw from Einstein's 1915 theory is the way ripples in the fabric of spacetime travel. Detailed calculations have shown that some revolving bodies, such as compact binary star systems, emit a weak gravitational field in the form of waves (see section 6.6). These waves propagate with the speed of light. Action at a distance is therefore avoided, and causality is restored.

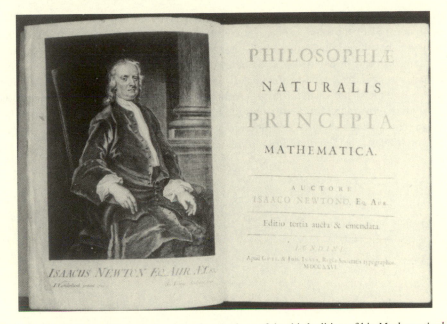

Figure 1.9 Sir Isaac Newton graces the frontispiece of the third edition of his *Mathematical Principles of Natural Philosophy*, published in London in 1726. Courtesy of Owen Gingerich.

observed trajectories of different comets. Yet, as was noted by Newton: "Hitherto we have explained the phenomena of the heavens and of our sea by the power of gravity, but have not assigned the cause of this power." Pointing out that he had made no unreasonable hypothesis, he further concluded: "And to us it is enough that gravity does really exist, and act according to the laws which we have explained, and abundantly serves to account for all the motions of the celestial bodies, and of our sea." Newton's theory is of great philosophical importance because it is the first successful attempt to explain the universe and its processes without reference to gods or animistic conceptions.

By the time geocentrism and terrestrial immobility had lost all credibility, it was widely accepted that the fixed stars were self-luminous bodies, the sun being one among the multitude of stars. However, because no stellar parallax could be detected from the earth's orbital motion round the sun, enormous arbitrary distances had to be assumed between the sun and the fixed stars. The problem was discussed by the Dutch polymath Christiaan HUYGENS (1629–1695) in his treatise *Cosmotheoros*; his efforts at measuring the distance of Sirius are particularly noteworthy. (The work was published posthumously in 1698.) Lacking any information about the global properties of stars, he assumed that his test star, Sirius, was exactly similar to the sun. If so, then, he correctly deduced that its distance from us should be as much greater than that of the sun as its apparent diameter was less than the diameter of the solar disk. In practice, to compare the sun and Sirius, he gradually lessened the sun's diameter in his telescope, closing one end of his twelve-foot tube with a plate that had a variable hole in its middle. Reducing the hole's diameter till the sun appeared of the same brightness as Sirius, he found that he had to make its diameter but the 27,664th part of the diameter of the observed solar disk. In our modern notation, he evaluated the distance of Sirius at about 27,664 AU whereas its distance is actually close to 546,000 AU. (Here AU denotes the astronomical unit, that is, the mean sun-earth distance.) Although we know that Huygens' evaluation was vitiated by an inadequate assumption about the global properties of Sirius, this observational work should not be ignored, because it is quite illustrative of the spirit and methods of the Scientific Revolution that changed our views of the universe.

Chapter Two

Three Centuries of Optical Discoveries: 1610–1910

Millimètre, Centimètre, Décimètre, Décamètre,
Hectomètre, Kilomètre, Myriamètre,
Faut t'y mettre, Quelle fêtre!
Des millions, Des billions, Des trillions,
Et des fraccillions!
— *L'enfant et les sortilèges*, a fairy opera with
COLETTE's libretto and Maurice RAVEL's music

Prior to the seventeenth century, our practical knowledge of the universe was based on visual observations made with the naked eye. In the 1610s, however, the entire course of astronomy was dramatically and permanently changed when Galileo and others raised their refracting telescopes toward the heavens, and so made observations possible beyond anything attainable before. Galileo's first series of telescopic discoveries were published early in 1610 in a twenty-four-page pamphlet entitled *Sidereus nuncius* (The sidereal messenger). Part of the little book discusses the revelation by the telescope of an immense number of fixed stars too faint for recognition by the naked eye. For example, he saw thirty-six stars in the Pleiades, which to the unaided eye consists of six stars only. Part of the Milky Way and various nebulous patches of light in Orion and other constellations were also discovered to consist of myriads of faint stars. However, it was noticed that the stars are not like planets, which appear as disks when magnified by the telescope, whereas the stars remain but points of light. This fact confirmed the idea that the stars are at huge distances from the sun, and it explained why even the best telescopes available in Galileo's day did not show any measurable annual motion of the pointlike stars on the celestial sphere. Actually, it was not until the 1830s that the first parallactic motion of a star was definitely detected, thereby proving once and for all the correctness of the heliocentric model developed by Copernicus and refined by Kepler.

Although Newton is primarily celebrated for his theoretical work in mechanics, it is appropriate at this juncture to recall two of his experimental achievements. In 1666 he positively established that by passing a beam of white light through a glass prism one does not produce color but merely separates the beam into a band of colors, like the rainbow, from red to violet. In 1668, he also constructed a one-inch reflecting telescope in order to overcome some of the problems with glass lenses. (In 1671 he built a larger instrument, which he presented to the Royal Society.) Owing to technical difficulties, however, about half a century elapsed before a reflecting telescope could equal in performance the best refracting telescopes of the time.

Progress in solar spectroscopy was also quite slow, for the first great discovery after Newton's spectrum studies was not made until the 1810s, when an almost countless number of dark lines were found in the spectrum of sunlight. However, for several decades to come, the role that these lines were destined to play in establishing the chemical composition of distant inaccessible bodies was not realized. Thus, in 1835 in his *Cours de philosophie positive*, Auguste COMTE (1798–1857) wrote of the celestial bodies that "never, by any means, will we be able to study their chemical composition, their mineralogic structure, and not at all the nature of organic beings living on their surface," noting also that "every notion of the true mean temperature of the stars will necessarily always be concealed from us." Two years after Comte's death, spectrum analysis arose and brought new, unexpected ideas about the nature of the sun and the stars.

In this chapter we shall outline the chief developments of solar and stellar astronomy from about 1610 to the 1910s, and present them in a form that is intelligible to anyone with an elementary knowledge of science. The emphasis here will be more especially on the basic observational data, setting the scene for subsequent chapters that will present more elaborate theoretical discussions. In terms of time, it is the period from 1860 to 1910 that saw the most rapid growth in the fields of solar and stellar research. Yet, for the most part, all this merely consisted of an accumulation of raw data and brute facts that were lacking a solid theoretical basis. Astrophysics did not become a full-fledged discipline until a new generation of physicists had developed the phenomena of radiation into a workable theory; as we shall see in chapter 4, this was achieved during the first decades of the twentieth century.

2.1 DISTANCES TO THE SUN AND THE STARS

During the second century B.C., Hipparchus estimated the distance of the moon to be between 59 and 67 ⅓ times the earth's radius (see figure 1.5). About three centuries later, Ptolemy obtained that distance by a parallax method that is quite identical with that still in use. If in figure 2.1 M denotes the moon, O the center of the earth, A the point on the earth at which the moon is overhead, and P any other point on the earth, then observers at points A and P see the moon in slightly different directions. Knowing the direction of the lines AM and PM, one can thus obtain the angle PMO, which depends on the distance of the moon, the earth's radius, and the position of the observer at point P. Ptolemy found that the mean distance of the moon was about equal to 59 times the earth's radius, in rough agreement with Hipparchus' result.

Until the first half of the seventeenth century, following Ptolemy's results, it was generally believed that the distance of the sun was about 1,200 times the earth's radius.[1] If so, then, the sun's parallax would at times be as much as 3′ and that of Mars, which in some positions is closer to the earth, proportionally larger. Unable to detect any parallax of Mars, Kepler correctly inferred that the distances of Mars and of the sun were much greater than had been hitherto assumed. Having no data

---

[1] See A. Van Helden, *Measuring the Universe* (Chicago: University of Chicago Press, 1985), 17–19.

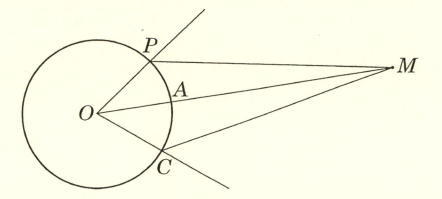

Figure 2.1 The parallax (see text).

at his disposal, he made the bold suggestion that the sun-earth distance should be about three times as great as its traditional value. This was still too short by a factor of seven.

The first reasonably accurate estimate of the distance of Mars and, hence, of the sun was made in the autumn of 1672, when Mars was in opposition and approaching the earth at a distance of 0.37 AU. The principle of the method is substantially identical with that used in the case of the moon (see figure 2.1, where M now denotes Mars). Thus, Jean RICHER (1630–1696) in Cayenne observed the direction of the line CM, while his colleagues Jean PICARD (1620–1682) and Giovanni Domenico CASSINI (1625–1712) in Paris observed the direction of the line PM. Since the line CP is known geographically, it is therefore possible to calculate the distance of Mars. They found that the sun's parallax was about 9.5″, corresponding to a distance of about 360 times that of the moon—that is, 22,000 times the earth's radius or 140,000,000 km. According to modern estimates, the astronomical unit, or AU, is a little less than 150,000,000 km.

Closely related to this measurement is the first evaluation of the *speed of light*. So far as we know, Empedocles—who lived during the fifth century B.C.—made the original suggestion that light travels and takes time to pass from one point to another (see section 1.3). However, the first successful measurements of the speed of light were made in Paris by the young Danish astronomer Ole RØMER (1644–1710). In 1676, studying the motions of Jupiter's bright satellites as they moved into the shadow of Jupiter, he found that when Jupiter was at its greatest distance from the earth, the eclipses occurred later by nearly 1,200 seconds than when Jupiter was closest to the earth. This he saw to be readily explained by the supposition that light travels through space at a finite speed, so that the time required by light to cross the diameter of the earth's orbit is about 1,200 seconds. Using the size of the earth's orbit as known at that time, one finds a speed of roughly 200,000 km s^{-1}. About fifty years later, in 1729, the English astronomer James BRADLEY (1693–1762) confirmed this result when he made the discovery of the *aberration of starlight* — an apparent displacement of the position of a star, which results from a combination of the earth's annual motion in its orbit around the sun with the velocity of light coming from the star.[2] Actually, this was the first direct demonstration that the heliocentric view of the universe was right, since aberration could not occur if the earth were stationary with the sun moving around it. Bradley's inferred value for the speed of light was approximately the same as Rømer's. The current value is a little less than 300,000 km s^{-1}.

Of great interest also is the work of Newton's friend Edmond HALLEY (1656–1742), who in 1718 called attention to the fact that Sirius, Aldebaran, Betelgeuse, and Arcturus had changed their angular distances from the ecliptic since Greek times, and that Sirius had even changed its position perceptibly since the time of Tycho Brahe. A similar conclusion was reached by the German astronomer Tobias MAYER (1723–1762) in 1756, from a comparison of star places recorded by Rømer in 1706 with his own observations. Thus certain stars could no longer be regarded as "fixed" since they had *proper motions* on the celestial sphere, relative to the general background of stars. Evidently, if a single star appears to move through space, the nearer it is to the observer the more rapid its motion appears to be. Apparent rapidity of motion, like apparent brightness, is thus a probable but by no means infallible indication of close proximity. This is of direct relevance to the problem of stellar distances, and it explains why the first definite detection of a parallactic motion was made for 61 Cygni — a star that is barely visible to the naked eye but is remarkable for its large proper motion, about 5″ per year. Actually, the Italian astronomer Giuseppe PIAZZI (1746–1826) was the first to discover the large proper motion of 61 Cygni and to point out, in 1806, that this star was a good candidate for parallax studies.[3] However, another thirty years would elapse before its parallactic motion could be detected.

The direct measurement of the distances of stars is basically a trigonometrical problem, and the method is the same as that used for measuring the distance of the sun. Indeed, as the earth describes an orbit around the sun, the location of a nearby star against the background of distant stars will appear to change slightly on the celestial sphere, since that star will be viewed from different positions over the year (see figure 2.2). The problem with the measurement of these *parallactic motions* is that all stars, including the nearest ones, are so far away in comparison with the radius of the earth's orbit. This is why it was not until the 1830s that astronomers finally had the required tools to measure the distances of the nearest stars. Thus, at the end of 1838 Friedrich Wilhelm BESSEL (1784–1846) in Königsberg (now Kaliningrad) reported a parallax of 0.31″ for the star 61 Cygni. (The best present determination is 0.292″, close to Bessel's estimate.) Early in 1839, the Scottish astronomer Thomas HENDERSON (1798–1844) announced a parallax of about one second of arc for the southern star α Centauri, which he had observed in 1832–1833 at the Cape of Good Hope; and in the following year Wilhelm STRUVE (1793–1864) in Dorpat (now Tartu) reported a parallax of 0.26″ for the star Vega that was monitored

---

[2] It is the same effect as that which causes a vertical shower of rain to appear to a moving observer as falling at an angle.

[3] Piazzi's results failed to win the attention of his fellow astronomers, however, until they were rediscovered by Bessel in 1812. The lack of recognition of Piazzi's 1806 paper has been discussed at great length by G. F. Serio in *J. History Astron.*, **21**, 275 (1990).

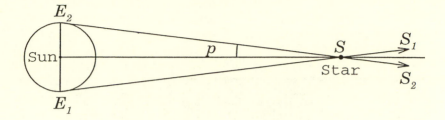

Figure 2.2 The stellar parallax. As the earth moves from E_1 to E_2, the star shifts its apparent
position from S_1 to S_2 by twice its parallax p.

over the years 1835–1838. (Modern estimates are $0.761''$ and $0.125''$, respectively.)
In subsequent years, many other parallaxes were measured, the largest one being
$0.763''$ for Proxima Centauri, a faint companion of the double star α Centauri.[4]

In figure 2.2 one readily sees that the distance d of a star, in astronomical units,
is related to its parallax p'', in seconds of arc, by the formula

$$d = \frac{206{,}265}{p''} \text{ AU,} \qquad (2.1)$$

since there are 206,265 seconds of arc in a radian. For convenience, one usually
defines the *parsec* (pc), which is the distance of a star with a parallax of $1''$. (One
parsec is equal to 206,265 AU or 3.086×10^{13} km.) One can also define the *light-
year*, which is the distance traversed by light in one year. One has

$$d = \frac{1}{p''} \text{ parsecs} = \frac{3.262}{p''} \text{ light-years,} \qquad (2.2)$$

since one parsec is equivalent to 3.262 light-years. The nearest star, Proxima Centauri, is then $3.262/0.763 = 4.27$ light-years distant from us. For comparison, the
sun's mean distance from the earth in terms of the speed of light is 498.5 light-
seconds, or 8.3 light-minutes, while the moon's distance from the earth is 1.3 light-
seconds. The average distance of Pluto from the sun is about 5.5 light-hours.

2.2 THE BEGINNINGS OF SPECTROSCOPY

Modern understanding of light began in the 1660s, when Newton convincingly
demonstrated that sunlight is composed of all the colors of the rainbow.[5] For this
purpose, with a glass prism he dispersed a beam of sunlight into its individual
components, one of which he isolated by means of diaphragms. He then allowed it to
fall upon a second prism and discovered that this component could not be divided any
further. From this experiment he correctly concluded that a prism merely separates

---

[4]See A. W. Hirshfeld, *Parallax: The Race to Measure the Cosmos* (New York: W. H. Freeman and
Company, 2001).

[5]For a detailed account of the contributions made by Newton's precursors, see V. Ronchi, *The Nature
of Light* (Cambridge, Mass.: Harvard University Press, 1970).

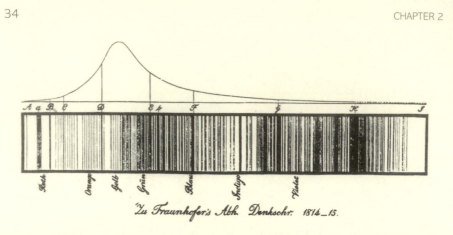

Figure 2.3 Fraunhofer's original drawing of the solar spectrum, to which are added the letters used to identify the more prominent absorption lines and a curve indicating his estimate of the relative luminosity of different parts of the solar spectrum. (Compare with Planck's black-body curves in figure 2.7.) From J. Fraunhofer, *Denkschriften der königlichen Akademie der Wissenschaften zu München für 1814 und 1815* (1817).

a beam of white light into a *spectrum* of colors, and that sunlight is always composed of all of them. In 1802, however, the London physician William Hyde WOLLASTON (1766–1828) was observing the spectrum of the sun when he noticed seven thin dark lines among the colors. These he took incorrectly to be the natural boundaries of the pure simple colors of the spectrum. Several years later, in 1814–1815, the Bavarian optician Joseph FRAUNHOFER (1787–1826) observed the sun's spectrum more carefully and discovered a total of six hundred such lines (see figure 2.3).

At first these so-called *Fraunhofer lines* were a complete mystery. But soon it was discovered that dark lines could be produced artificially in the laboratory by passing white light through bottles containing familiar gases, and that the spectra of certain flames were crossed by various bright lines. It was also found that each chemical element produces its own distinctive pattern of spectral lines among the colors, which — like a fingerprint — immediately identifies the element. For example, sodium when vaporized gives a spectrum characterized by two nearly coincident bright lines in the yellow part of the spectrum. Interestingly, these two lines agree in position with a pair of dark lines in the sun's spectrum.

The first satisfactory explanation of these observations was given in 1859 by Gustav KIRCHHOFF (1824–1887) of Heidelberg (figure 2.4), who at first collaborated with the chemist Robert Wilhelm BUNSEN (1811–1899). They noted that a glowing solid or liquid — or, as we now know, a hot dense gas — always emits a continuous spectrum; whereas a hot thin gas emits a spectrum consisting of bright lines, which depend on the particular chemical element and are characteristic of it. However, if a continuous spectrum radiated by a glowing solid or liquid was permitted to pass through the cooler vapor of any specific element, it was also found that the gas

Figure 2.4 The German physicist Gustav Kirchhoff as a young man. In 1859 he showed that
the Fraunhofer lines in the solar spectrum arise from the presence in the glowing
solar atmosphere of those substances which in a flame produce bright lines in
the same position. Since each chemical element has its own characteristic set of
spectral lines, Kirchhoff's empirical results therefore provided the Rosetta Stone
of solar and stellar spectroscopy. Courtesy of the Mary Lea Shane Archives of
the Lick Observatory.

absorbs radiation of the same wavelengths that it would emit if viewed alone, thus
producing dark lines in the spectrum.[6]

From these experimental results Kirchhoff explained the Fraunhofer lines in the
solar spectrum by the presence in the sun's atmosphere of those substances which in
a flame produce bright lines in the same position. Thus he concluded that sunlight
came from a red-hot sphere, solid or liquid, surrounded by cooler gases that have

---

[6]Some of these results were previously obtained in 1847 by John William DRAPER (1811–1882) in
the United States. Unfortunately, he incorrectly concluded from his experiments that hot thin gases gave
continuous spectra, but may have bright lines superposed. His error originated in his use of bright flames,
giving, in addition to the bright lines of the salt placed in the flame, the continuous spectrum of solid
carbon. Bunsen and Kirchhoff avoided this error by making use of the *Bunsen burner*, which gives a
nonluminous gas flame, in which chemical substances can be vaporized and a bright-line spectrum can
be obtained due to the luminous vapor only. See A. Berry, *A Short History of Astronomy* (London: John
Murray, 1898; New York: Dover Publications, 1961).

Figure 2.5 The Austrian physicist Johann Christian Doppler, discoverer of the effect known by his name, circa 1835. His earliest writings were on mathematics, but his name is associated with his work in physics. Lithograph by Franz Schier from a sketch by Antonin Machek. Courtesy of AIP Emilio Segrè Visual Archives.

absorbed light of the wavelengths corresponding to the Fraunhofer lines. By comparing these lines with the bright lines in the spectra of metals and other chemical elements in the laboratory, he was further able to show that, in addition to sodium, the elements iron, magnesium, copper, zinc, barium, and nickel exist in the sun's atmosphere. Kirchhoff's explanation of the Fraunhofer lines was truly epoch making since it provided astronomers with a method to discover what the sun and the stars are made of. However, it did not tell them how and why the spectral lines were formed.

Although the spectroscope came to be used extensively to study the chemical composition of celestial bodies, it was soon realized that it can also give direct evidence of stellar motion in a direction toward or away from us. The principle involved was first worked out for sound waves by the Austrian physicist Johann Christian DOPPLER (1803–1853; figure 2.5), who suggested in 1842 that the color of a luminous body must be changed by the relative motion between the source and the observer.[7] The principle was verified experimentally for sound by the

[7]From 1835 to 1847, Doppler was a professor at what is now the Czech Technical University in Prague. In 1992, on the occasion of the 150th anniversary of his discovery, that institution published an interesting book entitled *The Phenomenon of Doppler*, edited by I. Štoll. This book, which presents a detailed picture of the life and times of Doppler, also contains the reproduction of an oil painting and rare daguerreotypes of that too little known scientist.

Dutch meteorologist Christoph BUYS-BALLOT (1817–1890) in 1845. However, in 1848 the French physicist Hippolyte FIZEAU (1819–1896) showed that stellar radial velocities are much too small to cause an appreciable color change, but that this shifting may become noticeable through the examination of spectral lines. This phenomenon, which is known as the *Doppler effect*, was observed for the first time in 1868 by William HUGGINS (1824–1910), a wealthy English amateur, who was able to measure the minute shifts of the hydrogen lines in the spectrum of the star Sirius.

As a rule, if a source of light is approaching or receding from the observer, the lines in its spectrum are displaced, respectively, to shorter or longer wavelengths. One can show that this shift in wavelength, $\lambda - \lambda_0$, depends only on the relative speed of approach or recession. If v denotes that speed, c the speed of light, λ the measured wavelength, and λ_0 its usual position in the laboratory, the Doppler shift in wavelength is given by

$$\frac{\lambda - \lambda_0}{\lambda_0} = \frac{v}{c}, \qquad \text{when} \quad v \ll c. \qquad (2.3)$$

This formula, when solved for v, indicates the method by which the *radial velocity* of a luminous body may be obtained. According to this formula, therefore, velocities of approach should be counted as negative and velocities of recession as positive. For example, if a star is approaching us with a typical speed of 30 km s^{-1}, one readily sees that $v/c = -10^{-4}$. Hence, by virtue of eq. (2.3), a spectral line at $\lambda_0 = 5{,}000$ Å will be observed at $\lambda = 4{,}999.5$ Å in the star's spectrum. This is much too small to change the color of the star but the blueshift of the line is measurable.

Another important phenomenon was observed in the laboratory when light sources are operated in magnetic fields. This is the so-called *Zeeman effect*, the splitting up of spectral lines in a magnetic field, which was discovered by the Dutch physicist Pieter ZEEMAN (1865–1943; figure 2.6) in 1896. The theoretical explanation was first given by his fellow countryman Hendrik Antoon LORENTZ (1853–1928) soon after the effect was observed; Lorentz also predicted polarization effects which were verified later by experiment.

Thus far we have seen how spectral lines can be used to provide some essential information about the nature of the sun and the stars. Where then does the *continuous spectrum* come from? It is a matter of common knowledge that the intensity and color of light emitted by a hot body vary with its temperature. The question therefore arises, what is the spectral distribution of radiation for bodies at different temperatures? This problem was considered in the early 1860s by Kirchhoff and, independently, by Balfour STEWART (1828–1887) in Edinburgh. They found that for radiation of the same wavelength at the same temperature, the ratio of the emission and the absorption powers is the same for all bodies. For convenience, Kirchhoff defined a "perfectly black body" as one which at all temperatures completely absorbs all radiation falling upon it. Such a *black body* is also a "perfect radiator" in the sense that it emits radiation that depends only on its temperature; there is thus no dependence on the details of the chemical structure of this *ideal radiator*. As to the experimental realization of such a black body, Kirchhoff and Stewart suggested a closed box with black walls inside, kept at constant temperature, and having a tiny

Figure 2.6 The Dutch physicist Pieter Zeeman as a young man. He was awarded the Nobel
        Prize for physics in 1902 with Hendrik Antoon Lorentz for their work on the
        influence of magnetism on radiation. Photo by Gen. Stab. Lit. Anst.; courtesy of
        AIP Emilio Segrè Visual Archives.

pinhole through which radiation may pass from the inside to the outside. To a good
approximation, this outcoming radiation is the same as that of an ideal black body.

An empirical relation between the temperature and total brightness of radiation
from a black body was obtained first by the Austrian physicist Josef STEFAN (1835–
1893) in 1879. The resulting *Stefan's law* simply states that the total energy emitted
from each square centimeter of a black body in each second grows with the fourth
power of the temperature. This law was subsequently deduced by the Austrian
physicist Ludwig BOLTZMANN (1844–1906) in 1884 from thermodynamical considerations. We thus have

$$F = \sigma T^4, \tag{2.4}$$

where F is the total energy flux (in erg cm^{-2} s^{-1}) and $\sigma = 5.669 \times 10^{-5}$ erg cm^{-2}
s^{-1} K^{-4}. Hence, if we may approximate a star by a black body, the intrinsic or
absolute luminosity (in erg s^{-1}) of the star is just

$$L = 4\pi \sigma R^2 T^4, \tag{2.5}$$

since the surface area of a star of radius R is $4\pi R^2$.

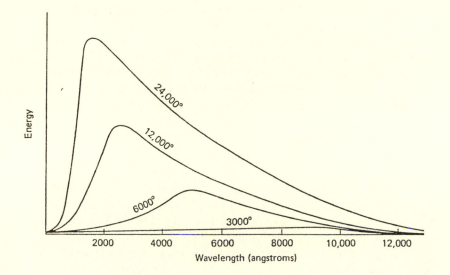

Figure 2.7 Black-body radiation. The curves are drawn according to Planck's radiation law, which gives the amount of energy emitted by an ideal radiator at different temperatures. (The curves are not plotted accurately to scale.) Violet light has a wavelength of about 4,000 Å; blue, 4,200 Å; green, 5,000 Å; yellow, 5,600 Å; and red, 6,800 Å. (One angstrom is equal to 10^{-8} cm.) From W. J. Kaufmann, *Astronomy: The Structure of the Universe* (New York: Macmillan, 1977).

Experiments made at the very end of the nineteenth century showed that for any given temperature the radiation emitted by a black body covers a wide range of wavelengths (see figure 2.7). The experimental curves indicated that for any given temperature there is a particular wavelength, λ_{max}, at which radiation is emitted with maximum intensity. It was also found that this particular wavelength depends inversely on the temperature. Thus, as the material gets hotter, the peak of radiation shifts toward the blue. The experimental results closely verified the law that the German physicist Wilhelm WIEN (1864–1928) had advanced on theoretical grounds in 1893. One has

$$\lambda_{max} T = 0.2898, \tag{2.6}$$

where λ_{max} is given in centimeters when T is given in kelvins. This is known as *Wien's displacement law*. From this law, we see what the observed color of a star can tell us about the star's surface temperature (see, however, section 4.1).

Several attempts were made during the nineteenth century to develop a theory that would explain the shape of the black-body curves obtained in the laboratory and which closely follow those depicted in figure 2.7. All of them were unsuccessful, however, because they were based on the idea that light behaves like a wave. In 1900 the German physicist Max PLANCK (1858–1947) resolved this dilemma by proposing that energy is radiated only in discrete packets called *photons* or *quanta*, whose energy depends on the wavelength of the light. Thence, he was able to prove both Stefan's and Wien's laws, as well as to calculate the intensity of radiation at each wavelength for a given temperature. The agreement between the measurements and *Planck's radiation law* was found to be nearly perfect. Although Planck had to make an assumption that seemed extremely strange to himself and his fellow scientists at the time, the value of his discovery was immense since it paved the way for *quantum mechanics*, a new field of physics whose early applications to solar and stellar spectroscopy were to be worked out between the two world wars.

2.3 THE SUN AS A STAR

The sun is the nearest of the stars, the only one whose disk we can distinguish, and hence the only one on which surface features can be seen easily. In view of its close proximity, more detailed information is therefore available about the sun than any other star. Here we shall briefly summarize some of its gross features, as they were known at the dawn of the twentieth century.

The *radius* of the sun, $R_\odot$, can be obtained from the apparent angular diameter of the solar disk — a little over half a degree — and its known distance from the earth — the astronomical unit. The result is

$$R_\odot = 696,000 \text{ km,} \tag{2.7}$$

or 109 times the earth's radius.

Another parameter of primary importance is the *mass* of the sun. It can be readily calculated from observations of relative motions. Consider the sun, of mass $M_\odot$, and a planet, of mass m, moving about it in a circular orbit of radius a. For reason of symmetry, the speed v of the planet must be constant but its direction is constantly changing. Such a changing velocity represents the centripetal acceleration which maintains the circular motion. The corresponding force, mv^2/a, must therefore equal the force of gravity, $GM_\odot m/a^2$, if the planet is to remain in its orbit. Hence we let

$$\frac{GM_\odot m}{a^2} = \frac{mv^2}{a}, \tag{2.8}$$

where G is the constant of gravitation. If P is the period of revolution of the planet, then one has $v = 2\pi a/P$. Solving eq. (2.8) for $M_\odot$, one readily sees that

$$M_\odot = \frac{4\pi^2}{G}\frac{a^3}{P^2}. \tag{2.9}$$

To obtain the mass $M_\odot$ we must therefore measure the value of the constant G in the first place.[8]

---

[8]The first quantitative estimate of the sun's mass is due to Newton. Since he did not know the value of the constant of gravitation, he compared the motion of a planet, Venus for example, around the sun with the motion of the moon around the earth. In this manner, knowing the orbital periods and the mean distances, he was able to determine the sun-to-earth mass ratio. In the 1687 edition of his *Principia*, Newton used too high a value for the solar parallax, and so found a sun-to-earth mass ratio that was too small by more than a factor of ten. In later editions, published in 1713 and 1726, he made use of a better estimate of the solar parallax and found the mass of the sun to be about 200,000 times greater than that of the earth, which is within a factor of two of the modern value.

This measurement was first performed in 1798 by the English physicist Henry CAVENDISH (1731–1810), who evaluated the force of attraction between two large fixed lead balls of equal mass and two smaller identical balls by observing the twisting of the torsion bar from which the small balls were hanging.[9] He found that $G = 6.7 \times 10^{-8}$ cm^3 g^{-1} s^{-2}. (The accepted value today is $G = 6.672 \times 10^{-8}$ in the same units.) Using the earth as the planet in eq. (2.9), we obtain

$$M_\odot = 2 \times 10^{33} \text{ g}, \tag{2.10}$$

which is about 330,000 times the mass of the earth. Note that this result does not depend on the choice made for the planet since the ratio a^3/P^2 is the same for all planets, as stated by Kepler's third law (see section 1.6).

Another important parameter is the *luminosity* of the sun, that is, the total amount of energy lost — or radiated — by the sun in unit time. This quantity is found by measuring the total radiation falling on a given area of the earth in a given time from direct sunshine, which is known as the *solar constant*. Original measurements were made in 1837 by Claude POUILLET (1790–1868) in France and, independently, by the English astronomer John HERSCHEL[10] (1792–1871) during his expedition to the Cape of Good Hope. Allowance being made for the solar heat stopped in the earth's atmosphere, an estimate was then formed of the total amount of energy radiated by the sun in all directions during one second of time. Making use of modern values for the solar constant, one has

$$L_\odot = 4 \times 10^{33} \text{ erg s}^{-1}, \tag{2.11}$$

which corresponds to a luminosity of 4×10^{23} kW. It follows at once that the energy flux per unit surface of the sun, $L_\odot/4\pi R_\odot^2$, is of the order of 6,500 W cm^{-2}.

Inserting next the values of $L_\odot$ and $R_\odot$ into eq. (2.5), we obtain the *effective temperature* of the sun, that is, the temperature at which a black body of the same surface area would radiate the same total energy as the sun. A straightforward calculation yields

$$T_\odot = 5{,}800 \text{ K}, \tag{2.12}$$

which is a reasonably adequate value of the *surface temperature* of the sun.

Kirchhoff considered the sun to be a hot liquid or solid body, surrounded by a less hot atmosphere containing terrestrial elements in a gaseous state, which produce the Fraunhofer lines. From the obviously high temperature of the sun, however, both

---

[9]The English physicist John MICHELL (1724–1793) was the original inventor of the torsion balance, but he did not live to put his method into practice; this was done by Cavendish, who gave proper credit to Michell.

[10]He was the son of the German-born musician and astronomer William HERSCHEL (1738–1822) who, among other achievements, had discovered in 1800 that there are infrared solar rays. Specifically, by measuring the temperature of various colors in the sun's spectrum, William Herschel had discovered the unequal distribution of heat among these colors, the heating being greatest below the red. Before him no one had suspected such an inequality. Appropriately, the English physicist Thomas Young (1773–1829), in his *Lectures* of 1807, says, "This discovery must be allowed to be one of the greatest that has been made since the days of Newton." In 1801 Johann Wilhelm RITTER (1776–1810) of Silesia made a discovery which was just as interesting, namely, that radiations beyond the violet end of the solar spectrum (the *ultraviolet rays*) produce fluorescence in silver chloride. See F. Cajori, *A History of Physics* (New York: Macmillan, 1929; Dover Publications, 1962), 178–184.

the Jesuit astronomer Angelo SECCHI (1818–1878) and John Herschel concluded in 1864 that the surface of the sun must represent a cloud layer lying on top of an otherwise mainly gaseous body. From eqs. (2.7) and (2.10) one readily sees that the *mean density* of the sun is only 1.41 g cm^{-3}— a value consistent with the fact that the sun is a centrally condensed sphere consisting mainly of gaseous components.

The visible disk of the sun, which is known as the *photosphere* or "sphere of light," is as far into the sun as we can see. (This will be discussed at greater length in section 2.4.) The gases above the photosphere constitute the sun's atmosphere. Its lower part, just above the photosphere, is called the *reversing layer*, and here the Fraunhofer lines are produced. Just above this layer, which is a few hundred kilometers deep, is the *chromosphere* or "sphere of color"; it appears as a red fringe due to hydrogen when seen around the dark disk of the moon during a total solar eclipse. The *corona*, the outer solar envelope, appears as a pearly halo during a total eclipse; it may extend more than a million kilometers from the sun.[11]

Photography was used to make pictures of the sun immediately after its discovery.[12] At the total eclipse of 1860 in Spain, this new technique was first effectively employed to ascertain the objective reality of the chromosphere and of those small pink objects, generally known as *prominences*, that were protruding at different points along the edge of the moon's disk. At the eclipse of 1868 in India, the spectrum of the chromosphere and the prominences was first obtained, showing that they were glowing masses of gas. The brightest lines were red and green emission lines of hydrogen (designated H_α and H_β), and a strong line in the yellow region that had never been observed on the earth. Immediately afterward, Jules JANSSEN (1824–1907), who was one of the observers, and Norman LOCKYER (1836–1920) independently presented a method whereby these lines could be made visible at the edge of the sun's disk in ordinary daylight, without waiting for a total eclipse.[13] The new bright yellow line was at first thought to come from a substance peculiar to the sun and hence was called "helium"; only in 1895 was that chemical element discovered by William RAMSAY (1852–1916) on the earth and found to be actually a new element.

At the eclipse of 1869, which was visible from North America, William HARKNESS (1837–1903) of the Naval Observatory (Washington, D.C.) and Charles Augustus YOUNG (1834–1908), then at Dartmouth College (Hanover, N.H.), discovered in the faint continuous spectrum of the solar corona a narrow green emission line of unknown origin. This and other coronal lines were long ascribed to a new element, called "coronium," until they were shown to come from the atoms of several common

---

[11] For centuries the corona could be observed only during total solar eclipses. It was not until 1930 that the French astronomer Bernard LYOT (1897–1952) invented an ingenious device, the *coronograph*, for recording the faint light of the corona outside eclipses.

[12] The first successful daguerreotype of the sun was made at Paris in 1845 by Hippolyte Fizeau and Léon FOUCAULT (1819–1868); it showed some spots. The use of high-resolution photography for the investigation of sunspots and the solar granulation was initiated by Jules Janssen in the 1870s.

[13] By a strange coincidence, both the Frenchman and the Englishman transmitted the news of their discovery to a meeting of the Académie des Sciences on the same day in 1868. To commemorate the event a medal bearing the portraits of both astronomers was struck by the French government in 1872. Both sides of the medal were reproduced by A. J. Meadows in vol. 4A of *The General History of Astronomy*, edited by O. Gingerich (Cambridge: Cambridge University Press, 1984), 7.

metals in a very high state of ionization (see section 5.8). At the eclipse of 1870 in Spain, Young made another important discovery during the few seconds of the eclipse when the chromosphere is visible. Having thus set the slit of his spectroscope nearly tangent to the sun's limb, he saw the flaring up of a very large number of bright lines. This was known as the *flash spectrum*, because it flashes into view into the spectroscope near the beginning of a total eclipse, reversing the Fraunhofer lines into bright lines. This is why the thin layer lying immediately above the photosphere was called the "reversing layer." In 1877, by then a recognized authority on solar spectroscopy, Young was appointed professor at the College of New Jersey (now Princeton University), a post he held until 1905.

The chromosphere, which owes its color to the hydrogen H_α line in the red part of the spectrum, cannot be directly observed outside a total eclipse. As mentioned, however, with the aid of a spectroscope the chromosphere can be seen just off the sun's limb at any time in full daylight. In the early 1890s, George Ellery HALE (1868–1938) in Chicago and Henri DESLANDRES (1853–1948) in Paris, independently and in slightly different ways, constructed an instrument, called a *spectroheliograph*, to observe the whole solar disk, not in white light, but in the light of a narrow wavelength band such as the strong H_α line or the calcium H and K lines. Photographs of the solar disk in hydrogen light were first obtained at the Yerkes Observatory (Williams Bay, Wis.) in 1903 using the blue and violet spectral lines of hydrogen. Five years later, at the Mount Wilson Solar Observatory (near Pasadena, Cal.), when methods for making photographic emulsions sensitive to red light had been developed, the H_α line was used. These photographs of the solar disk showed bright clouds, known as *plages* or *flocculi*, that were much greater in area than the sunspots around and above which they were always observed. However, it was also found that these plages may appear where no sunspots were visible on the photosphere. Long dark *filaments* were identified as prominences projected against the surface of the sun.

2.4 SOLAR ACTIVITY AND ROTATION

The study of solar activity and rotation began during the 1610s, when sunspots were observed for the first time through a refracting telescope. The first public announcement of an observation came in the fall of 1611 from Johann FABRICIUS (1587–1616/17) in Osteel, the son of Kepler's friend David FABRICIUS (1564–1617). From his observations the young Fabricius correctly inferred the spots to be parts of the sun itself, thus proving its axial rotation.[14] Unfortunately, his work was ignored by his closest competitors, Galileo in Florence and the Jesuit astronomer Christoph SCHEINER (1575–1650) in Ingolstadt. A controversy about the nature of sunspots made Scheiner a bitter enemy of Galileo and developed into a quarrel regarding their respective claims to discovery. Scheiner had been warned by his

---

[14]An English translation of his paper was published by W. M. Mitchell in *Popular Astronomy*, **24**, 149 (1916). Mitchell's paper also presents a brief discussion of the unpublished observations of sunspots made by Thomas HARRIOT (1560–1621) in 1610.

ecclesiastic superiors not to believe in the reality of sunspots because Aristotle's work did not mention them (see section 1.4), and so he announced his discovery in three letters written under the pseudonym Apelles. He explained the spots as being small planets revolving around the sun and appearing as dark objects whenever they passed between the sun and the observer. These views opposed those of Fabricius and Galileo, who claimed that the spots must be on or close to the sun's surface.

In 1613, Galileo made public his own observations in a pamphlet entitled *Istoria e dimostrazioni interno alle macchie solari* (History and demonstrations concerning the solar spots). In these three letters he refuted Scheiner's conclusions and, for the first time, publicly declared his adherence to the Copernican system, thus initiating the whole sad episode of his clashes with the Roman Inquisition (see section 1.6). As was observed by the young Fabricius, Galileo perceived that the changes in the motions of the spots across the solar disk were an effect of foreshortening, which would result if and only if the spots were near the solar surface. He also noticed that all spots moved across the solar disk at the same rate, making a crossing in about fourteen days, and that they all followed parallel paths. Obviously, these features would be highly improbable given the planetary hypothesis, which is also incompatible with the observed changes in the size and shape of sunspots.

The planetary hypothesis, championed by Scheiner among others, was thus convincingly refuted by Galileo. Eventually, Scheiner's own observations led him to realize that the sun rotates with an apparent period of about twenty-seven days. He also showed that the spots farther from the sun's equator moved with a slower velocity. He positively established, therefore, that the sun — at least in its surface properties — could not be a solid. To Scheiner also belongs the credit of the discovery of small bright patches on the surface of the sun, best seen near its edge, and from their brilliancy named *faculae*. Scheiner published his collected observations in 1630 in a volume entitled *Rosa ursina sive sol*, dedicated to the Duke of Orsini who sponsored the work. (The title of the book derives from the badge of the Orsini family, which was a rose and a bear.) This was truly the first monograph on solar physics.

No further advances of importance were made until the second half of the eighteenth century. As we shall see below, this lack of progress may be attributable to the fact that between 1645 and 1715 sunspots and other solar activity had all but vanished from the sun. Then, in 1774, Alexander WILSON (1714–1786) of Glasgow published some observations which seemed to show that sunspots were depressions below the normal surface of the sun. Specifically, he noticed that a regular round spot, on approaching the sun's edge, shows the penumbra to appear widest on the side nearest the sun's edge and narrowest on the opposite side—as if the spot were a saucer-shaped depression, of which the bottom formed the umbra and the sloping sides the penumbra. A similar observation was made at Slough (near London) by William Herschel, who in 1795 constructed an elaborate theory of the nature of the sun. He believed the interior of the sun to be a cold dark solid body, surrounded by two cloud layers, of which the outer was the photosphere and the inner served as a fire screen to protect the interior. The umbra of a spot was the dark interior seen through an opening in the cloud layers, whereas the penumbra corresponded to the inner cloud layer illuminated by light from above. This model was widely accepted during the first half of the nineteenth century, until thermodynamics showed the impossibility of a steady configuration with a cold interior surrounded by layers of hot and luminous clouds.

From the earliest solar observations, it was evident that the number of spots visible on the sun's surface varied from time to time. However, no law of variation was established until 1851, when Heinrich SCHWABE (1789–1875), an apothecary in Dessau, published in Humboldt's *Kosmos* the results of sunspot observations carried out during the previous quarter of a century. He showed that the number of spots visible increased and decreased in a period of about ten years. Later observations have confirmed this result, but the period was estimated as slightly over eleven years on the average. Shortly afterward, in 1852, three independent investigators — Edward SABINE (1788–1883) in England, Rudolf WOLF (1816–1893) and Alfred GAUTIER (1793–1881) in Switzerland — found a relationship between the periodic variations of sunspots and of various magnetic disturbances on the earth. In the 1870s Jules Janssen also noticed that the shape of the sun's outer corona changes with the sunspot cycle: The corona tends to be circular at spot maximum and seems to have axial symmetry near minimum. In the 1890s, however, Gustav SPÖRER (1822–1895) in Germany and Walter MAUNDER (1851–1928) in England called attention to the fact that the sunspot cycle and other signs of solar activity had varied considerably over the past centuries. The period of seventy years ending in about 1715, during which solar activity was almost nonexistent, is generally known as the *Maunder sunspot minimum.*[15]

For more than two centuries the problem of solar rotation was practically ignored, and it was not until the 1850s that any significant advance was made. Then, a long series of observations of the apparent motion of sunspots by Richard Christopher CARRINGTON (1826–1875), a wealthy English amateur, led to a clear recognition of the difference in the rate of rotation at different heliocentric latitudes. From his own observations made during the period 1853–1861, Carrington derived the following expression for the daily angle of solar rotation:

$$\Omega \text{ (deg/day)} = 14\overset{\circ}{.}42 - 2\overset{\circ}{.}75 \sin^{7/4} \phi, \qquad (2.13)$$

where ϕ is the heliocentric latitude. Somewhat later, however, the French astronomer Hervé FAYE (1814–1902) found that the formula

$$\Omega \text{ (deg/day)} = 14\overset{\circ}{.}37 - 3\overset{\circ}{.}10 \sin^2 \phi \qquad (2.14)$$

more satisfactorily represented the dependence of rotation on heliocentric latitude.

Carrington established the scarcity of spots in the immediate vicinity of the sun's equator, as well as their greater scarcity more than 35° from the equator. He was also the first to notice the migration of sunspots to lower latitudes in the course of the eleven-year cycle. That important finding was confirmed by Spörer, who made his own detailed study of the latitude drift of sunspots during the 1860s. This is known as *Spörer's law* (see figure 2.8).

---

[15]Perhaps because the results of Maunder and Spörer came as a surprise, not much attention was paid to them until their papers were rediscovered in the late 1970s. See J. A. Eddy, *Scientific American*, **236**, No. 5, 80 (1977).

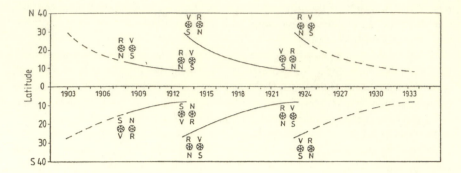

Figure 2.8 Schematic diagram showing the polarity rules for sunspots: N and S stand for north and south, respectively (R and V indicate the red and violet σ components of the Zeeman triplet). The curves show the migration in heliocentric latitude of the sunspot region. From G. E. Hale, *Nature*, **113**, 105 (1924).

Let us also note that on September 1, 1859 Carrington was engaged in his daily monitoring of sunspots when he was suddenly surprised at the appearance of two intensively bright patches of white light. These patches moved together over the surface of a very large spot and finally disappeared five minutes later. Similar observations of this exceptional event were made by his friend Richard HODGSON (1804–1872). Both of them reported that the magnetic instruments at the Kew Observatory were simultaneously disturbed to a great extent. Their papers present the first chance accounts of a white-light *solar flare* in the astronomical literature.[16]

The next step was taken by Hermann Carl VOGEL (1841–1907) in Potsdam. In 1871, by means of a new spectroscope devised by Friedrich ZÖLLNER (1834–1882), he showed that the solar rotation can be detected from the relative Doppler shift of the spectral lines at opposite edges of the solar disk, one of which is approaching and the other receding. Extensive measurements were made visually by Nils DUNÉR (1839–1914) in Lund and Uppsala, and then by Jacob HALM (1866–1944) in Edinburgh. They showed a rotation rate and equatorial acceleration that were quite similar to those obtained from the apparent motion of sunspots. They concluded that Faye's empirical law adequately represented the spectroscopic observations also, but their coverage of latitude was double that of the sunspot measurements. After these early visual observations made during the years 1887–1906, photography almost completely superseded the human eye. The first spectrographic determinations of solar rotation were undertaken in 1907 by Walter Sydney ADAMS (1876–1956) at Mount Wilson.

The spectrum of sunspots was also an object of many investigations. In 1892, Young noticed that some iron lines in the photosphere appeared double in a sunspot. Hale immediately invoked the Zeeman effect that had been recently discovered (see section 2.2). In 1908, at Mount Wilson, he investigated this effect further

---

[16]*Mon. Not. Roy. Astron. Soc.*, **20**, 13 and 15 (1860). Both papers were reprinted by A. J. Meadows in his book *Early Solar Physics* (Oxford: Pergamon Press, 1970).

with the new 60-foot solar telescope and a large spectroheliograph. Making use of a polarization analyzer, he found that the line components in sunspots were indeed polarized, thus confirming the presence of intense magnetic fields there. Hale and his associates also discovered that pairs of spots following one another in the solar rotation showed opposite magnetic polarities, and that the disposition of polarities was opposite for the northern and southern hemispheres. However, after the sunspot minimum of 1912, it appeared that the polarities of the two hemispheres had interchanged. A magnetic reversal occurred again after the minimum of 1923, and so it was concluded that, magnetically, the real period of the sunspot cycle was not eleven but twenty-two years (see figure 2.8). This was the beginning of several studies in solar magnetism.[17]

Spectrographic observations have also been used to measure the velocity fields found in sunspots. The largest components of these motions were discovered in 1909 by the English astronomer John EVERSHED (1864–1956), who was then working at the Kodaikanal and Madras Observatories in India. He observed that in the photospheric layers the motion in the penumbra of a spot is predominantly radial and outward, away from the central dark umbra. In the overlying chromospheric layers, however, one finds that the radial component of the motion is directed inward, toward the umbra. In 1913, at Mount Wilson, extensive work on the displacement of Fraunhofer lines in the spectra of sunspots was conducted by Charles Edward ST. JOHN (1857–1935), who confirmed Evershed's discovery concerning these inward and outward motions. This is known as the *Evershed effect*.

2.5 INTRINSIC PROPERTIES OF STARS

Hipparchus, who flourished during the second century B.C., is credited with preparing the first star catalog. For convenience, he divided all of the stars visible to the naked eye into six grades of brightness, or *magnitudes*, numbered from 1 to 6, with the first-magnitude stars the brightest. Later, in the second century A.D., Ptolemy adopted this system, which remained the dominant one until the middle of the nineteenth century. Then, photometric measurements showed that the average first-order magnitude star is about 100 times brighter than the average sixth-order magnitude star. In 1856 Norman Robert POGSON (1829–1891) at Oxford proposed that a standard ratio of $\sqrt[5]{100} = 2.512$ be adopted between any two adjacent magnitudes, so that a difference of five magnitudes would correspond exactly to a ratio in brightness of one hundred. However, to remain consistent with extant catalogs, Pogson's scale was adjusted to agree with the old one at the sixth magnitude, near the limit of visibility. The difference between the magnitudes of two stars is thus proportional to the logarithm of the ratio of their respective brightnesses. Specifically, if two stars have magnitudes m_1 and m_2, and if their brightnesses are l_1 and l_2, one has

$$\frac{l_2}{l_1} = 10^{0.4(m_1-m_2)} = 2.512^{(m_1-m_2)} \tag{2.15}$$

---

[17] See H. Wright, J. N. Warnow, and C. Weiner, eds., *The Legacy of George Ellery Hale* (Cambridge, Mass.: MIT Press, 1972); W. Sheehan, and D. E. Osterbrock, *J. History Astron.*, **31**, 93 (2000).

or[18]

$$\log_{10} \frac{l_2}{l_1} = -0.4(m_2 - m_1). \tag{2.16}$$

On this scale, the star Sirius has a magnitude of -1.4, the full moon of -12.5, and the sun of -26.8.

The apparent magnitude of a star depends on its intrinsic or absolute luminosity L and its distance d. Since we know that the illumination from a point source of light falls off inversely as the square of the distance from its source, we can thus calculate the absolute luminosity of a star, once we have measured its apparent magnitude m and its parallax p''. Following current practice, we shall define the *absolute magnitude* of a star as the magnitude it would have if seen at an adopted standard distance of 10 pc. If L is the absolute luminosity at a distance of $1/p''$ pc, then by the inverse-square law and eq. (2.15), we can write

$$\frac{L}{l} = \frac{(1/p'')^2}{10^2} = 10^{0.4(m-M)}, \tag{2.17}$$

where m and M are the apparent and absolute magnitudes of the star. Taking the logarithm of eq. (2.17) and rearranging terms, one finds that

$$M = m + 5 + 5\log_{10} p''. \tag{2.18}$$

As before, the ratio of the absolute luminosities for two stars is

$$\log_{10} \frac{L_2}{L_1} = -0.4(M_2 - M_1). \tag{2.19}$$

Since the sun's apparent magnitude is -26.8 and its parallax $206,265''$, it follows from eq. (2.18) that its absolute magnitude is 4.8. The sun would thus appear as a fifth-magnitude star if seen at a distance of 10 pc.

Before the advent of photography, the apparent brightness of stars was measured by eye, in conjunction with photometers designed for accurate comparison of brightness. These magnitudes came to be called *visual magnitudes*. Since the human eye is most sensitive to yellow-green light, visual magnitudes thus express the brightness of stars in that range of radiation. The early photographic plates recorded the brightness of stars in the blue-violet region of the spectrum. The magnitudes so determined were called *photographic magnitudes*. Following a suggestion made by the German astronomer and physicist Karl SCHWARZSCHILD (1873–1916; figure 2.9) in 1899, the *color index* of a star was generally defined as the difference between the photographic magnitude and the visual magnitude. The color index is then a numerical expression of the star's color. Because the magnitude scale is such that smaller numbers denote brighter stars, the color index is positive for red stars and negative for blue stars. As was noted in section 2.2, the color of a star depends on the temperature of its surface. Hence, the color index can be used directly to

---

[18] If the factor of -0.4 in Eq. (2.16) had been replaced by $+1$ or $+10$, the astronomical magnitude scale would have been defined exactly as the *bel* or the *decibel* in acoustics. But astronomers are a very conservative lot and neither of these solutions was deemed acceptable. See O. Lodge, *Nature*, **106**, 438 (1920).

Figure 2.9   Karl Schwarzschild, versatile German astronomer and physicist, who improved
photographic photometry, invented new optical systems, investigated the properties of stellar atmospheres, and explored Einstein's 1915 theory of gravitation.
At the outbreak of World War I, aged forty, he volunteered for the German army.
After serving in Belgium and France, he was sent to the Eastern Front, in Russia,
where he contracted a fatal skin disease. He wrote his last paper in a hospital,
just before his death in May 1916, applying Bohr's 1913 theory to spectral lines.
In 1960 the Berlin Academy honored him as the greatest German astronomer of
the last hundred years. Courtesy of Yerkes Observatory.

find the temperature at a star's surface by application of Wien's law, provided that
passage through interstellar dust has not reddened the starlight. It is a crude index
of temperature, however, because the stars show numerous absorption lines in their
spectra and so do not radiate exactly like black bodies (see also section 4.1).

Stellar spectra were first investigated visually in the 1810s by Fraunhofer, who
noticed that some of them displayed dark lines clearly different from those found in
the solar spectrum. Later, in 1863, Huggins and his friend William Allen MILLER
(1817–1870) of King's College, London, studied visually the spectra of a number
of bright stars and identified sodium, iron, calcium, and magnesium in these stars.
They therefore confirmed the widely held belief that the same elements are present
in the stars as in the sun and on the earth. The 1860s turned up also a variety of
rare objects with emission lines. As early as 1864 Huggins and Miller observed the

Figure 2.10 Edward C. Pickering, as director of Harvard College Observatory from 1877 to 1919, hired many women as human computers, cataloging and classifying thousands of stellar spectra. It was a reflection of the times that the group of women working at Harvard were nicknamed "Pickering's Harem." Courtesy of Dorrit Hoffleit, Yale University Observatory.

spectrum of a *planetary nebula* (NGC 6543 in Draco) and found that it has emission lines typical of a large mass of hot luminous gas.[19] This result was quickly followed by a similar conclusion for the great Orion nebula and several other bright diffuse nebulae. Especially noteworthy were a pair of unidentified green lines in their spectra and ascribed to the mystery element "nebulium." These strong nebulium lines presented a puzzle for many years, until they were shown to come from doubly ionized oxygen at very low densities (see section 5.8). In 1867 another type of object was found by the French astronomers Charles WOLF (1827–1918) and Georges RAYET (1839–1906), who reported finding three faint, eighth-magnitude

---

[19]The expression "planetary nebula" is a misnomer because these objects are not related to planets. In fact, they are expanding shells of rarefied gas around an extremely hot star. As in the Ring nebula in Lyra, a planetary nebula may appear as a ring because we are viewing it through the thin dimension on the front and rear parts of the expanding shell.

stars in Cygnus with very broad emission lines on a continuous background. These were the first examples of what are called *Wolf-Rayet stars*.

From 1863 to 1868, Father Secchi at the Specola Vaticana in Rome examined visually the spectra of over 4,000 stars and found that they could be classified into four groups that had features in common. Major progress was made in 1872, when Henry DRAPER (1837–1882) of New York City succeeded for the first time in photographing a stellar spectrum.[20] Later, in the 1880s, Edward Charles PICKERING (1846–1919; figure 2.10) instituted at Harvard College Observatory (Cambridge, Mass.) a photographic survey of stellar spectra. The spectra of the brightest stars were investigated by Antonia MAURY (1866–1952) who, starting with Secchi's four classes, found that she could arrange all spectra in a continuous series. The most original feature of her work was the further subdivision of the spectral classes, labeled from *a* to *c*, based on the appearance of certain lines, whether hazy or sharp. Annie Jump CANNON (1863–1941) continued this work, improved the classification, and developed the spectral types essentially in the form used today. (Maury's "collateral divisions" based on line width were not retained, however.) In 1922 the International Astronomical Union accepted the Harvard classification, named after Henry Draper, as the official system.[21]

Toward the end of the nineteenth century, the Harvard group classified the spectra of stars according to the strengths of the hydrogen absorption lines, and the spectral ordering was alphabetical, A through Q, omitting J. Some letters were eventually dropped and the ordering was modified because the letters originally assigned did not arrange in a physically continuous series. In the Harvard classification, as developed by Cannon and her associates (figure 2.11) around 1920, the *spectral types* are listed in order of decreasing temperature as follows: O, B, A, F, G, K, M. The vast majority of stars can be classified into one of these seven categories based on the appearance of the spectrum. The O-type stars are hot and blue, with temperatures around 30,000 K or more; the A-type stars look white to the eye, with temperatures around 10,000 K; and the M-type stars are cool and red, with temperatures around 3,000 K or less. Note that each spectral type is further subdivided into ten parts from 0 to 9 for finer distinctions. In that scheme, the sun is designated type G2, with a surface temperature of about 5,800 K.

So far we have discussed, admittedly in rather broad terms, the absolute luminosities, colors, spectral types, and atmospheric temperatures of stars. One of the most significant graphs ever plotted by astronomers is the *Hertzsprung-Russell diagram*, also called the H-R diagram, which is simply a plot of the absolute magnitudes

---

[20]Thirty years earlier, in 1842, the French physicist Edmond BECQUEREL (1820–1891) had successfully photographed the whole solar spectrum with nearly all the Fraunhofer lines. This experiment in photography led him to the discovery of numerous absorption lines in the ultraviolet region of the solar spectrum, which extended far beyond the visible limit. A few months later, John William Draper, Henry Draper's father, repeated Becquerel's work. His solar spectrum, which he published as a drawing in 1843, extended further into the red than Becquerel's.

[21]For a detailed historical account, see B. Z. Jones and L. G. Boyd, *The Harvard College Observatory* (Cambridge, Mass.: Harvard University Press, 1971). See also P. E. Mack, in *Women of Science: Righting the Record*, edited by G. Kass-Simon and P. Farnes (Bloomington: Indiana University Press, 1990), 72–116.

Figure 2.11 A group of staff members at the Harvard College Observatory circa 1925. Antonia C. Maury is first from the right in front, Annie J. Cannon is second from the left in the middle, and Cecilia H. Payne (later Payne-Gaposchkin) is second from the left in the rear. Courtesy of Harvard College Observatory.

of stars against their spectral types (see figure 2.12). Graphical plots were originally published in 1911 by Hans ROSENBERG (1879–1940) at Göttingen and Ejnar HERTZSPRUNG (1873–1967; figure 2.13) at Potsdam. These diagrams, the first unpublished version of which Hertzsprung had produced as early as 1908, depicted the relationship of apparent brightness to effective wavelength (as a measure of color) for stars in the Pleiades star cluster. (Since all the stars in a cluster are at about the same distance from us, it makes no difference whether one uses absolute or apparent magnitudes: only the origin of the magnitude scale is changed.) They found that the dots representing the stars did not scatter around the diagram, but in fact seemed to fall along a diagonal line. Hertzsprung's 1911 paper also gave the color-magnitude diagram for the Hyades star cluster, showing a "main sequence" of hot stars as well as four luminous red stars.[22] In 1914 the American astronomer Henry Norris RUSSELL (1877–1957) published a similar plot, this time of absolute magnitude versus spectral type for every star for which these two quantities were known.

Actually, the Irish amateur astronomer William MONCK (1839–1915) was the first to point out, in 1892, that stars of different spectral types have differences in proper

---

[22]See H. Rosenberg, *Astron. Nachr.*, **186**, 71 (1911); E. Hertzsprung, *Publ. Astrophys. Observ. zu Potsdam*, No. 63 (1911).

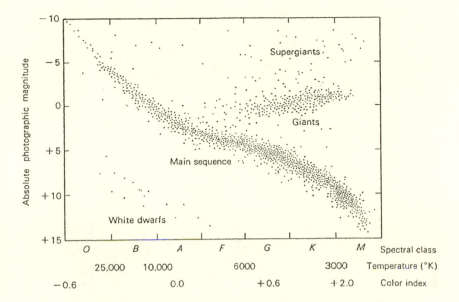

Figure 2.12 The Hertzsprung-Russell diagram. A typical observational H-R diagram of the
nearest and brightest stars, showing absolute magnitude versus spectral type,
temperature, or color index. Each dot represents a star. From G. O. Abell,
Exploration of the Universe, 3rd edition, ©1975. Reprinted with permission of
Brooks/Cole, a division of Thomson Learning: www.thomsonrights.com; FAX
800 730-2215.

motion, and hence also in absolute luminosity. In 1895 he further suggested that
there may be two different classes of yellow stars, one being dull and near us (like
α Centauri and Procyon) and the other bright and remote (like α Aurigae). About
ten years later, Hertzsprung, who was still an amateur astronomer while working
as a chemical engineer in Copenhagen, analyzed the proper motions of those stars
with unusually narrow lines and designed by Maury as "division *c*" in her pioneering classification. He found that the A-type and Orion-type stars all had small proper
motions and so were all of a uniformly high luminosity. For the later spectral types,
however, he noticed that Maury's division-*c* stars had very small proper motions and
parallaxes compared with other stars of the same color. He rightfully concluded,
therefore, that there were two different populations among the yellow, orange, and
red stars. This sharp dichotomy in the late-type stars between those of high and
low luminosity, as well as the marked spectral differences arising from luminosity, were properly established by Hertzsprung in 1907.[23] But his international repu-

---

[23] Soon afterward, as his local reputation grew, Hertzsprung was called to Göttingen as professor
of astronomy and later moved with his friend Karl Schwarzschild to Potsdam, when the latter became
the director there. After World War I, Hertzsprung moved to the Netherlands. There he was appointed
assistant director of the Leyden Observatory in 1919 and became director in 1935; he resigned in 1945.
He then retired to Denmark but continued measuring plates into his nineties.

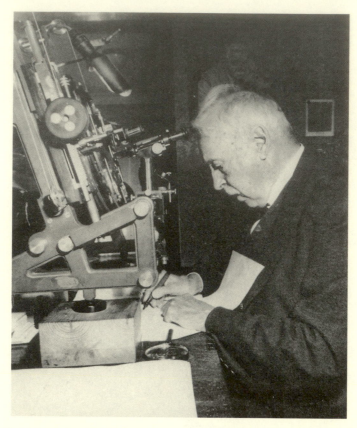

Figure 2.13 The Danish astronomer Ejnar Hertzsprung, in his later years, at work measuring a double-star photograph. In fact he was educated as a chemical engineer and never took an astronomy course. Yet he was the first to recognize, in 1907, the existence of giant and dwarf stars. This work was his springboard into professional astronomy, which brought him from Copenhagen to Göttingen, Potsdam, and Leyden; but he never forgot that he was a Dane, returning to his native country for two decades of retirement. Hertzsprung was primarily an observer, and his main preoccupation was searching for sound observational facts about stars. He has been compared to Tycho Brahe and Bessel. From A. V. Nielsen, *Centaurus*, **9**, 219 (1963); photo by H. Clausen, Roskilde.

tation was not yet established since, apparently oblivious of Hertzsprung's work, Russell from Princeton University was also led to conclude, in 1910, that the redder stars from spectral type G onward fall into two different populations of high and low luminosity stars. In any event, there is no doubt that the empirical relationship between the spectra or colors of stars and their absolute magnitudes was independently discovered by the Dane and the American well before it was ever plotted into a diagram. For many years, however, Hertzsprung's 1911 diagram for clusters continued to be relatively neglected, while Russell's 1914 plot of absolute magnitude versus spectral type became widely known as the "Russell diagram." In fact, it was not until the 1930s that several European astronomers popularized the term "Hertzsprung-Russell diagram," which is in common use today.[24]

At this juncture it is worth noting that the wide range in absolute luminosities for the late-type stars was also discussed in 1914 by Walter Adams and the German astronomer Arnold KOHLSCHÜTTER (1883–1969) at Mount Wilson. Among other results, they found that the strengths of certain lines in the spectra of stars correlated closely with the luminosities of those stars, thereby allowing their absolute magnitude, and hence their distance, to be deduced from the strengths of these lines. Admittedly, their new method of determining the distance of stars from the intensities of their spectral lines was purely an empirical one and had to be calibrated with stars of known trigonometric parallax and absolute magnitude. Yet, the *spectroscopic parallax*, as the new method was called, enabled the distances of stars to be determined that were far beyond the possibility of direct measurement. Actually, the range of these measurements was increased from about 10^2 pc — the upper limit of application of the trigonometric method — to about 10^3 pc.

Figure 2.12 shows a midcentury version of the H-R diagram based on many decades of observation. We note the remarkable tendency of stars to fall into three groups, designated as main-sequence stars, red giant stars, and white dwarf stars. The *main sequence* comprises the majority of all stars, including the sun. It extends from bright blue stars in the upper left-hand corner to dim red stars in the lower right-hand corner. The *red giants* are above and to the right of the main sequence. Since they are much brighter than main-sequence stars with comparable temperatures, eq. (2.5) implies that they are also much larger, with radii ranging from ten to several hundred times the sun's radius. These stars are giant in size as well as in brightness. The most extreme ones are called *supergiants*. As we shall see in section 4.5, the *white dwarfs*, which lie below the main sequence, are much smaller than the main-sequence stars. White dwarfs are typically about the size of the earth.

In summary, we have seen that the photospheric temperatures of the vast majority of stars range from 3,000 to 30,000 K. However, when taking the most extreme values into account, one finds that

$$2,000 \text{ K} \lesssim T \lesssim 100,000 \text{ K}; \tag{2.20}$$

the sun's value at 5,800 K is not near either extreme. The radii of ordinary stars range from $0.1 R_\odot$ up to hundreds of solar radii. Again taking extreme limits, one has

$$10^{-3} R_\odot \lesssim R \lesssim 10^{+3} R_\odot, \tag{2.21}$$

where $R_\odot$ is given in eq. (2.7). (Neutron stars, which were discovered during the twentieth century, are outside this range.) Deneb and Rigel, the most luminous stars in the sky, are almost 10^5 times as bright as the sun in absolute luminosity, and almost 10^{10} times as bright as the faintest star known. The full range of stellar luminosities is truly amazing:

$$10^{-6} L_\odot \lesssim L \lesssim 10^{+6} L_\odot, \tag{2.22}$$

---

[24]For a detailed historical review of the discovery of the H-R diagram, see D. H. DeVorkin, *Physics Today*, **31**, No. 3, 32 (1978). See also D. H. DeVorkin, *Henry Norris Russell* (Princeton, N.J.: Princeton University Press, 2000).

where $L_\odot$ is given in eq. (2.11). The masses and densities of the stars are discussed in the next section.

2.6 BINARY STARS AND STELLAR MASSES

One of the most notable of William Herschel's discoveries was a by-product of his unsuccessful search for stellar parallaxes. In the 1780s, following Galileo's original suggestion, he tried to measure the parallax of a star by observing the variations in its angular distance from some faint star close to it, which from its faintness was supposed to be much further off. William Herschel tended at first to regard double stars as *optical pairs*, i.e., as two independent stars seen along the same line of sight. His instruments were not sufficiently sensitive to detect parallactic motions, but in 1803 he found some double stars that showed a progressive change in the direction of the line joining their two components, clearly indicating that in each of these *visual binaries* the two components were revolving round one another. However, it was not until 1827, five years after William Herschel's death, that the French theoretician Félix SAVARY (1797–1841) had accumulated enough data to calculate accurately the orbit of the binary star ξ Ursae Majoris, showing that both components were revolving in ellipses about their common center of gravity with a period of 58 years.[25] In the 1830s William Herschel's son, John Herschel, set up a temporary observatory near the Cape of Good Hope, where he literally swept the southern skies and produced an extensive list of double and multiple stars. Meanwhile, in the northern hemisphere, many double stars were discovered and studied by several observers, among them Wilhelm Struve and his son Otto STRUVE (1819–1905), who cataloged thousands of visual binaries at the Pulkovo Observatory, near St. Petersburg. These stars are observed to have orbital periods ranging from about one year to thousands of years.

Some double stars have companions too faint to be detected with existing telescopes. Fortunately, the presence of an unseen companion can sometimes be detected by the gravitational effect that it has on the brighter star, causing the latter to move along a wavelike path in its proper motion across the sky. Bessel noticed in 1844 that both Sirius and Procyon had a slightly wavy motion superimposed on their linear proper motions. He correctly concluded that in both cases this secondary motion was caused by the action of an unseen companion, and estimated the orbital periods of both binaries at about 50 years. In 1862, while testing a new 18 ½-inch objective, the American optician Alvan Graham CLARK (1832–1897) discovered the predicted but hitherto unseen companion of Sirius. Several decades later, the unseen companion of Procyon was also detected visually. Such double stars are called *astrometric binaries*. The low-luminosity companions of Sirius and Procyon have proved to be of particular interest in astrophysics because they are the first two known examples of white-dwarf stars.

---

[25] In 1832 the German astronomer Johann Franz ENCKE (1791–1865) published a different method of solution which was somewhat better adapted to the needs of the practical astronomer; John Herschel quickly followed with a third method in 1833.

In a visual binary, the two components attract one another as point masses and move about one another in elliptical orbits whose orientation remains fixed in space. The importance of these binaries lies in the fact that they permit the determination of *stellar masses* when the parallax is known. If P is the orbital period, in years, and a the semimajor axis, in astronomical units, the sum of the masses, in solar units, is given by the relation

$$M_1 + M_2 = \frac{a^3}{P^2};$$ (2.23)

a simple geometrical argument shows that

$$a = \frac{a''}{p''},$$ (2.24)

where a'' is the angular semimajor axis of the orbit, in seconds of arc, and p'' is the parallax, in the same units. If the mass M_2 is much smaller than the mass M_1, one readily sees that eq. (2.23) reduces to Kepler's third law, as given in other units in eq. (2.9). In the general case, the sum of the masses, $M_1 + M_2$, is all that can be determined from eqs. (2.23) and (2.24). However, if the displacements of both components can be observed with respect to several nearby stars, then we can determine the individual masses, M_1 and M_2, since their ratio is inversely proportional to the ratio of the distances of the two components at any time from their common center of gravity. Sirius consists of a pair of stars for which the individual masses can be determined. Sirius A is an A-type main-sequence star of about two solar masses, and Sirius B is a white dwarf of about one solar mass.

In many cases the components of a double star are so close together that they cannot be seen as two distinct stars by a telescope. Yet they may sometimes be identified as a binary system by inspection of their combined spectrum, when the latter shows periodic Doppler shifts in its spectral lines as the two components approach and recede from us in the course of their revolutions. A binary system thus identified is called a *spectroscopic binary*. The brighter star of the visual pair Mizar, generally called Mizar A, and β Aurigae were the first two spectroscopic binaries to be detected. The binarity of Mizar A was found by Pickering in 1889, when he noticed that the spectral lines of this star were double in some photographs and single in others. The same year, 1889, Maury found a similar doubling in the spectrum of β Aurigae. They correctly deduced that the Doppler effect caused the doubling and they interpreted the changes in terms of orbital revolution. At a much later time, the orbital periods of Mizar A and β Aurigae were fixed at 20.5 days and 4 days, respectively. Typical periods for spectroscopic binaries range from hours to a few months. They are important because they include stars of high luminosity and permit the determination of masses for stars on the upper part of the main sequence.

Of particular interest is the special case in which the line of sight lies nearly or exactly in the orbital plane of a binary star, because then one component periodically eclipses the other as seen from the earth. The best known *eclipsing binary* is the bright star Algol, which undergoes periodic changes in its apparent brightness. Its variation in light had already been noted in 1669 by Geminiano MONTANARI (1632–1687) at Bologna, but the strict periodicity of its changes was first detected

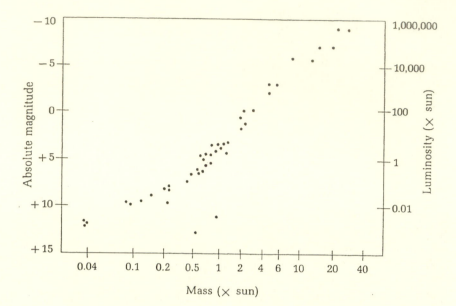

Figure 2.14 The mass-luminosity relation. A typical observational diagram showing that
there is a direct relationship between the masses of main-sequence stars and their
absolute luminosities. The two isolated dots represent white dwarfs. From W. J.
Kaufmann, *Astronomy: The Structure of the Universe* (New York: Macmillan,
1977).

in 1782 by the English amateur astronomer John GOODRICKE (1764–1786), who
was able to fix its period at 2.87 days. He also made the bold suggestion that Algol
might be partially eclipsed by a darker companion with this periodicity. Goodricke's
assumption was eventually confirmed in 1889, when Vogel found that Algol was
also a spectroscopic binary, with a period equal to that of its light variation.

One of the striking features of Algol is the sharpness of its light minimum, and
its nearly constant brightness between eclipses. In 1784 Goodricke found another
eclipsing binary, β Lyrae, whose brightness varies continuously between eclipses,
with a period of 12.94 days. The contrast between these two eclipsing systems is a
result of differences between their components. In β Lyrae the components are so
close together that mutual gravitational attraction distorts their figures into lemon-
shaped objects, whereas the components of Algol lie farther apart and so undergo
much less tidal distortion. Another most unusual case was originally discussed by
William Herschel in 1785; it is the star ι Bootis B, which exhibits light variations
with a period of 0.27 days. It is the first *contact binary* that was discovered, and
it belongs to the class of stars that we now call W Ursae Majoris systems. As was
originally shown by Russell in 1912, the eclipsing binaries are important because
they provide information on the densities, radii, and masses of the stars, as well as
on their internal structure.

In this section we have shown that it is possible to learn about the masses of stars
from observing binaries. Taking extreme limits, one has

$$0.05 M_\odot \lesssim M \lesssim 50 M_\odot \qquad (2.25)$$

for all stars, including the giants and white dwarfs. However, in examining the masses of main-sequence stars alone, one finds that their masses are directly related to their absolute luminosities: stars on the upper part of the main sequence have high masses, whereas stars on its lower part have low masses.[26] Figure 2.14 illustrates this *mass-luminosity relation*, masses ranging from about one-tenth to fifty times that of the sun. (As we shall see in section 5.5, the main-sequence components of contact binaries follow a distinct mass-luminosity relation.) Together with the H-R diagram, the mass-luminosity relation provided astrophysicists at the beginning of the twentieth century with the basic information they needed to start a meaningful study of stellar structure and evolution.

Having obtained the masses and radii of stars, we are now in a position to calculate the *mean density* of a star. In section 2.3 we showed that the sun's mean density $\bar{\rho}_\odot$ is 1.41 g cm^{-3}. One can show that stars along the main sequence have a general tendency to increase in mean density from O-type stars to M-type stars, ranging approximately from $0.05\bar{\rho}_\odot$ to $5\bar{\rho}_\odot$. In contrast, supergiants and giants have mean densities ranging from $10^{-7}\bar{\rho}_\odot$ to $10^{-4}\bar{\rho}_\odot$, whereas the mean density of a white dwarf may exceed $10^6\bar{\rho}_\odot$. Thus, although the stars have only a 1,000-to-1 range in mass, the range of radii is 10^6 to 1 and the range of mean density exceeds 10^{12} to 1.

2.7 VARIABLE AND UNUSUAL STARS

There are a number of stars in the sky that exhibit changes in their brightness.[27] A remarkable *variable star* is that known as Mira Ceti, "the wonderful." David Fabricius of East Friesland (Germany) saw it disappear in 1596, and Johann Phocylides HOLWARDA (1618–1651) of Friesland (the Netherlands) first clearly recognized its variable character in 1638.[28] The French astronomer Ismaël BOULLIAU[29] (1605–1694) in 1667 fixed its period at about 334 days, although it was found that its variations were somewhat irregular both in magnitude and in period. More importantly, Boulliau was the first to suggest that the variability of the light of Mira Ceti

---

[26]The existence of a correlation between the absolute magnitudes of the stars and their masses was originally suggested in 1911 by the German-born astronomer Jacob Halm, who was then working at the Cape of Good Hope Observatory, Cape Town (*Mon. Not. Roy. Astron. Soc.*, **71**, 610 [1911]). The mass-luminosity relation was further discussed by Russell in 1913 and by Hertzsprung in 1918.

[27]The search for variable stars was not made systematic until 1844, when the German astronomer Friedrich Wilhelm ARGELANDER (1799–1875) called for their observation. The number of known variable stars, which was less than 20 by 1844, was already reaching 4,000 in the 1910s, and was exceeding 20,000 in the 1970s. See, e.g., G. D. Roth, in vol. 1 of *Compendium of Practical Astronomy*, edited by G. D. Roth (Berlin: Springer-Verlag, 1994), 425.

[28]There is no ancient text that explicitly talks about variable stars. According to the Assyriologist Johann Schaumberger, however, the variability of Mira Ceti had probably already been noticed by Babylonian astronomers; some cuneiform texts speak of the constellation to which Mira Ceti belongs in terms which mean "to become visible and brighten up" and "fade away like a torch." See J. Schaumberger, *3. Ergänzungsheft zu F. X. Kugler, Sternkunde und Sterndienst in Babel* (Münster: Aschendorffschen Verlagsbuchhandlung, 1935), 350–352.

[29]Also known as BULLIALDUS or BOUILLAUD. See H.J.M. Nellen, *Ismaël Boulliau (1605–1694)* (Amsterdam: Holland University Press, 1994).

might be a direct consequence of axial rotation, the rotating star showing alternately its bright (unspotted) and dark (spotted) hemispheres to the observer. This idea was popularized by Bernard le Bovier de FONTENELLE (1657–1757) in his book *Entretiens sur la pluralité des mondes* — a highly successful introduction to astronomy that went through many revised editions and translations during the period 1686–1742. To be specific, he noted "that these fixed stars which have disappeared aren't extinguished, that these are really only half-suns. In other words they have one half dark and the other lighted, and since they turn on themselves, they sometimes show us the luminous half and then we see them sometimes half dark, and then we don't see them at all."[30] Boulliau's original idea was long forgotten when, in 1852, Rudolf Wolf at Bern noticed similarities between the newly discovered sunspot cycle and the irregular behavior of long-period variables; and so Wolf suggested again that their variability in light might be due to the rotational modulation of starspots. As we shall see in subsequent chapters, however, this explanation of the long-period variables did not withstand the passage of time.

The *long-period variables*, which are easily recognized by their large ranges of light, are the most common variable stars. About four hundred had already been recorded in the 1910s. Their large ranges of light variation average 5 magnitudes and may even exceed 10 magnitudes in visual brightness (see figure 2.15). Their periods also have a large spread, from 90 days up to 700 days; and their spectra are those of the coolest giant and supergiant stars. The physical nature of the long-period variables remained a topic of controversy well into the twentieth century, and it was not elucidated until stellar evolution theory had firmly taken shape.

Another important type of star that undergoes periodic variations in brightness is the *Cepheid variable*. The first known example was discovered in 1784 by Goodricke's friend Edward PIGOTT (1753–1825), who noticed that η Aquilae varied in brightness with a period of 7 days, and that it increased in brightness more rapidly than it decreased. Soon afterward, Goodricke found that δ Cephei was also a variable star, with a period of about 5 days, and that it exhibited light variations similar to those of η Aquilae. The Cepheid variables, which take their name from the prototype δ Cephei, are yellow supergiants. Their periods range from one day to several weeks and are most frequently observed around 5 days. The visual range of their light variation is often around 1 magnitude. At the end of the nineteenth century, however, Solon Irving BAILEY (1854–1931) of Harvard College Observatory observed regular brightness variations over periods less than one day in numerous stars belonging to globular clusters. Such stars were called *RR Lyrae stars* after the prototype that was discovered outside a globular cluster by his colleague Williamina FLEMING (1857–1911) in 1901. Six years later, Fleming published a study of 222 variable stars she had discovered. Observation of RR Lyrae stars in the same cluster shows that, unlike the Cepheid variables, they all have about the same mean absolute magnitude, which is close to +0.6 for all clusters. The RR Lyrae stars are blue giants varying in periods around half a day and with ranges up to 1 magnitude or more. They are the

---

[30]Bernard le Bovier de Fontenelle, *Conversations on the Plurality of Worlds*, translation of the 1686 edition by H. A. Hargreaves (Berkeley: University of California Press, 1990), 70. See also P. Brunet, *L'introduction des théories de Newton en France au XVIIIe siècle* (Paris: Albert Blanchard, 1931; Geneva: Slatkine Reprints, 1970), 223–228.

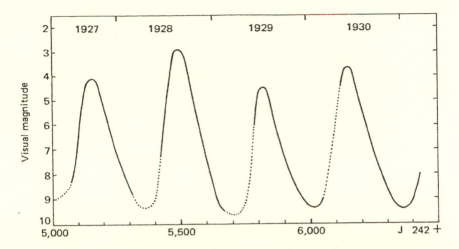

Figure 2.15 The visual light curve of Mira Ceti, "the wonderful." The lower scale represents Julian days (i.e., the number of days that have elapsed since Greenwich mean noon of an arbitrary zero day). The dotted portions are interpolations when the star is unobservable. Mira Ceti is a red supergiant at least ten times as massive as the sun, and its radius is seven hundred times the sun's radius. The average period of the light variation is 330 days. The greatest brightness ranges from the third to the fifth visual magnitude and the least brightness from the eighth to the tenth magnitude, where the star is invisible to the naked eye. From D. H. Menzel, F. L. Whipple, and G. de Vaucouleurs, *Survey of the Universe* (Englewoods Cliffs, N.J.: Prentice-Hall, 1970).

second most common variety of variable stars. By contrast, the Cepheid variables, which are much brighter than the RR Lyrae stars, are considerably rarer. As is the case for the long-period variables, progress in understanding the physical nature of these variable stars was quite slow, and it was not until the 1950s that an adequate explanation for their periodic changes in light was found.

The Cepheid variables are important because a relation has been found between their periods and their mean absolute luminosities. This relation, called the *period-luminosity relation*, was originally obtained by Henrietta Swan LEAVITT (1868–1921; figure 2.16) by studying Cepheid variables in the Small Magellanic Cloud, which was then known as a large aggregation of stars in the southern hemisphere. In 1908, using plates taken from the Harvard observation station in Peru, she found that the brightest stars had the largest periods. By 1912 she had increased the number of stars in her sample from sixteen to twenty-five with the same result. Since the cloud was very distant, she rightfully concluded that all the stars in the cloud were at very similar distances from us, so that differences in apparent magnitude were the same as differences in absolute magnitude. Leavitt had therefore demonstrated that the periods of the Cepheid variables are correlated with their mean absolute luminosities. Accordingly, once the absolute magnitudes of the closest Cepheid variables had been obtained, the period-luminosity relation could be used to determine the distance of any Cepheid variable from its measured period. Hertzsprung in 1913

Figure 2.16 Henrietta Swan Leavitt, who made the discovery in 1912 that a group of variable stars in the Small Magellanic Cloud show a correlation between their periods and apparent brightnesses. Leavitt's period-luminosity relation has become a prime tool in surveying the Milky Way and measuring the distance to all the nearer galaxies. Courtesy of Harvard College Observatory.

was able to calibrate the absolute magnitude scale of the Cepheid variables. This scale was recalibrated in the late 1910s by Harlow SHAPLEY (1885–1972), who was then on the staff of the Mount Wilson Observatory[31] (see, however, section 5.6). This enabled him to estimate the distance of globular clusters and hence to exhibit the true dimensions of the Milky Way, our galaxy. Later, in 1924, Edwin Powell HUBBLE (1889–1953) observed Cepheid variables in the Andromeda nebula using the Mount Wilson 100-inch reflector, and this enabled him to prove that it was an object outside the Milky Way. It was thus mainly through Leavitt's discovery that astronomers began to realize that the universe is really much larger than had been previously thought. Specifically, in the early 1910s the largest measured distances did not exceed one *kiloparsec* (i.e., 10^3 pc), whereas in the 1920s distances of one *megaparsec* (i.e., 10^6 pc) were already measurable.[32]

---

[31] The "Solar" was dropped from its name on January 1, 1918.

[32] For comparison, the diameter of the earth's orbit is of the order of 10^{-5} pc (see section 2.1).

At the dawn of the twentieth century, many other types of variable stars were already known, some of which were most intriguing and remained a mystery for a number of years. For example, in 1795 Pigott had discovered that R Scuti varied in brightness with a period of about 140 days, but tended to be more irregular than the Cepheid variables. This star is typical of the class of *RV Tauri stars*, which are yellow and reddish supergiants that form a sort of connecting link between the Cepheid variables and the long-period variables. The same year, 1795, William Herschel noticed that the brightness of α Herculis varied in irregular cycles lasting several months. As we now know, this star is typical of the class of *semiregular red variables*. More surprisingly, R Coronae Borealis was observed by Pigott in 1795 to suffer brightness changes in abrupt and unpredictable fashion. This is an *irregular variable* that exhibits no periodicity and spends most of the time at maximum brightness.

There is also a category of stars that apparently undergo some sort of outburst or explosion. A common type is the *dwarf nova*, such as U Geminorum and SS Cygni. U Geminorum was originally observed by John Russell HIND (1823–1895) in London to dim from magnitude 9 to below magnitude 13 in a few days at the end of 1855. Pogson found it again bright one hundred days later in March 1856. It was also observed by Eduard SCHÖNFELD (1828–1891) in Bonn to increase by 3 magnitudes in twenty-four hours in 1869. The dwarf novae spend most of the time at minimum brightness, but rise abruptly by up to 6 magnitudes, at irregular intervals of a few weeks or months, and then decline in a few days. A closely related group of stars, the *recurrent novae*, repeat their outbursts at intervals of a few decades. The *novae*, or "new stars," are faint stars that increase their brightness abruptly by about 10 magnitudes; they remain at maximum brightness for a couple of days, and then gradually decline during the next several months to their former faintness. The most intense nova since Kepler's new star of 1604 was discovered in 1901 by the amateur astronomer Thomas David ANDERSON (1853–1932) of Edinburgh.[33] Nova Persei 1901, as it was called, rose to apparent magnitude +0.1, as bright as Capella. Nova Aquilae 1918 reached magnitude −1.4, as bright as Sirius, and some years after the original explosion showed an expanding shell of gas that became large enough to be observed directly with a telescope. Another group includes the *supernovae*, which are exploding stars reaching enormous luminosities. The apparently brightest galactic supernova on record became as bright in the sky as Venus. It flared out in Cassiopeia in 1572 and was observed by Tycho Brahe and others, until after eighteen months it disappeared to the unaided eye. In 1604 another galactic supernova was observed by Kepler and his contemporaries in Ophiuchus. And, in 1054, the Chinese and Japanese chronicles recorded a "guest star" in Taurus, which was later identified with a supernova whose remnant is still visible as the Crab nebula. The first known extragalactic supernova, S Andromedae, was thoroughly monitored at Dorpat (now Tartu) by the German astronomer Ernst HARTWIG (1851–1923) in 1885. Shapley in 1917 estimated an absolute magnitude of −15 for this object if the Andromeda nebula were assumed to be an external star system, a figure which at that time seemed

---

[33]Anderson also discovered Nova T Aurigae, and announced it on February 1, 1892 by sending an anonymous postcard to the Astronomer Royal for Scotland.

impossible because it corresponded to about $10^8 L_\odot$. It was not until 1925 that Knut LUNDMARK (1889–1958) in Sweden was able to distinguish clearly between novae and supernovae.

2.8 THE RISE OF ASTROPHYSICS

In this chapter we have attempted to trace the development of solar and stellar physics through their principal directions, rather than in strict chronological order. For completeness, let us now briefly outline the main stages of this progress. Until the middle of the nineteenth century, astronomy was primarily concerned with counting the stars, producing star catalogs, and preparing ephemerides for the sun, moon, and planets. Fully aware of the limitations of their instruments, however, most (if not all) astronomers of that age were pessimistic about the prospects for ascertaining the physical and chemical composition of celestial bodies. In fact, most of them agreed with Bessel, the great expert in positional astronomy and celestial mechanics, who emphatically stated in his *Populäre Vorlesungen über wissenschaftliche Gegenstände* posthumously published in 1848 that "astronomy had no other task than to find rules for the motion of every star; its reason for being follows from this." It is not surprising, therefore, that almost all of the founders of *astrophysics*, or the "new astronomy" as it used to be called, were young physicists or amateur astronomers who stood completely outside the tradition of classical astronomy. Accordingly, the foundations of astrophysics did not come from astronomy, but from the conjunction of three physical methods which had undergone great advances in the first decades of the century: *spectroscopy*, *photometry*, and *photography*. It is primarily through the application of these new techniques that the constitution of the sun and stars has become an object of scientific inquiry.

Parenthetically, note that by a curious coincidence all the elements required for the growth of astrophysics were provided at about the same time in the late 1850s: Kirchhoff and Bunsen conclusively established the fundamental laws of spectroscopy in 1859; Pogson in 1856 laid down the fundamental relation that has since been adopted as the definition of the stellar magnitude scale while, three years later, Zöllner introduced the first modern photometer; and photography began about 1858 to produce the first results that really contributed to astronomical progress. Finally, Foucault in 1857 constructed the first telescope with a silvered glass mirror, which is ideally suited to the needs of spectroscopy and photography. This remarkable conjunction of events is of particular importance because, as was shown in sections 2.2–2.7, it led to the explosive development of observational astrophysics during the period from about 1860 to the outbreak of World War I.[34]

During the second half of the nineteenth century, there were only a few astrophysicists in Europe, among them Huggins and Lockyer in England, Janssen in France, Secchi in Italy, and Vogel in Germany. Despite this list of eminent scientists, one striking fact in the early development of observational astrophysics is that it was

---

[34]For further technical details, see especially G. de Vaucouleurs, *Discovery of the Universe* (London: Faber and Faber, 1957), 144–149.

already flourishing in the United States at the end of the century. Undoubtedly, no immutable tradition of the old astronomy had taken root there which could hamper the growth of this new area of research. Moreover, as was noted by Donald Oster-brock, a former director of the Lick Observatory, "American money and technology, applied at fine observing sites in the favorable climate of California, enabled the United States to overtake Germany and Great Britain, and become the world leader in observational astronomy. American physicists emphasized increasingly accurate quantitative measurements, and the physical conditions that could be drawn by them."[35] In the world at large, America had moved into the lead when, in 1888, a 36-inch refractor went into operation at the Lick Observatory, on Mount Hamilton, California. The size of this large instrument was actually surpassed in 1897 when, through the efforts of George Ellery Hale, a 40-inch refractor was installed at the Yerkes Observatory, on the shore of Lake Geneva, Wisconsin. After supervising the construction of this huge instrument (still the largest refractor in the world), Hale moved to southern California where he supervised at Mount Wilson the construction of a 60-inch reflector in 1908, soon surpassed by a 100-inch reflector in 1919. The large reflecting telescopes had finally come of age. The 100-inch Hooker reflector of the Mount Wilson Observatory, which was the most significant instrument in the world for thirty years, was surpassed only when the 200-inch Hale reflector at Palomar Mountain came on line in 1948.

And yet, in spite of these remarkable achievements in observational astronomy, prior to World War II America was much weaker than Europe in theoretical astrophysics. Except for Pickering at Harvard College Observatory and Russell at Princeton University, there were almost no theoretically inclined astrophysicists in American observatories and astronomy departments. Pickering and coworkers were essentially "classifiers," however, whereas Russell was neither an observer nor a theoretician, but a phenomenologist with a consummate flair for data analysis. In the mid-1930s the Russian-born American astronomer Otto Struve (1897–1963), who was director of the Yerkes Observatory from 1932 to 1947, clearly recognized the need for theoreticians, both as researchers and as teachers.[36] He thus brought several outstanding scientists to the Yerkes permanent staff from abroad, among them the "hard" theoretician Subrahmanyan Chandrasekhar (see section 4.5). Other universities and observatories soon followed Struve's lead and sought the best astronomers and theoreticians all over the world. This influx of foreign-born scientists, several of them refugees from Hitler's Germany, was the beginning of a golden age for theoretical astrophysics in the United States and, at the same time, the end of Europe's supremacy in that field. These matters are discussed further in chapter 5, following our presentation in the next two chapters of the theories about the sun and the stars as they were successively developed from the mid-1840s to the late 1930s.

---

[35] D. E. Osterbrock, *James E. Keeler: Pioneer American Astrophysicist* (Cambridge: Cambridge University Press, 1984), 2. See also D. E. Osterbrock, *Pauper & Prince: Ritchey, Hale, & Big American Telescopes* (Tucson: University of Arizona Press, 1993); J. Lankford and R. L. Slavings, "The Industrialization of American Astronomy, 1880–1940," *Physics Today*, **49**, No. 1, 34 (1996).

[36] D. E. Osterbrock, *Yerkes Observatory: 1892–1950* (Chicago: University of Chicago Press, 1997), 322–324.

Chapter Three

The Time of Pioneers: 1840–1910

For unto every one that hath shall be given, and he shall have
abundance; but from him that hath not shall be taken away
even that which he hath.
  —Matthew 25:29

Following Kepler's empirical derivation of the laws of planetary motion and
Galileo's deductions about falling bodies and projectiles, Newton virtually
created the new sciences of dynamics and celestial mechanics, which he discussed
in his *Principia* in 1687. In it, Newton made clear statements of the three basic laws
that govern the motions of material bodies, and explained the orbits of the celestial
bodies from their mutual gravitational attraction.

Specifically, *Newton's first law* states that a body will continue in a state of rest, or
of uniform motion in a straight line, if no force acts upon it. This law, which has its
roots in Galileo's theory of uniform motion, is of particular importance because, for
the first time, it was stated that a force was necessary, not for motion, but for a change
of motion (i.e., an acceleration). *Newton's second law* states that the rate of change of
momentum equals the applied force and is in the direction of that force. Thus we have

$$\mathbf{f} = \frac{d}{dt}(m\mathbf{v}), \tag{3.1}$$

where $\mathbf{f}$ is the applied force, m is the mass, $\mathbf{v}$ is the velocity, and $m\mathbf{v}$ is the momentum. If the mass does not change during its motion, eq. (3.1) reduces to the familiar
form $\mathbf{f} = m\mathbf{a}$, where $\mathbf{a}$ is the acceleration. *Newton's third law* states that to every
action there is always an equal and opposite reaction. The whole subject of classi-
cal mechanics grew up with these three basic assumptions about the nature of the
physical world (see also appendix F). The *law of gravitation* is to the effect that two
particles of masses m_1 and m_2 attract each other with a force of magnitude equal to

$$G\frac{m_1 m_2}{r^2}, \tag{3.2}$$

where G is a universal constant and r is the distance separating them, and that this
force acts in the direction of the line joining the two particles (see section 2.3).
Actually, the explicit form of the inverse square law of gravitation was first deduced
empirically from the observed motions of planets.

Combining this law for the gravitational force with his three basic laws of motion,
Newton employed the newly invented differential calculus to describe the motions
of the moon and the planets, and bequeathed to his successors the task of solving
the problem in greater detail. Surprisingly, whereas important observational work

was done in England during and after Newton's lifetime, his fellow countrymen contributed little to this complex theoretical problem during the eighteenth century. In fact, Newton's ideas were developed chiefly by a succession of able continental mathematicians, although not until nearly half a century after the publication of the *Principia*.[1] Between the 1740s and the early decades of the nineteenth century, in the absence of any knowledge of the exact distances and movements of the stars, mathematical analysis was thus applied almost exclusively to the study of planetary motions and the underlying problems of mechanics and gravitation. This work was summed up by Pierre Simon LAPLACE (1749–1827) in his monumental *Traité de mécanique céleste*, which appeared in five volumes between 1799 and 1825. Laplace also published in 1796 his *Exposition du système du monde* (Essay on the system of the world). In this semipopular book he presented his "nebular hypothesis," according to which the whole solar system had condensed out of a vast rotating mass, a huge nebula, that filled the bounds of the present solar system. This conception has remained an integral part of our views concerning the formation of the sun, the planets, and the stars.[2]

At the beginning of the nineteenth century, there was still little knowledge of the physical constitution of the sun and the stars, and there was none of their chemical composition. The first theoretical studies of the sun as a radiating body were published in the 1840s. By that time, however, investigations on the general structure of the universe fell naturally into two branches: *positional astronomy* and *celestial mechanics*. Since each of these fields already had a very long tradition, the great majority of observers and mathematicians engaged in astronomical studies were thus firmly entrenched into their respective specializations. Characteristically, the first lasting theoretical contributions to solar and stellar physics were made by scientists who had no professional affiliation with observatories. Some of them—such as Hermann VON HELMHOLTZ (1821–1894) and William Thomson, later Lord KELVIN (1824–1907)—were scientists of considerable repute who had already gained a certain influence in their own academic circles. Others were outsiders to the scientific community, and their pioneer work was essentially a lonely business: there were but a few at any one time during the period 1840–1890, each of them working in isolation. In this chapter we shall pay particular attention to these lesser known scientists and attempt to restore proper credit to their work.

3.1 THE PUZZLE OF THE SUN'S ENERGY

In the late 1830s the results of John Herschel and Claude Pouillet called attention to the enormous expenditure of the sun in the form of light and heat (see section 2.3). The questions naturally arise, What is the nature of the sun's heat? and How did this heat originate? It is now well established that in the seventeenth century several scientists saw more or less clearly that heat was a form of motion. This

---

[1]The Scottish mathematician Colin MACLAURIN (1698–1746), alone of Newton's British successors, ranked equal with the continental mathematicians of his day.

[2]Laplace's speculations were almost certainly independent of a somewhat similar but less detailed theory which had been suggested by the German philosopher Immanuel KANT (1724–1804) in 1755.

view was abandoned in the following century in favor of a materialistic theory of heat. Thus, while important progress was being made in celestial mechanics, heat was still described as an imponderable substance —called "caloric"—which was a very subtle, attenuated form of matter not capable of being weighed. During the first half of the nineteenth century, the caloric was progressively set aside and it became experimentally established that heat was due to molecular motion. At the same time, chemists with their sensitive balances established the law of the *conservation of mass*, and physicists recognized that heat was one form of what we now call energy and that it could be converted into other forms of energy. Then, in the 1840s, came the all-embracing principle of the *conservation of energy* which several scientists—Mayer, Waterston, Helmholtz, and Kelvin—sought to apply to the sun.[3]

Julius Robert MAYER (1814–1878; figure 3.1) was born at Heilbronn and trained for the medical practice. In 1840, while acting as ship's surgeon on a Dutch vessel trading to Java, he was impressed by the surprising redness of the venous blood of European sailors. Mayer surmised that a lower rate of metabolic combustion was needed to maintain the body heat in hot climates, so that the body was extracting less oxygen from the red arterial blood. This physiological observation led him to further consideration of those physical factors on which the phenomena of vitality depend. This, in turn, led him to make the bold hypothesis that body heat and muscular exertion were interconvertible manifestations of a single indestructible quantity, which we call energy, and that this quantity was conserved in any conversion. Upon his return to Heilbronn, where he took up a medical practice, he generalized this result to the phenomena of inorganic matter as well. Mayer's first paper on the conservation of energy appeared in 1842 in Liebig's *Annalen der Chemie*. It attracted no attention so that this principle was discovered again independently by other scientists. In fact, so unacceptable were his ideas that he had to pay for the publication, in 1848, of his paper on the source of the sun's heat.[4] Several personal disasters in 1849 and his failure to obtain recognition for his scientific work led to a mental collapse and his disappearance from the scientific scene for about ten years. Only after 1860 did Mayer gradually receive international recognition. He died in Heilbronn of tuberculosis in 1878.

The problem of the origin of the sun's heat was also investigated by the Scottish physicist John James WATERSTON (1811–1883; figure 3.2). From 1839 to 1857, Waterston held the post of naval instructor to the East India Company's cadets in Bombay. In 1857 he resigned his appointment in India and returned to Scotland, apparently having saved enough money to be able to devote his time to scientific work. After his death, an original manuscript on the kinetic theory of gases and solar radiation that he had submitted in 1845 to the Royal Society of London was discovered in the society's archives.[5] Had this important work not been rejected for publication in 1845, Waterston would have been recognized during his lifetime as

---

[3] For a detailed historical study of the theory of heat, which we now call thermodynamics, see D.S.L. Cardwell, *From Watt to Clausius* (London: Heinemann, 1971).

[4] *Beiträge zur Dynamik des Himmels in populärer Darstellung* (Heilbronn: Johann Ulrich Landherr, 1848). His major scientific papers were collected by J. J. Weyrauch in *Die Mechanik der Wärme*, 3rd edition (Stuttgart: Cotta, 1893).

Figure 3.1 The German physician Julius Robert Mayer, who in 1842 was the first to enun-
ciate the all-embracing principle of the conservation of energy, and who in 1848
speculated that the fall of meteoroids into the sun would lead to the conversion
of their mechanical energy into light and heat. From E. Dühring, *Robert Mayer*
(1904; reprint, Darmstadt: Wissenschaftliche Buchgesellschaft, 1972).

one of the leading figures of the theory of gases. Fortunately, at the 1853 meeting
of the British Association for the Advancement of Science he could communicate a
two-page paper in which he restated his ideas on the origin of the sun's heat that had
been censored in 1845.[6] Waterston also published several papers in the *Monthly
Notices of the Royal Astronomical Society*. In 1878 two of his papers were rejected
for publication in that journal; soon afterward he resigned, having been a member of
the Royal Astronomical Society since 1852. Waterston's case has become a classic
example of retreating after denial of recognition by established scientific institutions.

Attempts to solve the problem of the solar energy supply were originally made,
independently, by Mayer and Waterston in the 1840s. By this time, from fossils

---

[5]Thanks to Lord RAYLEIGH (1842–1919), Waterston's 1845 paper was eventually published in the
Philosophical Transactions in 1892. It is reprinted in *The Collected Scientific Papers of John James
Waterston*, edited with a biography by J. S. Haldane (Edinburgh: Oliver and Boyd, 1928).

[6]"On Dynamical Sequences in Kosmos," *The Athenaeum*, No. 1351, 1099–1100 (1853). (This work
is not reprinted in his *Collected Scientific Papers*.)

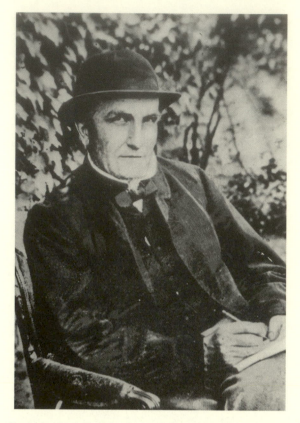

Figure 3.2 The Scottish physicist John James Waterston, aged forty-six, upon his return
from India. He was the first to advance, in 1845 and again in 1853, the solar
contraction theory and the accompanying change of potential energy into heat.
This is generally known today as the Helmholtz-Kelvin contraction theory. From
The Collected Scientific Papers of J. J. Waterston (Edinburgh: Oliver and Boyd,
1928).

found in rocks, it was well established that the age of the solar system was much
longer than the traditional 6,000 years,[7] and that the sun had been emitting roughly
the same amount of light and heat for a very long period of time. As was noted by
Mayer in 1848, the obvious explanation of the sun as a heated body cooling down
could be discarded because, without an internal source continuing to heat the body,
its temperature would drop severely in a few thousand years. Similarly, he pointed
out that the explanation of the sun as a furnace deriving its heat from coal burning
was also quite inadequate, as the sun could sustain its present luminosity for only a
few thousand years.

Clearly a more efficient source of energy had to be found, and this led Mayer
to advance in 1848 what rapidly became known as the "meteoric hypothesis" of
the sun's heat. Mayer speculated that meteoroids, falling under gravity with ever-

---

[7]Using the Old Testament as a base, James USSHER (1581–1656), archbishop of Armagh, calculated
with astounding certainty that God created the heavens and the earth in the year 4004 B.C.

increasing velocities on the sun, would build up a tremendous amount of kinetic energy which, either by friction with the solar gas or by impact on the sun's surface, would then be converted into light and heat—just as meteoroids do when they enter the earth's atmosphere and become visible as meteors, otherwise known as shooting stars. The idea that constant meteoroidic bombardment on a large scale may well account for the sun's energy was independently suggested by Waterston in 1853, and it was for a while advocated by Kelvin.[8] Very soon, however, it was realized that this hypothesis was incompatible with the known rate of meteor occurrence at the earth. Moreover, it was also observed that the required mass of meteoroids impacting the sun should increase the sun's mass at a rate that would be detectable in the motions of the planets Mercury and Venus. Since this was not observed, the Mayer-Waterston hypothesis was thus rapidly discarded.

It is not generally known that Waterston was the first to suggest, in 1845 and again in 1853, that the sun might be shrinking under the force of its own gravity and that heat and radiation would be produced by that contraction. Thus he wrote in his 1845 paper:

> The *vis viva* theory appears to harmonize well with the Nebular Hypothesis of Laplace. The intense activity of the molecules of the Sun's mass may be viewed as the result of, or to have been originally produced by, its centripetal force while condensing. The motion generated is not lost, as it is in appearance when inelastic bodies meet each other with equal momenta. The clashing together of the descending elastic matter is followed by equal recoil in the opposite direction, and molecular *vis viva* is generated. We see this take place on a minute scale when metals are hammered, or compressed, or rubbed. Friction and every other expenditure of mechanical force gives birth to heat or molecular *vis viva*, which is dissipated by radiation and conduction.

The results of his computations were as follows:

> If the Sun is supposed to contract uniformly throughout its mass so that its radius becomes 3 ⅓ miles less in consequence of the general increase of density, the force generated is sufficient to supply the solar radiation for about 9000 years.

This was a most promising conclusion, since the gravitational contraction needed to account for the whole annual output of solar heat was so small that it would be quite imperceptible to observation.

Unfortunately, as was noted above, Waterston's 1845 paper did not appear in print when it was presented. This led its frustrated author to publish a summary of his ideas on the sun's gravitational contraction in his 1853 communication to the British Association meeting. Helmholtz, who had attended the meeting, presented Waterston's contraction hypothesis in a public lecture delivered in 1854,[9] and then

---

[8] *Trans. Roy. Soc. Edinburgh*, April 1854. Reprinted in *Mathematical and Physical Papers*, by Lord Kelvin, vol. II (Cambridge: Cambridge University Press, 1884), 1.

[9] An English translation of that lecture is given by A. J. Meadows in his book *Early Solar Physics* (Oxford: Pergamon Press, 1970).

it became extremely popular. Specifically, he pointed out that the heat which the sun could have previously generated by its condensation from a primeval nebula would have been sufficient to account for the expenditure of solar heat at the present rate for no less than 22 million years. From the early 1860s to the 1890s, Kelvin became an ardent advocate of the gravitational contraction theory of solar energy.[10] And since it was fairly intuitive, the theory also enjoyed wide support among astronomers, who named it after Helmholtz and Kelvin.

By the end of the nineteenth century, however, the gravitational contraction theory was increasingly criticized by biologists and geologists, who pointed out that a span of about 20 million years was not sufficient to explain their observations. Since the theory was popular among astronomers, many of the proposals to explain the solar energy supply over a longer period of time came from people working in other disciplines. Thus, in 1868, the Scottish geologist James CROLL (1821–1890) suggested that the solar heat came from the kinetic energies of two cold bodies that collided long ago to form the sun. Or, in 1881, the German-born English engineer William SIEMENS (1823–1883) suggested that the sun was deriving all of its energy from chemical reactions maintained by a large-scale circulation in its atmosphere. Specifically, he hypothesized that the sun was continually ejecting chemically processed material at the equator and absorbing it again in rejuvenated form through the poles, that is, he assumed that in space the gases were broken back to their original form by the action of the sun's radiation. According to Siemens, such a cyclical process was driven by the rotation of the sun. As was shown by Kelvin, however, this was a most inefficient process because the rotational energy of the sun would suffice to drive the large-scale meridional flow for only a hundred years or so. Of great historical interest is also the book entitled *Stellar Evolution and Its Relations to Geological Time* that Croll wrote in 1889, and which contains what is perhaps the first modern attempt to explain the chemical composition of the stars. Following an original suggestion made in 1815 by the English physician William PROUT (1785–1850) and further developed by William CROOKES (1832–1919), Croll hypothesized that the atoms of the different elements were really aggregates of hydrogen atoms, and so were "manufactured" articles. He further argued that the gradual conversion of a given element into a heavier one required intense heat, and suggested that the heat was generated by originally cold matter coming together to form the sun. In its principles, this is a surprisingly modern overall picture of nucleogenesis.[11]

After 1896 studies of radioactivity further challenged the gravitational contraction theory, first by showing that the oldest rocks containing fossils of living creatures were very much older than 100 million years, and second by revealing that hitherto unknown energies were stored in atoms. These new developments, which truly belong to the twentieth century, will be discussed in sections 4.2 and 5.1.

---

[10] *Macmillan's Magazine*, March 1862. Reprinted by Lord Kelvin and P. G. Tait in their book *Principles of Mechanics and Dynamics*, formerly titled *Treatise on Natural Philosophy*, vol. II (New York: Dover Publications, 1962), 485.

[11] K. W. Siemens, *Proc. Roy. Soc. London*, **33**, 393 (1881); J. Croll, *Stellar Evolution and Its Relations to Geological Time* (London: Edward Stanford, 1889). See also N. J. Woolf, *Sky and Telescope*, **31**, 150 (1966); A. J. Meadows, *Early Solar Physics* (Oxford: Pergamon Press, 1970), 81–82.

3.2 THE FIRST SOLAR MODELS

The problem of the gravitational equilibrium of the sun was originally discussed by the American scientist Jonathan Homer LANE (1819–1880) in 1870. Lane, who graduated from Yale University in 1846, was primarily an engineer with a strong interest in physics, meteorology, and astronomy. From 1848 to 1857, he earned a living as an examiner in the Patent Office in Washington, D.C. Then, from 1857 to 1866, his course becomes obscure, but it is known that Lane went to Franklin, Pennsylvania in 1860 to live with his brother, a blacksmith. There, far from the turmoil of the Civil War, he did attempt to develop a low-temperature apparatus which was to utilize the expansion of gases for cooling. Lane went back to Washington in 1866 to join, three years later, the Office of Weights and Measures, the predecessor of the present National Institute of Standards and Technology. Although Lane remained something of an outsider in the Washington scientific community, he was acquainted with some of the most prominent American scientists of the time, especially with Simon NEWCOMB (1835–1909) who did much to make his work known abroad. Lane's work on the sun's interior is the only achievement for which he is now remembered.[12]

Lane's 1870 paper, "On the Theoretical Temperature of the Sun; Under the Hypothesis of a Gaseous Mass Maintaining Its Volume by Its Internal Heat, and Depending on the Laws of Gases as Known to Terrestrial Experiment," considers for the first time the equilibrium of a wholly gaseous, spherical fluid mass kept together by the mutual attraction of its parts.[13] In this paper, Lane thus made the assumption that the sun was entirely gaseous. (This was a rather bold assumption because, in the 1870s, there was a widespread belief that parts of the sun were liquid or even solid.) Hence he let

$$p = \mathcal{R}_g \rho T, \tag{3.3}$$

where p is the pressure, ρ is the density, T is the temperature, and $\mathcal{R}_g$ is a constant that depends on the nature of the gas. By virtue of eq. (3.3) there are thus only two independent variables determining the state of the material. Lane also formally recognized that two equilibrium conditions were needed to calculate these two variables: the first can be called the condition of "hydrostatic equilibrium"; the second was identified by Lane as a condition of "convective equilibrium."

A star is held together by its self-gravitation and supported against collapse by internal gas pressures that produce an upward thrust. *Hydrostatic equilibrium* ensues when the inward gravitational attraction exactly balances the outward pressure gradient at every point. Equation (3.2) expresses the gravitational attraction between two mass points. To find the gravitational attraction inside a star, we must sum the vectorial contributions due to each mass element of the body. Let r denote the radius vector, measured from the center of the star. For a spherically symmetric body, it can be shown that the shell of matter outside the sphere of radius r exerts no

---

[12] For a detailed study of the life and times of J. Homer Lane, see C. S. Powell, *J. History Astron.*, **19**, 183 (1988).

[13] *Amer. J. Sci. Arts*, 2nd ser., **50**, 57 (1870). This paper (except the blueprint on p. 69, here redrawn in figure 3.3) was reprinted by A. J. Meadows in his book *Early Solar Physics* (Oxford: Pergamon Press, 1970).

resultant attraction in its interior, while the entire mass inside the radius r—$m(r)$, say—attracts a mass element located on the sphere of radius r as would a point of mass $m(r)$ located at the center. The acceleration of gravity in a spherical star is, therefore, $Gm(r)/r^2$, where

$$m(r) = 4\pi \int_0^r \rho r^2 \, dr. \tag{3.4}$$

Hence, for hydrostatic equilibrium we should have

$$dp = -\frac{Gm(r)}{r^2} \rho \, dr, \tag{3.5}$$

expressing the increase of pressure as we descend in a column of fluid.

As was noted by Lane, however, if a gaseous sphere like the sun were left to itself in space, losing heat by radiation outward, it would give rise to convective currents—one set ascending with highly heated matter and the other descending with the material cooled by radiation. Granting that these heat-carrying currents were not likely to produce much disturbance to the hydrostatic balance (eq. [3.5]), Lane further assumed that such a stirring would go on until a condition of approximate equilibrium was reached. This is the concept of *convective equilibrium* that was originally broached by Kelvin in 1862.[14] In Kelvin's own words:

> The essence of convective equilibrium is that if a small spherical or cubic portion of the fluid in any position P is ideally enclosed in a sheath impermeable to heat, and expanded or contracted to the density of the fluid at any other place P', its temperature will be altered, by the expansion or contraction, from the temperature which it had at P, to the actual temperature of the fluid at P'.

The formulas that express this condition were first given by Siméon Denis POISSON (1781–1840) in 1823. They were then generally known as the equations of adiabatic expansion or contraction. For the ideal case of a perfect gas, one has

$$\frac{p}{p'} = \left(\frac{\rho}{\rho'}\right)^\gamma, \qquad \frac{\rho}{\rho'} = \left(\frac{T}{T'}\right)^{1/(\gamma-1)}, \qquad \frac{p}{p'} = \left(\frac{T}{T'}\right)^{\gamma/(\gamma-1)}, \tag{3.6}$$

where primed and unprimed quantities denote temperatures, densities, and pressures at any two places in the fluid, and γ is the ratio of the specific heats at constant pressure and constant volume.

Lane's mathematical formulation of the problem can be found in appendix A. Figure 3.3 illustrates his numerical results for the cases $\gamma = 1.4$ (as in common air) and $\gamma = 5/3$ (as in a monatomic gas). To specify the central densities ρ_c, however, he made the following assumptions: (i) the sun's atmosphere extends above the photosphere by 1/22 of the total radius, or about 20,000 miles, and (ii) within

---

[14]Lord Kelvin, *Mathematical and Physical Papers*, vol. III (Cambridge: Cambridge University Press, 1890), 235. As a matter of fact, Lane arrived at the notion of convective equilibrium from a consideration of the work of the American meteorologist James Pollard ESPY (1785–1860), who had previously discussed convective currents in his book *The Philosophy of Storms* (Boston: C. C. Little and J. Brown, 1841). Lane was also influenced by Faye's 1865 papers (see section 3.6).

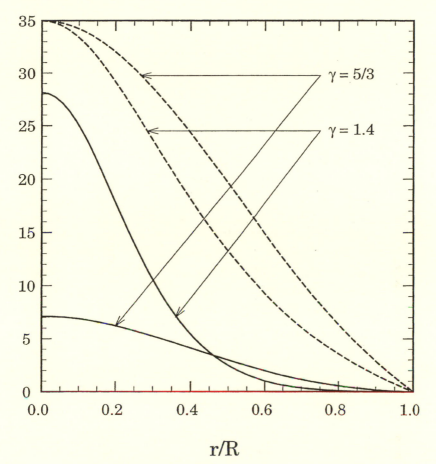

Figure 3.3 Density and temperature distributions inside the sun. The solid lines describe
absolute density distributions, in g cm^{-3}, for $\gamma = 1.4$ and $\gamma = 5/3$; the dashed
lines describe relative temperature distributions for $\gamma = 1.4$ and $\gamma = 5/3$. After
J. H. Lane, *Amer. J. Sci. Arts*, 2nd ser., **50**, 57 (1870).

the photosphere the mean density of the sun is 1/4 that of the earth, or 1.375 g
cm^{-3}. Given this choice, he found that $\rho_c = 28.16$ g cm^{-3} when $\gamma = 1.4$, and
$\rho_c = 7.11$ g cm^{-3} when $\gamma = 5/3$. Lane also calculated the temperature distribution
within his solar models but, as can be seen in figure 3.3, he did not calibrate them.
In fact, Lane's principal interest was to determine the temperature and density at
the surface of the sun. From the value of the solar constant that had been recently
obtained by John Herschel and Claude Pouillet, he estimated the surface temperature
to be 30,000 K. Thence, making use of his monatomic solar model, he found the
corresponding surface density to be 0.000363 g cm^{-3}. Although this was the first
theoretical investigation of the physical conditions in the sun's atmosphere, Lane is
chiefly remembered today as the author of the first paper on the internal structure
of the sun.

Table 3.1 Densities and Temperatures Inside See's Adiabatic Model for the Sun ($\gamma = 5/3$)

r/R	ρ (g cm^{-3})	T (K)	T (K)
1.00	1.2900E−5	6.0000E+3	1.2000E+4
0.95	6.4977E−2	3.7921E+5	3.5203E+6
0.90	1.9832E−1	7.9791E+5	7.4072E+6
0.70	1.3590E+0	2.8787E+6	2.6723E+7
0.50	3.5364E+0	5.4461E+6	5.0557E+7
0.30	6.2171E+0	7.9329E+6	7.3643E+7
0.10	8.1476E+0	9.5002E+6	8.8192E+7
0.00	8.4245E+0	9.7142E+6	9.0178E+7

After T.J.J. See, *Astron. Nachr.*, **169**, 321 (1905).

In 1905 his fellow countryman Thomas Jefferson Jackson SEE (1866–1962) remade the calculations for the case $\gamma = 5/3$, using much more accurate expansions than Lane's.[15] See essentially confirmed the soundness of Lane's results and found that in a monatomic gas sphere in convective equilibrium the central density is closely equal to six times the mean density. Letting $\bar{\rho} = 1.4026$ g cm^{-3} as in the sun, he thus obtained $\rho_c = 8.417$ g cm^{-3} in place of $\rho_c = 7.11$ g cm^{-3}. Evidently, this difference is due to the fact that Lane considered a polytropic model in which, by assumption, the mean density is equal to 1.375 g cm^{-3} multiplied by the factor $(22/23)^3$, or $\bar{\rho} = 1.20$ g cm^{-3}. Lane's convective model is basically sound, therefore, but it actually corresponds to a central density of $6 \times 1.20 = 7.20$ g cm^{-3}, instead of the value 7.11 g cm^{-3} given in his paper.

For completeness, See also calculated the temperature distribution within a monatomic gas sphere in convective equilibrium from the formula

$$T = T_S \left(\frac{\rho}{\rho_S}\right)^{2/3}, \tag{3.7}$$

where ρ_S and T_S are the surface density and temperature, respectively (see eq. [3.6]). Since the exact values of ρ_S and T_S for the sun were not known at the time, See made two sets of calculations so as to obtain upper and lower limits on the sun's internal temperatures. In one set he assumed that $\rho_S = 0.1$ of atmospheric air (or 0.000129 g cm^{-3}) and $T_S = 6,000$ K, which should give the minimum internal temperature at any point; in the other one he let $\rho_S = 0.01$ of atmospheric air (or 0.0000129 g cm^{-3}) and $T_S = 12,000$ K, which should give the maximum admissible temperatures. Table 3.1 illustrates the results of his numerical labor. They are of great historical interest because, perhaps for the first time, a sound theoretical calculation was showing that a temperature of millions of degrees could exist in the

---

[15]*Astron. Nachr.*, **169**, 321 (1905); **171**, 386 (1906) (erratum). Interesting analyses of See's controversial career can be found in the following papers: T. J. Sherrill, *J. History Astron.*, **30**, 25 (1999); W. Sheehan, *Mercury*, **31**, No. 6, 34 (2002).

Figure 3.4 August Ritter, versatile German structural engineer, who made the first systematic study of the equilibrium, oscillations, and stability of self-gravitating gas
spheres. He was also the first to suggest, in 1880, that most periodic variable
stars are pulsating in their lowest radial mode. Courtesy of Dr. Roland Rappmann, Rheinisch-Westfälischen Technischen Hochschule, Aachen.

central parts of the sun. (Modern values are close to 16 million kelvins for the central
temperature and 150 g cm^{-3} for the central density.) As was pointed out by See,
however, his result was somewhat puzzling because it strongly suggested that solar
heat and light might be supplied in the main not by convective currents but by direct
radiative transfer throughout the sun's interior. This was undoubtedly a minority
opinion in 1905 since See felt it necessary to note in his paper that "the views here
expressed depart widely from those heretofore held by leading authorities." This
will be discussed further in section 4.3.

In 1870 Lane made the first step toward a genuine understanding of the mechanical and physical conditions in the deep interior of the sun and the stars. The
next step was made around 1880 by the German structural engineer August RITTER
(1826–1908; figure 3.4). He was born in Lüneburg and graduated from Göttingen
University in 1853. Then, after holding various posts in Germany and Italy, he taught

mechanics and mechanical engineering from 1856 to 1870 at the Polytechnikum in Hannover. In 1870 he became professor of mechanics at the Polytechnikum in Aachen, where he remained until 1899. Ritter, who was a leading authority in the theory and calculation of iron bridges and roofs, had a keen interest in thermodynamics and stellar astronomy. But unlike Lane, who had very practical interests, Ritter was primarily concerned with the equilibrium, oscillations, and stability of self-gravitating gas spheres. His astronomical work extended over a period of six years, 1878–1883, during which time he published eighteen papers in Wiedemann's *Annalen der Physik und Chemie* under the general title "Untersuchungen über die Höhe der Atmosphäre und die Constitution gasförmiger Weltkörper" (Researches on the height of the atmosphere and the constitution of gaseous celestial bodies). Of particular importance is his original discussion of the equilibrium of self-gravitating gas spheres in which the pressure and density are connected by the relation

$$p = K\rho^{1+1/n},\tag{3.8}$$

where K and n are constants. A system that satisfies this relation is called a polytrope of index n.

In appendix B we present Ritter's original derivation of the fundamental equation governing the structure of a polytropic gas sphere. This equation, eq. (B.6), was first obtained in 1880 for a general index n.[16] However, the exact solution corresponding to $n = 1$, $y = (\sin x)/x$, was already familiar to Laplace,[17] whereas those corresponding to $n = 3/2$ and $n = 5/2$ were first discussed numerically by Lane, albeit in a somewhat different context (see figure 3.3). In fact, the importance of Ritter's investigations lies in the wealth of material that is contained in his work. Not only did he obtain solutions of eq. (B.6) for different values of n, but he also considered the equilibrium of composite systems consisting of incompressible cores and gaseous envelopes. He was also much interested in the applications of thermodynamics to the study of slowly contracting, self-gravitating bodies.[18] In particular, he calculated the relation between the amount of energy lost by radiation, ΔQ, and the work of gravitation, ΔW, during a slow contraction. He obtained the relation

$$\frac{\Delta Q}{\Delta W} = \frac{\gamma - 4/3}{\gamma - 1}.\tag{3.9}$$

For a monatomic gas, one thus has $\Delta Q/\Delta W = 0.5$, so that exactly one-half of the energy is radiated away, and the other half goes to elevating the internal temperature of the contracting body. Ritter was also the first to show that a system is dynamically unstable when $\gamma < 4/3$.

In 1887, unaware of the work done by Lane and Ritter, Kelvin discussed the homologous contraction or expansion of a polytropic gas sphere. Among other re-

---

[16] *Ann. Phys. Chemie*, **11**, 332 (1880).

[17] *Traité de mécanique céleste*, vol. V (Paris: Bachelier, 1825), 49.

[18] Ritter was the first to publish the law $rT(r) = constant$ governing the uniform expansion or contraction of a gas sphere. However, this transformation is often called "Lane's law" because, according to Newcomb's testimony, such a transformation was already known to Lane, although his 1870 paper contains no explicit reference to it. For a detailed historical account of "Lane's law," see S. Chandrasekhar, *An Introduction to the Study of Stellar Structure* (Chicago: University of Chicago Press, 1939; New York: Dover Publications, 1957), 177.

sults, he properly established that the central temperature T_c of a polytrope varies inversely as the $2/(n-1)$ power of its outer radius R; e.g., $T_c \propto 1/R^4$ in a homologously contracting or expanding monatomic gas sphere in convective equilibrium. Kelvin returned to the subject only twenty years later in a paper that was published posthumously in 1908.[19]

Unlike Ritter, who was active in Aachen at a time when the astronomical community had little interest in the internal structure of stars, the Swiss physicist Robert EMDEN (1862–1940) flourished in Munich when astrophysics was slowly being recognized as a branch of astronomy. Emden, who married Karl Schwarzschild's sister Klara in 1907, was also interested in the applications of thermodynamics to geophysics and astrophysics. He taught physics and meteorology at the Technische Hochschule in Munich and, later, astrophysics at the university. Soon after his dismissal when Hitler and the Nazis came into power in 1933, Emden returned to his native Switzerland, where he died in 1940. He was one of the founders of *Zeitschrift für Astrophysik*, which he edited until 1936. Although Emden did not initiate the subject of self-gravitating gas spheres, he was instrumental in developing Ritter's theory into a comprehensive survey, using numerical tables and figures that caught the eye of the astronomers. His epoch-making book *Gaskugeln* published in 1907 thus became the standard reference on polytropic gas spheres for a whole generation of scientists. And so, despite the fact that Emden gave proper credit to Ritter's work in his book, eq. (B.6) and its solutions became generally associated with the names of Lane and Emden. It is unfortunate that the value of Ritter's astronomical papers has never been adequately recognized.

3.3 THE PULSATION THEORY OF VARIABLE STARS (I)

As was noted in section 2.7, Boulliau originally suggested in 1667 that the variability of Mira Ceti might be due to its rotation around its axis, the spinning body showing alternately its bright (unspotted) and dark (spotted) hemispheres to the observer. Boulliau's concept of rotational modulation of starspots was revived in the 1850s by Rudolf Wolf for the long-period variables, and again independently for the Cepheid variables by Pickering in the 1880s. Some astronomers, however, thought that the variable stars were eclipsing binaries, like Algol, but the shape of their light curves was then very difficult to explain. In 1880 Ritter made the bold suggestion that the variable stars might be self-gravitating fluid masses undergoing radial oscillations.[20] Although this new theory was ignored for several decades after its inception, it is worth discussing here because, as we know, it is this theory that now seems to be the most plausible.

Ritter's mathematical discussion of the adiabatic oscillations of a gaseous star is presented in appendix C. For a purely radial motion with frequency σ, he found

$$\sigma^2 = (3\gamma - 4)\frac{GM}{R^3} = \frac{4}{3}(3\gamma - 4)\pi G\bar{\rho}, \qquad (3.10)$$

---

[19] Lord Kelvin, *Mathematical and Physical Papers*, vol. V (Cambridge: Cambridge University Press, 1911), 184 and 254.

[20] *Ann. Phys. Chemie*, **8**, 157 (1880).

where $\bar{\rho}$ is the mean density of the star. This formula is exact only when the density is uniform throughout the system, but it is a reasonably good approximation for centrally condensed stars. The corresponding period of oscillation, $P = 2\pi/\sigma$, is

$$P = \pi \left[\frac{3}{(3\gamma - 4)\pi G \bar{\rho}} \right]^{1/2}.$$ (3.11)

One readily sees that this period varies inversely as the square root of the mean density. Hence, given two stars with mean densities $\bar{\rho}$ and $\bar{\rho}'$, the ratio of their corresponding periods is

$$\frac{P}{P'} = \left(\frac{\bar{\rho}'}{\bar{\rho}} \right)^{1/2}.$$ (3.12)

Since one has $P \approx 6$ hours for the sun, Ritter rightfully concluded in his paper that the variable stars should have very small mean densities indeed.

Astronomers of the time apparently had difficulties with the physical concept of purely radial pulsations. In 1909 the American theoretician Forest Ray MOULTON (1872–1952) therefore suggested that nonradial oscillations of the kind previously discussed by Kelvin could give rise to the observed phenomena in the variable stars.[21] Consider an incompressible fluid sphere (of constant density ρ) in a stable figure of equilibrium. If this configuration is slightly distorted and left to its own gravitation, it will oscillate around the spherical shape, the actual period of oscillation depending upon the geometrical character of the deformation. As is well known, any slight deformation of a sphere can be represented by a sum of spherical harmonics. The problem of determining the period of oscillation of the harmonic of order l, $P_l = 2\pi/\sigma_l$, was originally solved by Kelvin (figure 3.5) in 1863.[22] We have

$$\sigma_l^2 = \frac{2l(l-1)}{2l+1} \frac{GM}{R^3} = \frac{4}{3} \frac{2l(l-1)}{2l+1} \pi G \rho,$$ (3.13)

where $l = 2, 3, 4, \ldots$. Again we note that the periods of oscillation are inversely proportional to the square root of the density. The period corresponding to the second-order harmonic $l = 2$ is

$$P_2 = 2\pi \left(\frac{5}{4} \right)^{1/2} \left(\frac{R^3}{GM} \right)^{1/2}.$$ (3.14)

This is equal to about 3 hours for a homogeneous mass having the mean density of the sun. Very much like Ritter, Moulton was thus led to the conclusion that the observed periods of the Cepheid variables and longer-period variables demanded very low densities in these stars.

At the dawn of the twentieth century, although the exact nature of the oscillations to be considered was still very uncertain, the idea that certain classes of variable stars might owe their variability to small-amplitude oscillations was slowly gaining

---

[21] *Astrophys. J.*, **29**, 257 (1909).

[22] Lord Kelvin, *Mathematical and Physical Papers*, vol. III (Cambridge: Cambridge University Press, 1890), 384.

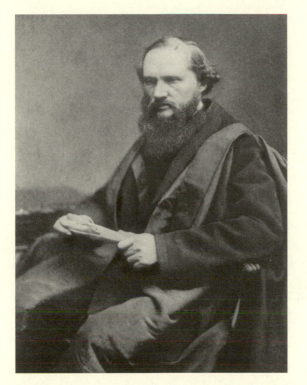

Figure 3.5 The Scottish physicist William Thomson, who became Baron Kelvin of Largs
in 1892, was the most respected British scientist of his day. Among his many
interests was the application of mathematical and physical principles to the study
of the sun and stars. In astronomy, he is chiefly remembered today for the so-
called Helmholtz-Kelvin contraction and for the Kelvin modes of oscillation of a
star. The portrait was taken in 1870. From S. P. Thompson, *The Life of William
Thomson, Baron Kelvin of Largs* (London: Macmillan, 1910; New York: Chelsea
Publishing Company, 1976).

acceptance. In fact, there was a general feeling that nonradial oscillations were more
plausible because it was thought that a nonradial mode belonging to the second-order
harmonic $l = 2$ could be more easily excited than a radial mode. As we shall see in
section 4.6, however, from the 1920s onward the emphasis was laid mainly on purely
radial pulsations, almost to the exclusion of the nonradial modes of oscillation.[23]

3.4 THE DOUBLE-STAR PROBLEM

In section 2.6 we pointed out that a large proportion of stars are members of gravitationally bound double (or multiple) systems. Binary stars range from close pairs
in which the components are almost in contact (with centers 10^{-2} AU apart) to wide

---

[23]For a more detailed review of the literature, see P. Ledoux and Th. Walraven, *Handbuch der Physik*,
51, 353 (1958).

pairs with major semiaxes up to 10^4 AU. We may therefore question whether all double stars are objects of a single genus, or whether we must look for a separate mode of formation for the wide and close binary systems. Three principal theories have been proposed to explain the formation of the double stars: (i) simple capture, (ii) formation by separate nuclei, and (iii) fission and subsequent separation.[24]

The *capture theory* was apparently first advanced by the Irish physicist George Johnstone STONEY (1826–1911) in 1867.[25] It is based on the hypothesis that two stars, originally independent, might approach each other under such conditions that they would change their initial paths and be forced to revolve about a common center of gravity. Very few people accepted this theory, however, since it requires a third body (i.e., a third star or a resisting medium) to absorb the excess energy liberated during the capture of one star by another to form a gravitationally bound double star.

The *separate nuclei hypothesis* was originally suggested by Laplace in 1796. In his *Exposition du système du monde*, he wrote (in translation):

> Such groups (as the *Pleiades*) are a necessary result of the condensation of nebulae with several nuclei, for it is plain that the nebulous matter being constantly attracted by these different nuclei must finally form a group of stars like the *Pleiades*. The condensation of nebulae with two nuclei will form stars in very close proximity, which will turn one around the other similar to those double stars whose relative motions have already been determined.

This theory apparently affords sufficient latitude for the explanation of any binary (or multiple) system, except perhaps the very close double stars. It supposes that, in some manner, nuclei of condensation developed that began to go round each other under conditions not much different from those observed in double stars. It was criticized by Moulton and others as not constituting an explanation but as explaining the little known by the less known. In any case, because this approach does not account for the very close binary stars, it should be supplemented by some other mechanism that can produce very close systems. The process of fission might be regarded as such.

The *fission theory* postulates conditions such that a star in its primal stage (or perhaps at a later one) divides into two masses which would at first revolve in surface contact and in circular orbits. To the best of our knowledge, fission was explicitly discussed for the first time in the 1883 edition of the *Treatise on Natural Philosophy* by William Thomson (Lord Kelvin) and Peter Guthrie TAIT (1831–1901).[26] As a matter of fact, their approach to fission has its roots in the theory of rotating fluid masses in equilibrium.

In 1742 Maclaurin showed that axially symmetric ellipsoids, or spheroids, represent the simplest figures of equilibrium for self-gravitating fluid masses in uniform rotation. Later, in 1834, the German mathematician Carl JACOBI (1804–1851)

---

[24] R. G. Aitken, *The Binary Stars* (New York: McGraw-Hill Book Company, 1935; Dover Publications, 1964).

[25] *Proc. Roy. Soc. London*, **16**, 25 (1867).

[26] *Principles of Mechanics and Dynamics*, formerly titled *Treatise on Natural Philosophy*, vol. II (New York: Dover Publications, 1962), 334. See also S. G. Brush, *Physics Today*, **33**, No. 3, 42 (1980).

opened up a new aspect of the subject by showing that, given a sufficiently rapid rotation, an ellipsoid with three unequal axes can be a figure of equilibrium. The classical formulation of the fission problem considers the slow evolution of a uniformly rotating fluid mass that contracts under its own gravitation while preserving its total angular momentum. Friction is supposed to be large enough to keep the slowly contracting body in uniform rotation. If so, the system can move along the sequence of Maclaurin spheroids and, at some point, bifurcate to the sequence of Jacobi ellipsoids. Kelvin and Tait argued that fission into two masses would follow. Soon afterward, in 1884–1885, the problem was taken up, independently, by Aleksandr LYAPUNOV (1857–1918) in St. Petersburg and Henri POINCARÉ (1854–1912) in Paris. They demonstrated the existence of a sequence of pear-shaped figures that branch off from the Jacobi ellipsoids—just as Jacobi's sequence branches off from the Maclaurin spheroids. The Lyapunov-Poincaré figures of equilibrium gave rise to extensive investigation, for it was then conjectured that they might eventually come apart into two detached masses rotating around each other. In 1905 Lyapunov showed that the pear-shaped figures are never stable, so that the body would necessarily undergo very large deformations followed by catastrophic disruption. The dynamical instability of the pear-shaped figures was subsequently confirmed, in 1924, by the French mathematician Élie CARTAN (1869–1951). The fission theory was also discussed by the English theoretician James Hopwood JEANS (1877–1946), who became its most ardent upholder during the 1910s and 1920s.[27]

Of direct relevance to the double-star problem is the search for equilibrium configurations of two detached bodies (of masses M and M') revolving round one another in such a way that the whole system remains at rest relative to a system of axes rotating with the Keplerian angular velocity Ω. The problem was originally tackled by the French mathematician Édouard Albert ROCHE (1820–1883; figure 3.6), whose fame rests upon two major achievements: his discussions of the equilibrium and stability of a synchronously rotating ellipsoid that is distorted by the tidal action of an external point of mass M', and of the geometry of the equipotential surfaces which surround a rotating gravitational dipole of masses M and M' (see eqs. [5.9]–[5.11]). In 1850 he showed that no ellipsoidal figure of equilibrium is possible if the angular velocity Ω exceeds the limit

$$\frac{\Omega^2}{\pi G \rho} = \frac{M'}{\pi \rho d^3} \leq 0.090093. \tag{3.15}$$

The lower limit to the mutual distance d set by this inequality is called the *Roche limit*. The dipolar model, which he developed in 1873, is called the *Roche model*, and it has become an essential tool for studying the internal structure and evolution of

---

[27] In 1982 modern computers were used to calculate a sequence of pear-shaped figures that branch off from the Jacobi ellipsoids and have finite deformations. Contrary to earlier widespread expectations, however, it was found that this sequence of incompressible configurations does not lead to fission but terminates at a critical point where mass is shed from the equator. See Y. Eriguchi, I. Hachisu, and D. Sugimoto, *Progress Theor. Phys.*, **67**, 1068 (1982). For an overall review of these recent developments, see R. H. Durisen and J. E. Tohline, in *Protostars & Planets II*, edited by D. C. Black and M. S. Matthews (Tucson: University of Arizona Press, 1985), 534.

Figure 3.6 Oil painting of the French mathematician Édouard Albert Roche circa late 1850s
(artist unknown). His fame in stellar astronomy rests upon the so-called Roche
model (1873), which adequately approximates the structure of a close binary star
as a rotating gravitational dipole. Since the 1940s, "Roche lobes" and "Roche-
lobe overflow" are household terms in the astronomical literature. Courtesy of
Henri Reboul, Service des collections, © Université Montpellier II–Astronomie.

close binary stars. In retrospect, it would seem that both achievements were too far
ahead of their time to be appreciated by his contemporaries. Indeed, when Roche's
name was put forward in 1883 for election to full membership of the Académie des
Sciences, out of fifty-six votes cast he received only one. Two days later, Roche died
of pneumonia in his hometown of Montpellier, hardly aware of the contemptuous
attitude of the Parisian establishment.[28]

3.5 EARLY VIEWS OF STELLAR EVOLUTION

Interest in the internal structure and evolution of stars began in the 1860s with the
first attempts at cataloging stellar spectra (see section 2.5). Two basic schemes of

---

[28]Z. Kopal, *The Roche Problem* (Dordrecht: Kluwer, 1989).

stellar evolution have emerged from these pioneering studies. The simplest one was originally proposed in 1865 by Zöllner, who argued that stars begin their life as hot blue stars and then contract and cool to become small red stars. The other one, which was developed by Ritter in 1883, rests upon the idea that stars start off as large cool red stars, heat during their gravitational contraction to become small blue stars, and finally cool to become small red stars.[29]

Zöllner, who was a professor at the University of Leipzig, had outstanding talents for constructing scientific instruments. Among his main achievements was the design of a new spectroscope that was used by his younger colleague Vogel to determine the sun's rotation rate (see section 2.4). Zöllner was also one of the first to examine systematically the luminosities, spectra, and colors of stars. His observational results led him to develop, in 1865, what is probably the first phenomenological theory of stellar evolution. In Zöllner's scheme, a star begins as a gaseous planetary nebula. (This was suggested by the work of Huggins and Miller who had shown in 1864 that some nebulae have emission lines typical of a large mass of hot luminous gas.) This hot cloud then contracts and cools to become a liquid sphere with a solid crust. This crust fractures from time to time to release hot interior liquids so that the star becomes variable in brightness. Eventually, the star's crust thickens so much that it can fully confine its liquid interior. At this stage the star cools, changes color from blue to yellow to red, and finally disappears from sight. In Zöllner theory, therefore, blue stars are young whereas red stars are old. (This was the origin of the usual expressions "early type" and "late type.") His evolutionary scheme was further discussed by Vogel, who created in 1874 a system of spectral classification based upon the idea that stars cool with age. However, Vogel modified Zöllner's original scheme by assuming that stars must pass through an early heating phase as they evolve from the initial nebula, but argued that such a phase was so short that no stars could be observed passing through it.

Unlike Zöllner and Vogel, Ritter approached the problem of stellar evolution not as a phenomenologist, but as a theoretician, proposing a simple model and then comparing the model to the observational data.[30] His theory is largely based on the behavior of a perfect gas sphere in convective equilibrium (see section 3.2). Specifically, he calculated the increase in surface temperature during a gravitational contraction. The result is quite simple: as the outer radius is reduced from R_1 to R_2, the ratio of the corresponding surface temperatures, T_1 and T_2, is

$$\frac{T_2}{T_1} = \left(\frac{R_1}{R_2}\right)^{(2-\gamma)/\gamma}, \qquad (3.16)$$

where γ is the adiabatic exponent. To this increase in surface temperature corresponds of course a variation in the total luminosity of the contracting body (see eq.

---

[29]For a detailed presentation of these early evolutionary schemes, see D. H. DeVorkin, in vol. 4A of *The General History of Astronomy*, edited by O. Gingerich (Cambridge: Cambridge University Press, 1984), 90.

[30]*Ann. Phys. Chemie*, **20**, 137 (1883). At the request of Huggins, who found this paper particularly interesting, Hale had it translated and published in the *Astrophysical Journal* (see *Astrophys. J.*, **8**, 293 [1898]).

[2.5]). Making use of Stefan's 1879 law, eq. (2.4), we obtain

$$\frac{L_2}{L_1} = \left(\frac{R_1}{R_2}\right)^{(8-6\gamma)/\gamma}.$$ (3.17)

Since a state of equilibrium is stable only if $\gamma > 4/3$, as shown in eq. (C.11), it follows from eq. (3.17) that the total luminosity of a perfect gas sphere necessarily decreases as the body contracts. Ritter also discussed the dependence of surface temperature on mass in a contracting star. Letting M_A and M_B denote the masses of two distinct stars, he found

$$\frac{T_B}{T_A} = \left(\frac{M_B}{M_A}\right)^{(\gamma+1)/3\gamma} \left(\frac{\bar{\rho}_B}{\bar{\rho}_A}\right)^{(2-\gamma)/3\gamma},$$ (3.18)

where $\bar{\rho}$ is the mean density. For monatomic gas spheres ($\gamma = 5/3$), eq. (3.18) reduces to

$$\frac{T_B}{T_A} = \left(\frac{M_B}{M_A}\right)^{8/15} \left(\frac{\bar{\rho}_B}{\bar{\rho}_A}\right)^{1/15},$$ (3.19)

so that the ratio of the surface temperatures can be considered as nearly independent of the mean densities. Thus, to a good approximation, the surface temperatures of two stars are to each other nearly as the square root of their masses.

Given these theoretical results, Ritter distinguishes three phases in the evolution of a star. In the first phase, the material density is assumed to be so low that the contracting body is almost transparent to radiation. If so, then the total luminosity can depend solely on temperature, steadily increasing with the rise of temperature caused by gravitational contraction. In the second phase, when the stellar material has become sufficiently opaque to radiation, Ritter assumes that the star behaves as a perfect gas sphere in convective equilibrium. Hence, by virtue of eqs. (3.16) and (3.17), the star that started as a huge red star does not grow in brightness anymore; instead, it becomes hotter and dimmer as it contracts to become a small blue star. Ritter's third phase starts when, at some point during the contraction, the star ceases to behave as a perfect gas. It is then conjectured that the actual surface temperature must begin to fall so that the small blue star cools, via a small yellow star, to become a small red star. As was shown by Ritter, however, the maximum temperature that a star can reach depends on its mass, with the heaviest stars reaching the highest surface temperatures (see eqs. [3.18] and [3.19]). This theory, which was subsequently advocated by Lockyer and Russell, is often referred to as the "giant-and-dwarf theory" of stellar evolution.

Norman Lockyer (figure 3.7), an amateur-turned-professional based in London, was a pioneer in solar and stellar spectroscopic studies. He also created in 1869 the scientific journal *Nature*, which he edited for the first fifty years of its existence. Lockyer was the first to propose in 1878 that, with increased temperature, atoms in stellar atmospheres would break down to form what he called protoelements.[31]

---

[31]This was a bold suggestion indeed, since it was not until 1897 that Joseph John THOMSON (1856–1940) positively established the existence of the electron.

Figure 3.7 Sir Norman Lockyer as a young man. A self-made entrepreneur in astronomy
           and laboratory physics, he contributed to many branches of the nascent sciences
           of astrophysics and atomic physics. Helium, it will be remembered, was inde-
           pendently discovered in the sun in 1868 by Jules Janssen and Norman Lockyer,
           who gave it the name "helium." Courtesy of the Mary Lea Shane Archives of the
           Lick Observatory.

This was the "dissociation hypothesis," an extension of his laboratory observations
of spectral changes brought about by increased temperature. This idea, like many of
Lockyer's speculations, met with almost universal opposition because most physicists of the time were strongly entrenched in the belief that atoms were indestructible.
(Eventually, in the twentieth century, it was shown that these protoelements were,
in fact, ionized atoms.) Perhaps the most far-reaching of Lockyer's ideas was his
"meteoritic hypothesis" that he proposed in 1887 for the birth of stars.[32] In his own
words: "All self-luminous bodies in the celestial spaces are composed of meteorites,
or masses of meteoritic vapor produced by heat brought about by condensation of
meteor-swarms due to gravity." Within the next few years, he developed and con-

---

[32] *Proc. Roy. Soc. London*, **43**, 117 (1887). See also A. J. Meadows, *Science and Controversy: A
Biography of Sir Norman Lockyer* (Cambridge, Mass.: MIT Press, 1972).

stantly improved a theory of stellar evolution based on the idea that a star begins as a vast extended swarm of meteoritic material contracting under the action of gravity. As condensation proceeds in the cloud, the star then gets hotter, causing its color and spectrum to vary from red to blue. Eventually, the heat generated by collisions of meteoritic material is no longer sufficient to balance the heat lost by radiation, and so the star starts cooling, progressively changing color from blue to red, and hence passes to extinction. Unlike Ritter, who had suggested that only the most massive stars can attain the highest surface temperatures, Lockyer remained firmly guided by his concept of dissociation and assumed that all stars must follow the same evolutionary path.

The giant-and-dwarf theory of stellar evolution aroused much opposition because it assumes that a star begins as a cold body, rises to a maximum temperature, and then cools down again, so that the coolest stars observable would belong to two groups, young and old. In fact, most astronomers favored the idea that stars were born hot and cooled continuously during their lives, and the succession of stellar spectra was held to support it. The problem was cleared up over the years 1905–1910 by Hertzsprung and Russell, who found that there were indeed two different populations of high- and low-luminosity red stars (see figure 2.12). Russell, from Princeton University Observatory, therefore suggested in 1914 that the whole process of stellar evolution starts with red giants, which contract and become hotter, heated as perfect gas spheres, to become bluish-white B- and A-type stars, at densities too great for the perfect gas condition. The stars then contract and cool, progressively changing color from bluish-white to yellow to orange to red, and ending their lives as red dwarfs. Russell properly acknowledged his debt to Ritter and Lockyer, but noted that his scheme had the advantage of differentiating those stars on the heating branch of the temperature curve from those on its cooling branch.

Russell also noted that the problem in interpreting the main sequence in the color-magnitude diagram as the cooling branch of the temperature curve was that the average masses decreased along the main sequence from blue to red (see figure 2.14). This observational result led him to suggest that only the most massive stars reached the hottest temperatures on contraction, as was originally pointed out by Ritter in 1883. Less massive stars would peak at lower temperatures on the giant branch and then join the main sequence as dwarfs at a lower point. In this manner, only the most massive stars would join the main sequence at the top; yet the average mass would still diminish down the main sequence, as observed, since its lower parts would be enriched with the inclusion of more stars of lower masses. All stars were still assumed to follow the main sequence down as they contracted and cooled. For reasons explained in section 4.4, the giant-and-dwarf theory was abandoned in the mid-1920s.

3.6 OUTLINE OF SOLAR ACTIVITY AND ROTATION

From observations of sunspots made prior to 1630 by Father Scheiner, it was apparent that spots near the sun's equator had shorter periods of revolution than those at higher heliocentric latitudes. However, it was the long series of observations published by Carrington in 1863 that led to the first clear recognition of solar differential rotation (see eqs. [2.13] and [2.14]). The extreme internal mobility betrayed by his observations led to the inference that the matter composing the sun was mainly or wholly gaseous. In 1864, from the obviously high surface temperature of the sun, Father Secchi and John Herschel made a similar suggestion about its internal structure, but their views first gained general acceptance through a phenomenological theory of the sun that was proposed by Faye in 1865.

Hervé Faye, who taught geodesy in Paris and astronomy in Nancy, was primarily a theoretician with a strong leaning toward phenomenology. He had a brilliant career, became a member of the Académie des Sciences at the age of thirty-three, and published more than two hundred research papers in its *Comptes-rendus hebdomadaires des séances*. His contribution to solar physics is important because he was the first to propose a coherent scheme of the solar constitution covering the whole range of surface phenomena.[33] Like most of his contemporaries, Faye believed that the sun could no longer be regarded as a cool body surrounded with a cocoon of fire, as was suggested by William Herschel in 1795, but that instead it was a vast heat-radiating machine. In his view, the sun was pervaded by extensive convective currents bringing up successive portions of the intensely heated interior mass. This retarded its surface rotation since the rising currents were continually left behind as the sphere widened in which they were forced to move. The effect was less at the equator than at higher heliocentric latitudes, however, because Faye further assumed that the vertical distance ascended by the currents went on increasing from the equator to the poles, following a pattern similar to that of the surface rotation rate. This was plainly an arbitrary expedient of the theory.

Another novelty introduced by Faye consisted in regarding the photosphere no longer as a defined surface, in the mathematical sense, but as a vast region in the sun's atmosphere. In Faye's 1865 theory, sunspots were regarded as simply breaks in the photospheric clouds, where the ascending currents of gas had strength to tear them into pieces. Through the sunspots, it therefore became possible to look down into the sun's deep interior, where the gases were too hot to emit visible light. In other words, the darkness of sunspots was attributed to a deficiency of emissive power. This explanation was criticized at once on the ground that it ran counter to the well-known Kirchhoff-Stewart relationship for the radiation emitted and absorbed by a hot body (see section 2.2). Indeed, since emissive and absorptive power are strictly correlative, it was concluded that the supposed lack of visible light from the sun's interior would be exactly compensated by an increase of its transparency. Accordingly, the light from the far side of the photosphere would shine across the whole body of the sun, so that no sunspots at all would remain visible to the naked eye on the near side of the photosphere.

---

[33] *Comptes-rendus Acad. Sci. Paris*, **60**, 89 and 138 (1865); **61**, 1082 (1865); **75**, 1664 and 1793 (1872); **76**, 509 (1873). This and other theories of the sun were discussed by J. Bosler in his book entitled *Les théories modernes du soleil* (Paris: Octave Doin et fils, 1910). See also A. M. Clerke, *A Popular History of Astronomy during the Nineteenth Century*, 4th edition (London: Adams and Charles Black, 1908); A. J. Meadows, *Early Solar Physics* (Oxford: Pergamon Press, 1970).

In the early 1870s, Faye abandoned the idea that sunspots were regions of ascending currents and proposed an entirely new approach. He again explained the solar differential rotation as due to large-scale convective currents. However, he suggested now that the latitude variations in the surface rotation rate led to the formation of a multitude of circular gyratory motions around a vertical axis extending to a great depth, as in our rivers and in the upper currents of our atmosphere. These eddies, which follow the large-scale motion of the photosphere, then develop into full-scale cyclones that carry downward the cooler materials of the upper layers, and thus produce in their center a decided extinction of light and heat as long as the cyclonic motion continues. Although Faye's second theory of sunspots accorded better with the observations than his first theory, it was objected that only a small fraction of all sunspots showed a cyclonic motion. Moreover, those which did resemble circular vortices had either clockwise or counterclockwise motion in the same hemisphere, whereas in Faye's theory the circular gyratory motion should always be in the same direction. It was also pointed out that the relative drift at different heliocentric latitudes was generally too small to generate the full-scale cyclones that were attributed to it.

Although Faye's theories of solar activity and rotation did not withstand the passage of time, we still find that several ideas of permanent value were embodied in his sketch of the solar constitution. The principal of these are that the sun is predominantly gaseous, that several surface phenomena are due to convective currents, that the solar differential rotation results from a complex interaction between rotation and convection, and that the photosphere is a region of the sun's atmosphere rather than a definite surface. Throughout the end of the nineteenth century, however, the nature of the sunspots remained a much debated question.

A common picture of a sunspot was that of a shallow depression in the solar surface filled with a darker and more absorbing mass. For example, in the early 1870s Secchi suggested that sunspots were formed by masses of absorbing vapors brought out from the sun's interior by violent eruptions. These vapors, which he identified with faculae and prominences, then cooled and fell back into the photosphere, forming a sort of cavity or basin filled with cool, dense gases. These shallow depressions he identified with sunspots. Later, in 1886, Lockyer also advocated the idea that sunspots were regions of descending currents of gas.[34] In his all-embracing theory, unlike Secchi's, the descending currents were the direct consequence of a large-scale circulation in the sun's atmosphere, with material flowing out of the polar regions and showing up as polar rays during eclipses. If so, then, the eleven-year sunspot cycle was a direct effect of this atmospheric circulation, while the presence of spots at middle latitudes resulted from the fall of matter from above descending on the photosphere. Specifically, at high latitudes there were no spots because the height from which the matter would fall back on the sun's surface was too small to cause any sizable disturbance. At lower latitudes, however, the matter would fall back from very great heights so that spots could be produced when the condensed matter was hitting the surface. Simultaneously, the material thrown up by the impact would then rise into the sun's atmosphere as a prominence. This large-scale atmospheric

---

[34] *Proc. Roy. Soc. London*, **40**, 347 (1886).

circulation also explained the differential rotation of the sun since, as was noted by Lockyer, the matter falling from greater heights near the equator would also bring a larger amount of angular momentum and, hence, accelerate the sun's rotation in the equatorial belt, as is observed. Unfortunately, Lockyer's large-scale circulation has never been observed in the sun's atmosphere.

Although many other hypothetical explanations of the development and evolution of sunspots were put forth in the second half of the nineteenth century, it is worth noting that considerable effort was expended during that period on the study of planetary influences on the eleven-year sunspot cycle.[35] To the best of our knowledge, Rudolf Wolf in 1859 was the first to contemplate a causal relationship between planetary motions and sunspot number variations. By the early 1890s, however, he came to the conclusion that none of the attempts, by himself and by others, to fit these variations with models based on planetary influences had produced truly satisfactory results. The planetary influence thesis enjoyed a short-lived revival at the turn of the twentieth century, but its downfall was precipitated by Hale's discovery, in 1908, of the magnetic nature of sunspots. In the following years, he and his colleagues at Mount Wilson also established that the sunspot cycle is the manifestation of an underlying magnetic cycle having twice the period of the sunspot cycle, therefore suggesting that solar activity is a hydromagnetic phenomenon operating within the solar interior (see figure 2.8). These and related concepts will be discussed in sections 4.8 and 5.8.

3.7 RETROSPECT: THE NINETEENTH-CENTURY ADVANCES

This and the preceding chapters have contained some account of progress made in observational and theoretical astrophysics from about 1610 to the 1910s. As was shown in section 2.3, by 1810 astronomers already possessed a fair knowledge of several of the sun's properties—that is, its distance, size, mass, mean density, and rate of rotation. There was little positive knowledge of the nature of the stars, however, since their distances had not yet been measured. In fact, it was not until the 1860s that astronomers could conclusively establish that stars were self-luminous objects, basically similar to the sun but seen as small points of light in the night sky by virtue of their large distance. This event, of great conceptual significance in astronomy, was actually the result of two major instrumental achievements. One was the measurement of the first stellar parallaxes and so of the distances of stars in the late 1830s. The other was the development of spectroscopy, which led to the discovery that stellar spectra were basically absorption-line spectra, and that some (but not all) stars had absorption lines in their spectra that were similar to those observed in the solar spectrum. Solar and stellar physics proper began when in 1859 Kirchhoff[36] showed "that the dark lines of the solar spectrum, which are

---

[35]For a close scrutiny of these solar cycle models, see P. Charbonneau, *J. History Astron.*, **33**, 351 (2002).

[36]*Monatsberichte Berliner Akad.*, S. 662 (Oktober 1859).

not caused by the terrestrial atmosphere, arise from the presence in the glowing solar atmosphere of those substances which in a flame produce bright lines in the same position." Since each chemical element has its own characteristic set of spectral lines, the existence of many elements, known terrestrially, could therefore be established for the sun and the stars.

As of 1810, there was also little knowledge of the internal structure and evolution of the sun. Following a suggestion made by William Herschel in 1795, it was generally believed that the interior of the sun was composed of solid rock, surrounded by a luminous atmosphere, and that sunspots gave glimpses through this superficial layer to the sun's dark interior; life on the sun was considered possible. In the 1850s, following recent developments in the new science of thermodynamics, astronomers abandoned Herschel's picturesque theory and came to regard the sun as made of compressible gases that through progressive compression generated the heat and light that it was radiating into space. Little known is the fact that this mechanism, which is known as the *Helmholtz-Kelvin contraction*, was originally suggested by Waterston in 1853. In the late 1890s, however, this idea was also discarded because, if it were true, the sun could not be more than about 20 million years old, less than the geologists were claiming for the oldest rocks containing fossils of living creatures. Obviously, the astronomers of that period did not have the theoretical tools for studying the physical constitution of the sun. In fact, much later in the twentieth century they would realize that the gravitational contraction mechanism was capable of sustaining the energy radiated during the early phases of a star's history, when its central temperature was not yet hot enough to permit the transmutation of hydrogen into helium.

The period from about 1860 to the 1910s witnessed the acquisition of an unprecedented amount of detailed knowledge of the sun and the stars. Indeed, thanks to the creation of solar and stellar spectroscopy in the 1860s, astronomers had good grounds for thinking that the atmosphere of the sun contained iron and other metals in the form of vapor; that the photosphere, which gave the continuous part of the solar spectrum, was certainly hotter; and that the same thing was probably true in general of the stars, even though their spectra showed considerable variations in the lines they contained and in their relative intensities. Two obvious paths of research thus lay open. One was the investigation of spectra of individual heavenly bodies with the aim of identifying chemical elements; the other was the classification of stellar spectra with the aim of discovering relations between them. After preparatory work by Father Secchi in the 1860s, stellar spectra were eventually classified in the extensive *Henry Draper Memorial Catalogue* of the Harvard College Observatory, which was published by Pickering in 1890.[37] Thus it was conclusively established that a large fraction of stellar spectra could be arranged into one continuous sequence, essentially according to decreasing temperature. Later advances in photometry showed that another parameter—absolute luminosity—was necessary for stellar classification. The discovery in the 1910s of the *Hertzsprung-Russell diagram*, or temperature-luminosity diagram, was the culmination of these pioneer-

---

[37] Its spectral classification represents the forerunner of the system for the *Henry Draper Catalogue*, which was published between 1918 and 1924 in nine volumes of the *Harvard Annals*.

ing years (see figure 2.12). Leavitt's period-luminosity relation for the Cepheid variables also belongs to that period. Also by 1910, as the result of several decades of research, solar physicists possessed a fair knowledge of solar activity and rotation. Broadly speaking, they had firmly established that the rotation period of the sun was minimum at the equator and increased gradually toward the poles, that the production of sunspots followed an eleven-year cycle, that chromospheric activity and coronal shape varied along with the sunspot cycle, that sunspots were the seat of strong magnetic fields, and that the variations of terrestrial magnetism followed with extreme closeness the variations of the sunspot numbers. And yet, despite the wealth of observational material thus accumulated, what the solar physicists of 1910 most needed was a solid theoretical framework for investigating the causes of solar activity, the origin of the sun's heat and light, and ultimately the sun's place in the evolution of the stars and nebulae. Real progress would not begin until the 1920s and 1930s, however, with the advent of modern atomic physics, nuclear physics, and magnetohydrodynamics.

Investigations of a somewhat different character—of the equilibrium, oscillations, and stability of self-gravitating masses—were also carried out during the period from 1870 to the 1910s. The first quantitative model of the solar interior was obtained by Lane in 1870 under the assumption that the sun was a polytropic gas sphere in convective equilibrium. Thus the sun was envisioned as a perfect gas, and the outward flow of energy was assumed to occur as a result of convection. Although Lane's model fitted well with the idea that gravitational contraction was the source of energy for the sun, his work was given little credence, because in the 1870s there was no evidence that any star existed for which the theory of a perfect gas would be applicable. Lane's prescient paper was followed and amplified in the 1880s by investigations on similar lines by Ritter and Kelvin, culminating in Emden's 1907 book *Gaskugeln*. Little known is the fact that Ritter in 1880 was the first to suggest that variable stars might be self-gravitating masses undergoing radial pulsations (see appendix C). To the same period belong the pioneering works of Roche, who developed in 1873 a deceptively simple representation of a double star, the so-called *Roche model*. Both Ritter and Roche were far ahead of their time since it was not until the late 1910s that a few theoreticians developed a genuine interest in stellar pulsations and binary stars. By then, thanks to Karl Schwarzschild's 1906 paper, it was widely accepted that in stellar conditions the main process of transfer of energy was by radiation and that the other modes of transfer—conduction and convection—may be neglected. This was the start of modern studies in solar and stellar physics.

Chapter Four

The Formative Years: 1910–1940

We may perhaps be glad that our lives have fallen in the
beginning, rather than at the end, of this great stretch of time.
We may well imagine that if man survives to the end of it, he
will have infinitely more knowledge than now, but one thing he
will no longer know—the thrill of pleasure of the pioneer who
opens up new realms of knowledge.
  —Sir James JEANS

The nineteenth century saw the completion of the science of dynamics founded two centuries earlier by Galileo and Newton. This theory not only provided a successful explanation for all mechanical phenomena known at that time but also laid the foundations of the kinetic theory of gases. Classical statistical mechanics, in turn, provided a sound base for thermodynamics, which had become an exact science with the enunciation of the all-embracing principle of the conservation of energy. The development of electrodynamics also made it possible to understand the phenomena of light, electricity, and magnetism in terms of a single set of basic physical principles. This theory, which was originally developed by James Clerk MAXWELL (1831–1879) in 1865, provided a most convenient explanation for the wave propagation of light. At the end of the nineteenth century, however, it became increasingly evident that many phenomena, especially those connected with radiation, defied explanation on classical principles.

   As was noted in section 2.2, one of the first disturbing failures of classical physics resulted from attempts to explain the shape of the black-body curves, as depicted in figure 2.7, on the basis of the wave theory of light. The first development that led to the resolution of this dilemma was Planck's introduction of the quantum hypothesis. Specifically, he approached the problem of *black-body radiation* in a cavity by treating the interactions of a set of harmonic oscillators with the radiation field. By assumption, each oscillator that is covering the walls of the cavity radiates an exact integral multiple of the basic quantum $\epsilon = h\nu$, where ν is the frequency of the oscillator and h is the quantum action, now known as Planck's constant. (It has the numerical value $h = 6.62 \times 10^{-27} \, \mathrm{g \, cm^2 \, s^{-1}}$.) The next important development in quantum mechanics was the work of Albert EINSTEIN (1879–1955), who suggested in 1905 that not only the oscillators but also the radiation field itself might be quantized, so that energy can be absorbed from it only in quanta of size $h\nu$. This corpuscular theory of light led Einstein to his explanation of certain properties of the *photoelectric effect*—an experimentally observed phenomenon in which electrons are emitted from certain illuminated metals. The success of Einstein's approach clearly demonstrated the corpuscular nature of radiation, thereby establishing the puzzling wave-particle duality of light.

Nowhere, perhaps, is the inadequacy of classical physics revealed more strikingly than in *spectroscopy*. Indeed, it has long been known that there are some remarkable regularities in the spectrum of hydrogen. As early as 1885, Johann Jacob BALMER (1825–1898) in Basel showed that the wavelengths λ of many spectral lines are well represented by the empirical formula

$$\frac{1}{\lambda} = R \left(\frac{1}{n^2} - \frac{1}{m^2} \right), \qquad (4.1)$$

in which n and m are integers, and R is a constant, the so-called Rydberg constant.[1] (It has the numerical value $R = 109,677.6\,\mathrm{cm}^{-1}$.) If, for example, one takes $n = 2$ and $m = 3, 4, 5, \ldots$, eq. (4.1) gives the wavelengths of the lines in what is called the Balmer series. The choices $n = 1, 3, 4,$ and 5 yield, respectively, the Lyman, Paschen, Brackett, and Pfund series, after their discoverers. Note that these series are seen to approach a limit as m approaches infinity.

The next major development in the quantum theory was due to the Danish physicist Niels BOHR (1885–1962) who proposed, in 1913, a theory of the hydrogen atom that explained the regularities summarized in eq. (4.1). By that time there was no longer any doubt as to the validity of the solar-system model of the atom proposed two years earlier by Ernest RUTHERFORD (1871–1937), with the nucleus and the electrons playing, respectively, the role of the sun and the planets. However, Maxwell's theory of electromagnetism unequivocally predicted that an electron revolving around a positive charge would necessarily radiate energy and eventually fall into the nucleus. This difficulty led the young Dane to postulate that atomic electrons obey quantum laws that select for them certain specific energies and force them to move in definite orbits, in defiance of the laws of classical electrodynamics. Moreover, when an electron jumps from one quantum state of energy E_1 to another of energy E_2, the difference $E_1 - E_2$, if positive, is emitted in the form of a monochromatic wave of frequency ν such that $E_1 - E_2 = h\nu$. Similarly, for an electron to move to a higher state of energy, the atom must absorb a quantum of exactly the correct energy. In appendix E we develop Bohr's picture of the hydrogen atom and show that it gives amazingly accurate agreement with the observed hydrogen spectrum.

The application of this theory to systems with more than one electron proved difficult, however. Bohr's simple picture of the atom was therefore superseded by more elaborate models, but the notion remains that spectral lines exist because the bound energy states of an atom are quantized. As a matter of fact, the Bohr atom is one of the major concepts in the historical development of quantum mechanics—a development which, in turn, paved the way for the quantitative analysis of stellar spectra in the 1920s.

---

[1]This constant was named after the Swedish physicist Johannes Robert RYDBERG (1854–1919), in recognition of his work in sorting out many series in the measured spectra of various elements.

4.1 THE BEGINNINGS OF QUANTITATIVE ASTROPHYSICS

As was noted in section 2.2, it was in 1859 that Kirchhoff laid the foundations of astronomical spectroscopy by showing how the chemical composition of the sun and stars could be determined from dark lines in their spectra. Later, in 1868, Anders Jonas ÅNGSTRÖM (1814–1874) at Uppsala University published a comprehensive table of solar spectrum wavelengths which for a long time served as a standard. By 1897 the work of Henry Augustus ROWLAND (1848–1901) at Johns Hopkins University (Baltimore, Md.) had led to the identification of thousands of solar absorption lines with spectral lines observed in the laboratory. It was also known that most stars had spectra that resembled the solar spectrum in that they consisted of a bright continuous band of radiation crossed by absorption lines, but there were many more that were markedly different. Specifically, the continuous spectra of certain stars were richer in blue light or in red light than the spectrum of the sun, and there were also enormous differences in their absorption spectra. Some of them showed practically nothing except the lines of hydrogen and helium; others, like the solar spectrum, had very strong lines of iron and other metals; still others were even known to have molecular bands corresponding to diatomic molecules, such as CH, CN, and TiO.

Not surprisingly, when these remarkable spectroscopic differences were first observed, astronomers concluded that stars differed widely in chemical composition, namely, that some consisted principally of hydrogen or helium, some of metals, some of simple chemical compounds. In the 1870s important progress was made by Lockyer when he showed experimentally that the spectrum of a single substance changed as the temperature rose. Thence, he advanced the bold idea that, just as the transition of band spectra into line spectra may be explained by the dissociation of molecules into atoms, so the spectral changes brought about by increased temperature may be due to the breakup of atoms into still more elementary particles, which he called protoelements (see section 3.5). In the 1890s, at a time when it was realized that stellar spectra can be arranged into a continuous sequence, Lockyer correctly suggested that, with increased temperature, the spectra of stars should show different spectral lines that he termed "enhanced," meaning that they were stronger at the higher temperatures; in his opinion, therefore, it should be possible to estimate qualitatively the relative temperatures of stars from the particular pattern of lines of each element observed in their spectra.

Another important discovery was made in 1896 by Pickering, who found that ζ Puppis, a Wolf-Rayet star, had a spectrum that was "very remarkable and unlike any other as yet obtained."[2] It contained six lines that alternated with the Balmer lines of hydrogen, having half-integer values ($m = 5/2, 7/2, \ldots$), instead of the integral values ($m = 3, 4, \ldots$) associated with the Balmer lines (see eq. [4.1]). Pickering concluded that these new lines, instead of being due to some unknown element as was supposed at first, were probably due to hydrogen under conditions of temperature and pressure as yet unknown. Soon afterward, Lockyer attributed

---

[2]*Astrophys. J.*, **4**, 369 (1896); **5**, 92 (1897). See also N. Robotti, *Hist. Studies Phys. Sci.*, **14**, 123 (1983).

Pickering's new spectral lines to protohydrogen and suggested that ζ Puppis was one of the highest-temperature stars known. These enhanced lines took some years more to be identified, however, in part because of the difficulty in reproducing them in the laboratory. It was not until 1912 that Alfred FOWLER (1868–1940), one of Lockyer's students, was able to show that the Pickering series lines could be reproduced in the laboratory using a mixture of hydrogen and helium at high temperature. The following year, 1913, Bohr showed that these lines were indeed due to *ionized* helium, whose atomic structure closely resembles that of hydrogen apart from having twice the nuclear charge and four times the mass (see appendix E). With this work, a new framework for spectroscopy emerged within which Bohr's ionized atoms replaced Lockyer's protoelements and transformed his enhanced lines into the expression of a well-defined subatomic phenomenon.

The first quantitative measurements of stellar temperatures were made in 1909 by Johannes WILSING (1856–1943) and Julius SCHEINER (1858–1914) at Potsdam, and by Charles NORDMANN (1881–1940) in Paris. Broadly speaking, they compared the dependence of a star's brightness on color (or wavelength) with the theoretical predictions of Planck's black-body radiation curves, using data from visual spectrophotometry. In 1914 Hans Rosenberg at Tübingen undertook photographic spectrophotometry for the same purpose. Although large systematic errors were made, mainly due to the departure of the energy distribution of stellar radiation from black-body curves,[3] the important aspect of all these temperature determinations was the systematic trend that earlier spectral types in the Harvard classification were hotter than later types. This was clear indication that it was the temperature that determined the spectral type, and not, as was originally suggested, the chemical composition.

The birth of quantitative astrophysics actually began in 1920, when Meghnad SAHA (1893–1956; figure 4.1), an Indian physicist working in London, applied the concept of ionization equilibrium to astronomy.[4] In his analysis Saha, seeing that the ionization of atoms is analogous to the dissociation of molecules, was able to show how the degree of ionization of an atomic species depends on temperature and pressure. Saha initially applied his theory to the problem of ionization in the solar chromosphere. In particular, he calculated the degree of ionization of calcium as a function of temperature and pressure in the reversing layer, which was believed to produce the flash spectrum (see section 2.3). Saha assumed the reversing layer to

---

[3]Particularly noticeable on spectrograms of Vega and stars of similar type is the sudden fall in the intensity of radiation beyond the end of the Balmer series in the ultraviolet. This phenomenon, which is known as the *Balmer jump*, was studied extensively during the 1930s by the French astronomers Daniel BARBIER (1907–1965) and Daniel CHALONGE (1895–1977) from observations made at the Jungfraujoch (Switzerland) at the altitude of 3457 m. (The high altitude of the site was ideally suited to ultraviolet spectrophotometry.) By 1941 they could then propose a new, accurate system of spectral classification based on two parameters, that is, the magnitude of the Balmer jump and the wavelength of the midpoint of the intensity decrease. For a succinct presentation of this and similar classifications, see J. B. Hearnshaw, *The Analysis of Starlight: One Hundred and Fifty Years of Astronomical Spectroscopy* (Cambridge: Cambridge University Press, 1986), 277.

[4]*Phil. Mag.*, **40**, 479 (1920); *Proc. Roy. Soc. London*, **99A**, 135 (1921); *Zeit. Phys.*, **6**, 40 (1921). For a detailed study of the life and times of Meghnad Saha, see D. H. DeVorkin, *J. History Astron.*, **25**, 155 (1994); see also S. Chatterjee, *Indian J. History Sci.*, **29**, 99 (1994).

Figure 4.1 The Indian physicist Meghnad Saha, who is noted for his development of the thermal ionization equation and for his application of it to the interpretation of stellar spectra. From G. Gamow, *A Star Called the Sun* (New York: Viking Press, 1964). Courtesy of the Gamow Estate.

be at 6,000 K but with a pressure decreasing outward from 10 atm to only 10^{-12} atm in its outermost layers. He found that the degree of ionization of calcium was only 2% in the deeper levels but, because of the falling pressure, was almost complete at the level where a pressure of 10^{-4} atm was reached. The great strength of the H and K lines of ionized calcium in the solar chromosphere could thus be explained, whereas the 4,227 Å neutral calcium line was strongest in the deeper layers.

Within a year after this work, Saha applied his ionization theory to the Harvard sequence of spectral types. By considering the temperatures at which different spectral lines should show marginal appearances, he was able to deduce a temperature scale for all the spectral types, from O to M. Comparing next his results with the color temperatures obtained by Wilsing and others, he found good agreement between his temperature scale and theirs. He therefore concluded that the Harvard sequence was indeed a temperature sequence, with spectral differences being caused by differences in excitation rather than differences in chemical composition. Noting that at constant temperature the degree of ionization of a chemical element increases

Figure 4.2 Edward Arthur Milne (*right*) with Henry Norris Russell in 1939. Milne first
became internationally known for his pioneering work on radiative transfer and
stellar atmospheres. All this work was completed in the 1920s and, when he left
the field in 1930, he moved on to stellar interiors and to cosmology. His later
work was not as successful, but it stimulated important work by others. Milne
was appointed the first Rouse Ball professor of mathematics at Oxford University
in 1929; he remained at Oxford for the remainder of his career, apart from service
on the Ordnance Board in 1939–1944. Courtesy of Meg Weston Smith, Milne's
daughter.

with decreasing pressure, Saha further pointed out that the observed correlation between line intensity and absolute magnitude for late-type stars of apparently the
same color, and hence temperature, was probably due to differences in atmospheric
density. This theoretical prediction was soon to be confirmed by observers, who
showed that giants and supergiants do indeed have lower densities and pressures
than main-sequence stars of the same spectral type.

By 1921 Saha had thus realized that differences in spectral type could be attributed to differences in temperature and pressure. This important work was further developed in 1923 by Ralph Howard FOWLER (1889–1944) and Edward Arthur
MILNE (1896–1950; figure 4.2) who were then collaborating at Cambridge University. Combining Saha's theory of ionization with Boltzmann's theory of excitation,
they showed that the number of atoms of a chemical species responsible for the
production of a spectral line could be estimated from the line intensity, once the
temperature and pressure in the star's atmosphere were known. Clearly, the path
was now open for a detailed comparison between theory and observation. This was
accomplished by Cecilia PAYNE (1900–1979), a former undergraduate student of

Milne at Cambridge, who went to Harvard and there wrote her famous book, *Stellar Atmospheres* (1925), for which she was awarded the degree of Ph.D. from Radcliffe College, Harvard's sister institution.[5] Using the new ionization-excitation theory and the Harvard collection of stellar spectra, she established the cosmic abundances of the chemical elements and showed that temperature, not composition, played the dominant role in fixing a star's spectral type. Payne was also the first to demonstrate the great preponderance of the lightest elements, hydrogen and helium, in stars. For the remaining elements, however, she found a close agreement between the composition of the earth's crust and the stars, thus corroborating the idea of abundance uniformity throughout the Harvard sequence of spectral types.

Three years later, at a time when the presence of large amounts of hydrogen in stars was still regarded as spurious, Albrecht UNSÖLD (1905–1995) in Kiel showed that hydrogen was also the most abundant element in the sun by analyzing the intensity profiles of the darkest lines. The following year, 1929, a much more detailed spectroscopic analysis of the sun by Russell, in which the abundances of fifty-six elements were obtained, fully confirmed the very high abundance of hydrogen in the sun's atmosphere. Russell's work also led to the establishment of a reliable table of atomic abundances in the sun.

In the 1930s important progress was made in our understanding of physical processes in stellar atmospheres, including the formation and broadening of spectral lines and the development of model atmospheres (see also section 4.7). The first *curves of growth* for spectral lines, which relate line strength to element abundance for a given star, were constructed by the Flemish-born astronomer Marcel MINNAERT (1893–1970) and his colleagues in the Netherlands. At about the same time, in 1931, *model atmospheres* were calculated by William McCREA (1904–1999) in Edinburgh for hot stars with neutral atomic hydrogen opacity. In 1939 the failure of this opacity source to model the observations for cooler stars led Rupert WILDT (1905–1976), a German émigré who was then working in Princeton, to suggest that the rather unstable negative hydrogen ions, H⁻, are the main source of opacity for these stars. Soon afterward, the first solar models that incorporated the new H⁻ opacity were constructed by the Danish astronomer Bengt STRÖMGREN (1908–1987). All these new techniques greatly improved the development of stellar spectral analysis and eventually led to the result that most of the stars have atmospheres that are very close in composition to the sun's. However, as we shall see in chapter 5, there are many other stars in the galactic halo that greatly depart from the usual solar abundances.

4.2 THE STELLAR-ENERGY PROBLEM

The origin of the sun's heat and radiation was first discussed on a scientific basis in the nineteenth century. At that time only *gravitational contraction* was consid-

---

[5]By her own account, Cecilia Payne (later Payne-Gaposchkin) moved to the United States because she felt that there was no future for her in England other than teaching, but soon she found out that academic advancement at Harvard was still denied to women. In 1956 she was finally made a full professor, the first woman at Harvard to be promoted to this rank. See C. Payne-Gaposchkin, *Cecilia Payne-Gaposchkin: An Autobiography and Other Recollections*, edited by K. Haramundanis (Cambridge: Cambridge University Press, 1984).

ered to be capable of sustaining the energy radiated by the sun. As explained in section 3.1, Waterston was the first to show, in 1845 and again in 1853, that the sun's progressive contraction might maintain its heat for a long period of time by transforming gravitational potential energy into thermal energy. The solar contraction theory and the accompanying change of potential energy into heat was also advanced by Helmholtz in 1854, and it was soon ably advocated by Kelvin until the end of the 1890s.

The Helmholtz-Kelvin theory, as it became known, played a prominent role in the controversy between physicists and astronomers who argued that the theory satisfied all the requirements of stellar astronomy proper, and geologists and planetary cosmologists who thought that the theory was inadequate because it did not allow the sun and the earth-moon system an age of more than twenty million years. In 1899, amid the heated controversy between Kelvin and his opponents, a little-known but important remark was made by the American geologist Thomas Chrowder CHAMBERLIN (1843–1928).[6] In his own words:

> In the first place, without questioning its *correctness*, is it safe to assume that the [Helmholtz-Kelvin] hypothesis of the heat of the sun is a *complete* theory? Is present knowledge relative to the behavior of matter under extraordinary conditions as obtain in the interior of the sun sufficiently exhaustive to warrant the assertion that no unrecognized sources of heat reside there? What the internal constitution of the atoms may be is yet an open question. It is not improbable that they are the complex organizations and the seats of enormous energies.

And he went further, adding:

> The [Helmholtz-Kelvin] theory takes no cognizance of latent and occluded energies of an atomic or ultra-atomic nature. A ton of ice and a ton of water at a like distance from the center of the system are accounted equivalents, though they differ notably in the total sum of their energies. These familiar latent and chemical energies are, to be sure, negligible quantities compared with the enormous resources that reside in gravitation. But is it safe to assume that this is true of the unknown energies wrapped up in the internal constitution of the atoms? Are we quite sure we have yet probed the bottom of the sources of energy and are able to measure even roughly its sum-total?

To the best of our knowledge, this was the first explicit suggestion that *subatomic processes* could liberate energy within the sun and hence sustain its present luminosity for a long enough period of time.

This controversy came to an abrupt end with the discovery of radium's tremendous heating effect in 1903, when many physicists and astronomers became willing to recognize that the radiation of the sun and the stars might originate in *radioactivity*, that is, the spontaneous breaking down of one form of atom into another. During the next decade, however, calculations by Rutherford and others showed that there is far too little uranium, thorium, or any of the elements heavier than lead in the sun

---

[6]*Science*, **9**, 12 (1899).

for appreciable amounts of energy to be generated by this mechanism. Meanwhile, in 1904 Jeans suggested that an enormous store of energy could be derived from the total *annihilation of matter*, with positive and negative charges occasionally colliding and so annihilating one another, with their mutual energy being set free as radiation. This hypothesis was further explored in the late 1910s by the English astrophysicist Arthur Stanley EDDINGTON (1882–1944), who concluded that only in specially favorable cases will annihilation result and that, as the lighter elements are gradually consumed, this supply of energy will become exhausted.

The next step toward the resolution of the stellar-energy problem came in 1915, when William Draper HARKINS (1873–1951) of the University of Chicago began work on the structure and the reactions of atomic nuclei. At that time the only nuclear reactions that had been studied were the decomposition reactions of radioactive nuclei, for which Einstein's 1907 equation relating mass and energy predicted the observed energies (see appendix F). With this equation Harkins showed the enormous amount of energy produced by the conversion of hydrogen into helium, with the attendant mass loss; he also identified this reaction as the source of stellar energy.[7] Four years later, a similar suggestion was independently made by the French physical chemist Jean PERRIN (1870–1942),[8] who wrote (in translation):

> Imagine that the tiny dust particles or isolated molecules that constitute the primeval nebula were made of light atoms such as hydrogen, nebulium, or helium. Knocking each other about with the large velocities they perhaps already had in coming from outer space, or that they acquire when falling toward the center of the nebula, these particles radiate heat and the mean temperature gradually increases. Thus, on the one hand, with the nebula contracting, encounters between light atoms capable of assembling into heavy atoms will become more and more numerous; on the other hand, with the temperature rising, the intensity of the radiation that can induce these encounters continues increasing. For these two reasons, the formation of heavy atoms becomes more and more important, being accompanied by ultra-hard X-rays which, for the most part, do not escape from the celestial body, whose temperature becomes gigantic: the star has begun to shine.
>
> It has begun to shine by a mechanism definitively comparable to that by which the combustion of a blazing mass kindles when brought to a sufficient temperature . . . . In short, one recovers the old idea of combustion but, instead of a combustion in which the atoms barely penetrate each other as in the formation of carbonic gas, there is deep penetration that mixes the atomic nuclei in a much more intimate manner.

Instead of being concerned with the heaviest of the elements, Harkins and Perrin were therefore suggesting that it is the building up of complex chemical elements from simpler ones that maintains the enormous expenditure of the sun and the stars

---

[7] See especially W. D. Harkins and E. D. Wilson, *Phil. Mag.*, **30**, 723 (1915).
[8] *Annales de physique*, **xi**, 90–91 (1919).

in the form of heat and radiation. The process involved became known as *nuclear fusion*.

In February 1920 Perrin restated his original suggestion while making it more specific.[9] His proposal was that if four atoms of hydrogen could be converted into one atom of helium, energy would be released because the mass of the helium atom is less than the sum of the masses of four hydrogen atoms. As explained in appendix F, this energy release is a general consequence of *special relativity*: a change of energy δE in any system involves a change in its mass equal to $\delta E/c^2$, where c is the speed of light. Specifically, since the atomic weight of hydrogen is 1.008 and that of helium 4.002, there is a loss of mass in the fusion process amounting to about 0.03 (the unit being the sixteenth part of the mass of the oxygen atom), and this is the mass of radiation discharged in space when helium is made out of hydrogen.

The same suggestion was made a few months later by Eddington[10] (figure 4.3) who, apparently unaware of Perrin's work, was led to the following conclusion: "If 5 per cent of a star's mass consists initially of hydrogen atoms, which are gradually being combined to form more complex elements, the total heat liberated will more than suffice for our demands, and we need look no further for the source of a star's energy." The year was 1920, however, and the details of how four atoms of hydrogen could be converted into one atom of helium were still lacking. Eddington therefore remarked:

> So far as the immediate needs of astronomy are concerned, it is not of any great consequence whether in this suggestion we have actually laid a finger on the true source of the heat. It is sufficient if the discussion opens our eyes to the wider possibilities. We can get rid of the obsession that there is no other conceivable supply besides contraction, but we need not again cramp ourselves by adopting prematurely what is perhaps a still wilder guess.

As we now know, the development of quantum mechanics and nuclear physics was to provide a resolution of this difficulty, but the solution did not come until the late 1930s, when nuclear physics had become a mature science (see section 5.1). Throughout the 1920s and 1930s, therefore, astrophysicists had to exercise their ingenuity in developing sound, but necessarily incomplete models that had to be adjusted to the observational data. As we shall see in section 4.3, this strategy led them to make some progress in understanding the conditions under which subatomic energy is liberated in the sun and the stars.[11]

4.3 THE INTERNAL STRUCTURE OF STARS

The first attempt at a scientific theory of stellar structure was the theory of polytropic gas spheres, initiated by Lane in 1870 and further developed by Ritter in 1880. As

---

[9] *La revue du mois*, **xxi**, 158–164 (février 1920).

[10] *Nature*, **106**, 14 (September 1920).

[11] For a quasiexhaustive analysis of the literature, see K. Hufbauer, *Hist. Studies Phys. Sci.*, **11**, 277 (1981). Surprisingly, this paper does not mention Jean Perrin's contributions.

Figure 4.3 Sir Arthur Eddington, who is particularly known for his fundamental researches
on the internal constitution of the stars. In 1913 he was elected to the Plumian professorship of astronomy and experimental philosophy at Cambridge University
as successor to Sir George Darwin, and in the following year he became director
of the university observatory. Concurrently with his stellar studies, Eddington
was deeply interested in Einstein's 1915 theory of gravitation. From 1930 onward he worked on a unification of general relativity and quantum mechanics,
but this work has not won general acceptance. He was also extraordinarily gifted
in popular exposition. Courtesy of Yerkes Observatory.

was noted in section 3.2, this early work was expanded and codified in Emden's
1907 book *Gaskugeln*. The sun and the stars were thus envisaged as a perfect gas of
uniform chemical composition, and the energy transport from their central regions
to the surface was believed to occur as a result of convection. Although the exact
nature of stellar matter was still a highly debatable question, this model fitted well
with the idea that gravitational contraction was the source of energy for the sun
and stars. As early as 1893, however, the British astronomer Ralph Allen SAMPSON
(1866–1939) had already pointed out that the transport of heat by radiation must
far exceed the transport of heat by large-scale convective currents, but his analysis
could not be developed fully without the more recent progress of thermodynamics.
In 1905 the German-born English physicist Arthur SCHUSTER (1851–1934) applied

Sampson's idea to energy transfer in an atmosphere and obtained solutions of the basic equations. The following year, 1906, Karl Schwarzschild brought this notion into prominence in his epoch-making paper concerning the radiative equilibrium of the solar atmosphere.[12] Then, in 1916, Eddington began his classical work on the radiative equilibrium of stars, which was ultimately published in his book *The Internal Constitution of the Stars* (1926). During this short span of time he gave convincing evidence that the sun and the stars are gaseous bodies with central energy sources, probably of subatomic origin, and that the energy liberated is transported radially outward by radiation.

The Mathematical Formulation

Eddington's basic assumptions, in addition to the usual assumption of a perfect gas, were those of radiative heat transfer and partial support by radiation pressure. The total pressure p is thus made up of gas pressure p_g and radiation pressure p_r, so that

$$p = p_g + p_r. \tag{4.2}$$

For a perfect gas, one has

$$p_g = \frac{\mathcal{R}}{\mu}\rho T, \tag{4.3}$$

where ρ is the density, T is the temperature, μ is the mean molecular weight in terms of the hydrogen atom, and $\mathcal{R}$ is the universal gas constant with a value $\mathcal{R} = 8.317 \times 10^7$ erg g^{-1} K^{-1}. One also has

$$p_r = \frac{1}{3}aT^4, \tag{4.4}$$

where a is the radiation pressure constant with a value $a = 7.564 \times 10^{-15}$ erg cm^{-3} K^{-4}.

By making use of eqs. (4.2)–(4.4), we readily see that any one of the three variables p, ρ, T can be calculated from the other two. Two equilibrium conditions are therefore needed to calculate these two independent variables: (i) the condition of *hydrostatic equilibrium*, which was already encountered in section 3.2, and (ii) the condition of *thermal equilibrium*, which requires that the star is in perfect energy balance, that is to say, the rate at which energy is being liberated in any small region of the star must equal the rate at which energy is carried away from that region. Following Eddington, we shall abandon Lane's hypothesis of convective equilibrium and assume instead that all the energy liberated is carried from the interior to the surface of the star by radiation. This special case of thermal equilibrium is known as *radiative equilibrium*.

---

[12]A. Schuster, *Astrophys. J.*, **21**, 1 (1905); K. Schwarzschild, *Nachr. Königl. Gesell. Wiss. Göttingen*, **195**, 41 (1906). Both papers were reprinted (the latter in translation) in *Selected Papers on the Transfer of Radiation*, edited by D. H. Menzel (New York: Dover Publications, 1966).

As explained in section 3.2, for a spherically symmetric star, the condition of hydrostatic equilibrium is

$$\frac{dp}{dr} = -\frac{Gm}{r^2}\rho, \tag{4.5}$$

where m is the mass inside a sphere with radius r. By virtue of eq. (3.4), we also have

$$\frac{dm}{dr} = 4\pi r^2 \rho. \tag{4.6}$$

This is the mass equation.

The energy-balance equation can be derived in a manner analogous to the mass equation. Let ϵ be the net release of energy per gram and per second, and define the luminosity $L(r)$ as the net amount of energy crossing a sphere with radius r. Conservation of energy implies

$$L(r) = 4\pi \int_0^r \rho\epsilon r^2 \, dr. \tag{4.7}$$

The rate of change of the function $L(r)$ with the radius is, therefore,

$$\frac{dL(r)}{dr} = 4\pi r^2 \rho\epsilon. \tag{4.8}$$

Let us now admit that the only mode of energy transport is radiation. In stellar interiors, where the radiation field is nearly isotropic, radiative equilibrium requires that the force due to the gradient of radiation pressure be equal to the momentum absorbed from the radiation field in passing through matter. As was shown by Eddington in 1916, this balance can be expressed as

$$\frac{dp_r}{dr} = -\frac{\kappa\rho}{c}\frac{L(r)}{4\pi r^2}, \tag{4.9}$$

where κ is the absorption coefficient per gram and c is the speed of light. The quantity $(\kappa\rho)^{-1}$ is the approximate distance that photons travel before absorption. The coefficient κ is thus a measure of the opacity of a star, that is to say, a measure of the resistance to the outward flow of radiation from the star. Combining next eqs. (4.4) and (4.9), one readily sees that the temperature gradient necessary to drive the radiation flux is

$$\frac{dT}{dr} = -\frac{3}{4ac}\frac{\kappa\rho}{T^3}\frac{L(r)}{4\pi r^2}. \tag{4.10}$$

This is the fourth basic equilibrium condition.

There are now four differential equations: eqs. (4.5), (4.6), (4.8), and (4.10); they have to be integrated subject to the boundary conditions that p and p_r simultaneously become negligible at the star's surface, and that $m(0) = L(0) = 0$ at the center. In his first attack on the problem, Eddington assumed uniform chemical composition, constant opacity, and constant rate of generation of energy per gram. Although these must be regarded as convenient assumptions, they are worth discussing here because they led to the so-called *standard model*, which played an important role in subsequent developments. From eqs. (4.5) and (4.9), we have

$$\frac{dp_r}{dp} = \frac{\kappa\epsilon}{4\pi Gc} = \text{constant},\tag{4.11}$$

since $\epsilon = L(r)/m(r)$ when $\epsilon = $ constant (see eq. [4.7]). Integrating eq. (4.11) and using the surface boundary conditions, one readily sees that the ratio p_r/p of the radiation pressure to the total pressure is a constant throughout the model. And since eqs. (4.2)–(4.4) can always be written in the form[13]

$$p = \left[\left(\frac{\mathcal{R}}{\mu}\right)^4 \frac{3}{a} \frac{1-\beta}{\beta^4}\right]^{1/3} \rho^{4/3},\tag{4.12}$$

where $\beta = p_g/p$, it follows at once that for Eddington's standard model we have the relation

$$p = K\rho^{4/3},\tag{4.13}$$

where

$$K = \left[\left(\frac{\mathcal{R}}{\mu}\right)^4 \frac{3}{a} \frac{1-\beta}{\beta^4}\right]^{1/3}\tag{4.14}$$

is a constant since, by assumption, β is a constant. We have thus recovered the condition for a polytropic gas sphere of index $n = 3$ (see eq. [3.8]).

Some Preliminary Results

Making use of Emden's 1907 tables of polytropic functions, Eddington was able to derive some general properties of his model. In particular, he found that in the range of masses covered by actual stars, the ratio p_r/p increased from small values to nearly unity. Eddington concluded that stars cannot exist with masses greater than about one hundred times the sun's mass because with too great a mass the gradient of radiation pressure would exceed the gravitational force that holds the star together, and the star would then explode. He also found that for the giant stars, which have substantial radiation pressure, the total luminosity of a star is determined almost completely by its mass. In 1924, making use of the opacity law $\kappa \propto \rho T^{-7/2}$ derived by the Dutch physicist Hendrik KRAMERS (1894–1952) the preceding year, he obtained very similar results. When he came to compare his results with observations, he unexpectedly found that his theory described the mass-luminosity relation law of the main-sequence stars as well (see figure 2.14). However, to get agreement he had to use empirical data from the star Capella to calibrate his theoretical mass-luminosity relation. He took this agreement as confirming that the main-sequence stars are ideal classical gases, essentially because the high temperature in these stars causes high ionizations that make the effective size of atoms much smaller than normal ones. Later, in 1932, Strömgren studied models of stars with

---

[13] See, e.g., S. Chandrasekhar, *An Introduction to the Study of Stellar Structure* (Chicago: University of Chicago Press, 1939; New York: Dover Publications, 1957), 74.

the assumed law $\epsilon \propto \rho^\alpha T^\nu$, where α and ν are positive constants, and deduced that the total luminosity depended mainly on mass but slightly on radius also, with a weak dependence on the concentration of the energy sources toward the center (see eq. [4.16]). It was also found that models made of light elements ($\mu \approx 1$) gave a better fit to the observational data than those made of heavy elements ($\mu \approx 2$).

In retrospect, it is quite evident that in the 1920s and 1930s the astrophysicists had a very imperfect knowledge of the physical laws that govern the structure and evolution of stellar interiors and atmospheres. Not unexpectedly, this led to very strong differences of opinion between Eddington, who was a theoretician with a strong leaning toward phenomenology, and his closest competitors—Jeans and Milne—who were more mathematically inclined.[14] Both of them criticized Eddington's work on the ground that he could not possibly deduce a unique mass-luminosity relation from the incomplete equations he was using. Eddington disagreed: by making use of various approximations, he had shown that luminosity depended only slightly on quantities other than the mass. One cannot but feel that Eddington had some luck but his intuition was right. Strictly speaking, however, they were right and he was wrong because a true mass-luminosity relation can be derived only when the law of energy release is known. This is embodied in the work of the German astrophysicist Heinrich VOGT (1890–1968; figure 4.4) who pointed out in 1926 that, given the mass and composition, four differential equations and four boundary conditions, and expressions for pressure, opacity, and energy release, there is a unique model for a star.[15]

Another important advance in the study of stellar structure was made in the 1930s through the revival of interest for convection. The condition for ascending and descending motions to be present in a star was given by Karl Schwarzschild in 1906. As we know, convective equilibrium is defined by the condition $p \propto T^{\gamma/(\gamma-1)}$, where γ is the ratio of specific heats (see eq. [3.6]). In a spherically symmetric star in which convective equilibrium prevails, the temperature gradient is, therefore,

$$\left(\frac{dT}{dr}\right)_{ad} = \frac{\gamma - 1}{\gamma} \frac{T}{p} \frac{dp}{dr}. \tag{4.15}$$

Schwarzschild's criterion states that if the radiative temperature gradient (eq. [4.10]) is steeper than the adiabatic lapse rate (eq. [4.15]), convective currents must arise.[16]

In Eddington's standard model, which is a polytrope of index $n = 3$, convective motions cannot arise if $\gamma > 4/3$. Since in stellar material the adiabatic exponent is normally closely equal to $5/3$, it follows that convection cannot appear in such

---

[14]For a firsthand account of these contentions, see T. G. Cowling, *Quart. J. Roy. Astron. Soc.*, **7**, 121 (1966). See also R. J. Tayler, *ibid.*, **37**, 355 (1996); L. Mestel, *Physics Reports*, **311**, 295 (1999).

[15]This result was independently derived by Russell in 1927 and is therefore called the Vogt-Russell theorem. It is not a theorem in the mathematical sense of the term, however, because stellar models can be constructed that violate the theorem; see, e.g., J. P. Cox and R. T. Giuli, *Principles of Stellar Structure* (New York: Gordon and Breach, 1968), 569.

[16]As was noted in section 3.2, the fact that convective currents arise when the temperature gradient is steeper than the local adiabat was already familiar to Espy and the meteorologists in the 1840s. However, Karl Schwarzschild was the first to apply this criterion to stars, which explains why it is generally named after him in the astronomical literature. It was not until 1947, however, that Paul Ledoux considered the general case where the chemical composition varies with radial distance (*Astrophys. J.*, **105**, 305).

Figure 4.4 The Heidelberg astrophysicist Heinrich Vogt in the late 1950s. He is best known
for the important theoretical contributions he made in 1925–1926 to the internal
structure of stars. Unfortunately, because Vogt was working in Germany while the
study of stellar interiors was largely an Anglo-Saxon endeavor, he never received
full credit for his original work. In fact he was the originator of both the Vogt-
Russell theorem and the Eddington-Vogt currents. Courtesy of Dr. Gerhard Klare,
Landessternwarte Heidelberg.

a model. As was shown by Unsöld, however, convection could appear in the layer
just below the surface of cool stars, in which hydrogen is becoming ionized. A
superadiabatic gradient may also occur in the central regions of a star, if the sources
of energy are strongly concentrated near its center. This was investigated by Ludwig
BIERMANN (1907–1986) at Göttingen, who showed in 1932 that even very slight
mass motions are sufficient to reduce a superadiabatic temperature gradient to a
steady one that differs little from the adiabatic lapse rate (eq. [4.15]). Three years
later, the English astrophysicist Thomas George COWLING (1906–1990) constructed
the first model with a convective core and a radiative envelope. This model, which
is generally known as the *Cowling point-source model*, is essentially the opposite
of the uniform energy source model. It is characterized by having a law of energy

release that is so temperature sensitive that all energy generation takes place within the convective core. This type of model is important from a historical standpoint because of its complete independence of energy sources within the convective core, since the only effect of these sources is to produce a negligibly small superadiabatic lapse rate there. It was also the first successful attempt at modeling main-sequence stars more massive than the sun.

4.4 PRE-1938 VIEWS OF STELLAR EVOLUTION

All stars (except the sun) are too far away to show visible disks. In 1890, however, the American physicist Albert Abraham MICHELSON (1852–1931) pointed out that it should be possible, by using the interference between the waves of light coming from opposite sides of the disk of some of the largest and nearest stars, to measure their angular diameters in seconds of arc. If the distance of the star is known, this measure can then be converted into actual dimensions. Because only a few stars are large enough and also near enough for their angular diameters to be measurable, actual observations for giant stars could not be made until after 1920. Then, Michelson and his associates—John August ANDERSON (1876–1959) and Francis Gladheim PEASE (1881–1938)—succeeded in measuring stellar diameters with an interferometer attached to the 100-inch telescope at Mount Wilson Observatory. Mira Ceti gave a diameter of 0.056″, Betelgeuse 0.047″, and Antares 0.040″. These were equivalent to diameters of about 700, 650, and 800 times that of the sun, respectively. This was the first direct evidence for the existence of giant stars which had been deduced earlier by Hertzsprung and Russell (see section 2.5).

The existence of giants and dwarfs is at the core of the evolutionary scheme developed by Russell in 1914. As was seen in section 3.5, in this approach a star starts as a red giant, which contracts and becomes hotter as it ages, heated as a perfect gas sphere. This contraction is arrested when the star reaches the main sequence, where the densities become too high for a star to behave like an ideal classical gas. The star then cools and contracts, descending progressively along the main sequence and ending its life as a red dwarf.

Now, in 1924 Eddington showed that his theoretical mass-luminosity relation was consistent with the observational fact that the average mass of main-sequence stars decreases with diminishing luminosity. Russell's theory was inconsistent, therefore, since a star could not go down the main sequence unless it lost a sizable amount of mass along the way, which seemed most unlikely. Moreover, Eddington had shown that stars would still follow the perfect gas law on the main sequence, and would thus heat up, rather than cool down, as they descended along the main sequence. Obviously, Russell's 1914 evolutionary scheme was no longer viable.

The following year, 1925, Russell (figure 4.5) came out with a new version of the life history of a star, based on the acceptance of the hypothesis that the source of stellar energy was the transformation of matter into radiation. Russell observed that, according to Eddington's theory, all main-sequence stars had approximately the same central temperature of about 30 million degrees. He also noted that the central temperature of the giants was much lower, of the order of 1 million degrees. He

Figure 4.5 Henry Norris Russell, a former student of Charles Augustus Young and later director of Princeton University Observatory from 1912 to 1947, circa late 1920s. A
phenomenologist of remarkable versatility, he contributed to all branches of stellar astronomy and helped make atomic physics central to astrophysical practice.
Courtesy of Yerkes Observatory.

therefore suggested that there were *two* energy sources, each involving a different
kind of active matter, each one being kindled as the central temperature of the star
rose to the required value. He adopted the words "dwarf stuff" for the main-sequence
stars and "giant stuff" for the giants. By assumption, the active material constitutes
but a small fraction of the star's mass, so that all through its history the mass, and
hence the absolute magnitude of the star, will remain nearly the same. A star thus
evolves almost horizontally, from right to left, in the H-R diagram (see figure 2.12).
The whole process starts with cold red giants, which contract rapidly under gravity
and exhaust most of their giant stuff while crossing the giant-sequence band. As
the stars become smaller, the pressure increases and stops their rapid contraction
phase when they meet the main sequence. At that point, dwarf stuff provides energy
until it has been exhausted. Proton-electron annihilation was conjectured as the
process during the main-sequence phase. The process responsible for the giant stars
remained vague, however. Russell's second theory did not withstand the passage of
time. Yet it is worth mentioning because, for the first time, an evolutionary scheme

was proposed that described how subatomic processes could sustain the energy radiated by a star as it moved across the H-R diagram.[17]

The next important progress was made in the 1930s, and it combined the analysis of observations with theoretical studies. Observations carried on Hertzsprung's 1911 analysis of the Hyades star cluster, in which the first color index–absolute magnitude diagram appeared. Over the years 1917–1920 Shapley published several H-R diagrams for *globular clusters*, and found that for some of them the upper part of the main sequence was absent. Robert Julius TRUMPLER (1886–1956) of the Lick Observatory followed by making a detailed investigation of fifty-two *galactic clusters* in 1925. He found that the population density of stars at various places on the main sequence and on the giant branch was significantly different for these clusters compared with that in the H-R diagram for bright stars in the solar neighborhood. Some clusters had virtually no stars cooler than type F5, while others had virtually no stars hotter than F0. All clusters had main-sequence stars, but some had no giant stars at all. Figure 4.6 shows the superposition of H-R diagrams for several clusters.

Trumpler made the plausible assumption that all stars belonging to a cluster were formed at nearly the same time, and interpreted the different population densities of stars on the H-R diagram as being due to the facts that some clusters comprised more massive stars than others, and that some clusters were older than others. Of course, in 1925 the evolution of a star was still believed to proceed from the red giants to the left in the H-R diagram, and then down along the main sequence. Trumpler was thus led to the conclusion that the galactic clusters were already of considerable age, for otherwise we would not find the lower main sequence so well formed in all cases, nor could we explain the general scarcity or total absence of giant stars. Shortly after Trumpler's investigation, however, a major reversal occurred in the ideas concerning the life history of a star. This major change was made in Copenhagen by Strömgren, who gave convincing evidence in 1932 that a star on the main sequence would move toward the giant branch, at the top right of the H-R diagram. If so, then the galactic clusters must be viewed as relatively young groups of stars.

The change in theory concerned the chemical composition of the stars. In Eddington's earlier work it was assumed that all stars were made of heavy elements and had the same mean molecular weight ($\mu \approx 2$). Once it had been established in the late 1920s that, in the solar atmosphere, hydrogen atoms were far more abundant than all other atoms together, Strömgren could make the first tentative steps toward developing a quantitative theory of stellar evolution based on Eddington's standard model.[18] He began with stars located on the main sequence and followed their global properties in a theoretical H-R diagram, assuming that hydrogen was gradually transformed into helium, even though the details of the fusion process were not yet known. Specifically, assuming further that the stars remain well mixed throughout their evolution, for the standard model one has

$$ L = \text{constant} \times \frac{M^{7/2}}{R^{1/2}} \, \mu^{15/2}, \tag{4.16} $$

---

[17]For an exhaustive discussion of the life, work and times of Russell, see D. H. DeVorkin, *Henry Norris Russell* (Princeton, N.J.: Princeton University Press, 2000).

[18]*Zeit. Astrophys.*, **4**, 118 (1932); **7**, 222 (1933).

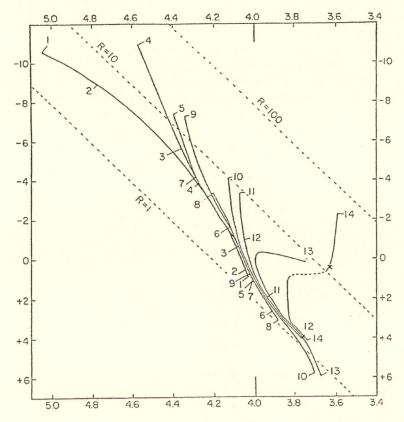

Figure 4.6 Superposition of Hertzsprung-Russell diagrams for several clusters (see text). The abscissae are $\log T_{\text{eff}}$; the ordinates are absolute magnitudes. From G. P. Kuiper, *Astrophys. J.*, **86**, 185 (1937).

where L is the total luminosity, M the mass, R the radius, and μ the mean molecular weight. Making use of eq. (2.5), we can also write

$$L = \text{constant} \times R^2 T_{\text{eff}}^4, \qquad (4.17)$$

where T_{eff} is the effective temperature, that is, the temperature at which a black body of the same surface area would radiate the same total energy as the star considered. Eliminating R from eqs. (4.16) and (4.17), one readily obtains a relation between the quantities L, T_{eff}, M, and μ. This can be depicted as a series of curves in a theoretical L-T_{eff} diagram, each curve corresponding to a different hydrogen content (i.e., a different value of μ). The curve corresponding to 42% hydrogen (by weight) came remarkably close to the observed main sequence, while the curves corresponding to lower hydrogen content lay above and to the right of the main sequence. Thus, with variable chemical composition, luminosity was no longer a function of mass only, but also of hydrogen content. According to Strömgren, as a star burns its hydrogen it thus moves to the right and upward in the H-R diagram, from the main sequence toward the giant branch.

In the late 1930s the Dutch-born American astronomer Gerard Peter KUIPER (1905–1973), who initially worked at the Lick Observatory and later at the Yerkes Observatory, combined Trumpler's observational results and Strömgren's theoretical results. He found that the various sequences in figure 4.6 agreed approximately with Strömgren's curves of constant hydrogen content. He therefore concluded that clusters such as 12 Monocerotis and S Monocerotis (numbers 1 and 2) consisted of hydrogen-rich stars, while the brighter stars in clusters like Coma Berenices and NGC 752 (numbers 13 and 14) contained less hydrogen. This lent further support to Strömgren's original proposal that hydrogen depletion brought a star from the main sequence to the giant branch. Even though the standard models used by Strömgren were much too simple to model actual stars, here a most plausible course of evolution was sketched out, which opened new avenues of research for the 1940s and 1950s.[19]

4.5 WHITE DWARFS AND NEUTRON STARS

The discovery by Adams at Mount Wilson in 1915 that the faint companion of Sirius was white, not red, brought to the foreground a quite unexpected problem— the existence of high-density stars. As was noted in section 2.6, this faint star was in fact detected before it was seen, for its existence was proved by Bessel in 1844 from the wavy path described by Sirius on the sky. This led to the conclusion that Sirius is really a double star, with an orbital period of about 50 years, and that its slightly irregular proper motion is the result of the gravitational pull exerted by an unseen companion. In 1915 the paths in which the two stars revolved around their common center of gravity were well known. From their distance—about 8 ½ light-years—it was therefore possible to derive their masses and absolute luminosities. Sirius A proved to be a very luminous white star, much heavier than the sun, but its faint companion—Sirius B—turned out to give only 1/360 part of the light of the sun, and to have a mass about equal to that of the sun. But then, as was shown by Adams, Sirius B had a spectrum only slightly different from the A0-type spectrum of Sirius A. Its high surface temperature in combination with its low luminosity thus meant that Sirius B had to be very small—only about the size of the earth (see eq. [2.5]). This implies an enormous density, of the order of 10^5 g cm^{-3}, which was at first considered absurd.

Ten years later, Eddington showed that there was nothing inherently absurd about the high densities of Sirius B and other *white dwarfs*. He argued that in the hot stellar interiors atoms collide violently with each other, stripping themselves of their electrons. Soon, therefore, all the atoms become completely ionized and the matter inside the stars consists of bare nuclei floating among the free electrons. An independent check of the large densities of white dwarfs was suggested by Eddington in 1924

---

[19]For a penetrating review of the changing interpretations of the Hertzsprung-Russell diagram during the period 1910–1940, see B. W. Sitterly, *Vistas in Astronomy*, **12**, 357 (1970).

and attempted the following year by Adams. According to Einstein's 1915 theory of gravitation, the radiation of a certain wavelength emitted by a compact body will be observed at a slightly longer wavelength, with the corresponding redshift being proportional to the mass and inversely proportional to the radius. Using Einstein's formula, Eddington predicted that the gravitational redshift for Sirius B was equivalent to about 20 km s^{-1}, if it was interpreted as a consequence of radial motion (see eq. [2.3]). After making allowance for the relative orbital motion of the two stars, Adams found a mean value of 21 km s^{-1}. This was taken as a confirmation of Einstein's general theory of relativity and as a proof of the existence of extremely dense stars.[20]

Some difficulties remained, however, and Eddington was quick in pointing out a paradox in the theory of condensed matter. Indeed, if we assume that white-dwarf material behaves more or less like an ideal classical gas, there is a perfectly definite relation between the energy and the temperature. So long as the star contains matter at high temperature, radiation of energy must presumably go on. But then, according to Eddington, there may come a time when the stellar material will have radiated so much energy that it has less energy than the same matter made of atoms at ordinary density and absolute zero temperature. This paradox was resolved by Ralph Fowler (figure 4.7), who pointed out in 1926 that the correct relation between energy and temperature in a white dwarf can be found from the new quantum statistical mechanics developed the previous year by Enrico FERMI (1901–1954). This theory is essentially based on the *exclusion principle* of Wolfgang PAULI (1900–1958), which prescribes that in a dense gas the number of electrons with momenta within a definite range cannot exceed a certain maximum. Making use of these new principles, Fowler showed that individual electrons at absolute zero temperature still have a kinetic energy comparable with the thermal energy of particles in an expanded gas whose temperature is as large as 10 million degrees. He was also able to show that the pressure of the electron gas in a white dwarf is unaffected by its temperature, with the pressure being thus a function of the density.

In a completely degenerate electron gas, the density is high enough so that all the available electron states having energies less than some maximum energy are filled. If the corresponding maximum momentum p_M is much less than $m_e c$, where m_e is the electron mass and c is the speed of light, one finds that the electron pressure p_e at mass density ρ is proportional to $\rho^{5/3}$. As the electron density is increased, the maximum momentum grows larger, and eventually a density is reached where the most energetic electrons become relativistic (see eq. [F.6]). In the limit $p_M \gg m_e c$, one can show that the electron pressure becomes proportional to $\rho^{4/3}$. The exponent $4/3$ means that the electron pressure force ($dp_e/dr \propto M^{4/3}/R^5$) supporting a spherically symmetric white dwarf against gravity ($\rho g \propto M^2/R^5$) grows no faster than the increasing gravitational force as the star contracts, with the result that there exists a limiting mass above which the electron pressure force cannot support the star against collapse.

---

[20]This excellent agreement between theory and observation was purely fortuitous, however, since both Adams and Eddington had derived incorrect values for the gravitational redshift of Sirius B. Recent work has shown that there is indeed good agreement between theory and observation, with modern values being more in the range 70–100 km s^{-1}. See F. Wesemael, *Quart. J. Roy. Astron. Soc.*, **26**, 273 (1985); J. L. Greenstein, J. B. Oke, and H. Shipman, *ibid.*, **26**, 279 (1985).

Figure 4.7 Sir Ralph Fowler, who in 1926 was the first to realize that the white-dwarf stars are made of degenerate matter. Courtesy of AIP Emilio Segrè Visual Archives.

The first explicit estimate of the maximum mass for a spherically symmetric white dwarf was made in 1930 by a student of Ralph Fowler, Edmund Clifton STONER (1899–1968), who found a critical mass of about $1.1 M_\odot$ based on a uniform distribution of density in the star. The following year, making use of the theory of polytropic gas spheres in conjunction with the relation $p_e \propto \rho^{4/3}$, the Indian-born astrophysicist Subrahmanyan CHANDRASEKHAR (1910–1995; figure 4.8) derived a critical mass of about $0.91 M_\odot$. Then, in 1932, the Soviet physicist Lev Davidovich LANDAU (1908–1968) showed that those stars composed of a degenerate electron gas whose mass exceeds about $1.5 M_\odot$ must, because of their increasing gravitational forces, collapse without limit into a point. In modern language, a too massive star must eventually become a *black hole* (see section 5.4).

A more precise value of this upper mass limit was obtained by Chandrasekhar, who derived numerically the exact equilibrium configurations in which degenerate electron gases support their own gravity. This work was accomplished with the help of a mechanical calculator and was completed by the end of 1934 (see figure 4.9). He found that the mass of a completely degenerate white dwarf must necessarily comply with the condition

Figure 4.8 Subrahmanyan Chandrasekhar in 1936. Courtesy of Yerkes Observatory.

$$M < \frac{5.84}{\mu_e^2} M_\odot = M_3 \text{ (say)}, \qquad (4.18)$$

where μ_e is the mean molecular weight per electron ($\mu_e \approx 2$ for all elements other than hydrogen). Chandrasekhar submitted his results for presentation at the January 11, 1935, meeting of the Royal Astronomical Society, London. After his talk, however, Eddington was invited to present a paper on "Relativistic Degeneracy." First he noted that Chandrasekhar's results predicted that a white dwarf with mass in excess of $1.4 M_\odot$ would continue to radiate and shrink until it disappeared. Then he went on to declare that "there should be a law of Nature to prevent a star from behaving in this absurd way," and that Chandrasekhar's calculations merely showed that the theory of relativistic degeneracy was incorrect. Although knowledgeable physicists of the time considered Eddington's 1935 assertions to be nonsense, they were unwilling to enter the controversy. At the Paris meeting (figure 4.10) it was clear to all that the consensus view was moving toward Chandrasekhar's analysis, but no one, not even Russell, was willing to press this point on Eddington. As a result, confusion prevailed among astronomers, and more than two decades passed before Chandrasekhar's theory was completely accepted. Chandrasekhar left Cambridge University and moved to the Yerkes Observatory of the University

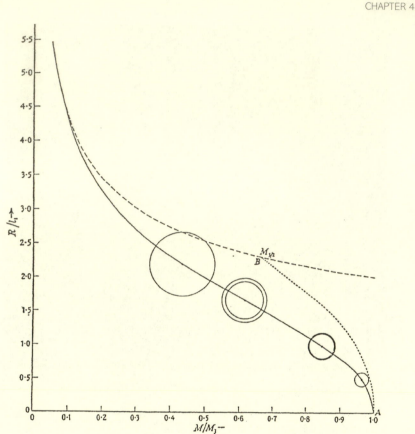

Figure 4.9 Chandrasekhar's original mass-radius relation for completely degenerate white dwarfs. The solid line represents the exact solution. The dotted lines represent approximate solutions, using $p_e \propto \rho^{5/3}$ when $M \to 0$ and $p_e \propto \rho^{4/3}$ when $M \to M_3$. From S. Chandrasekhar, *Mon. Not. Roy. Astron. Soc.*, **95**, 207 (1935).

of Chicago in 1937, but lived to see his vindication.[21] For the next fifty-nine years, the University of Chicago provided Chandrasekhar with his scientific home—at the Yerkes Observatory until 1964 and then at the Chicago campus. Equation (4.18) is known as the *Chandrasekhar limit*.

Neutron Stars

The existence of stars with densities much greater than those encountered in white dwarfs was originally suggested by Walter BAADE (1893–1960) and Fritz ZWICKY

---

[21] For a detailed study of the confrontation between Chandrasekhar and Eddington, see K. C. Wali, *Chandra: A Biography of Chandrasekhar* (Chicago: University of Chicago Press, 1984). The relevant scientific papers about white dwarfs are: R. H. Fowler, *Mon. Not. Roy. Astron. Soc.*, **87**, 114 (1926); W. Anderson, *Zeit. Phys.*, **54**, 433 (1929); E. C. Stoner, *Phil. Mag.*, **7**, 63 (1929); **9**, 944 (1930); S. Chandrasekhar, *Astrophys. J.*, **74**, 81 (1931); L. D. Landau, *Phys. Zeit. Sowjetunion*, **1**, 285 (1932); S. Chandrasekhar, *Mon. Not. Roy. Astron. Soc.*, **95**, 207 (1935).

Figure 4.10 Conference on Novae, Supernovae, and White Dwarfs in Paris (July 1939).
Except for Henri Mineur, the photograph shows *all* invited participants in the
conference. *Left to right*: (*front*) Frederick Stratton, Cecilia Payne-Gaposchkin,
Henry Norris Russell, Amos Johnson Shaler (the conference's secretary), Arthur
Eddington, Sergei Gaposchkin; (*back*) Carlyle Beals, Bengt Edlén, Pol Swings,
Gerard Kuiper, Bengt Strömgren, S. Chandrasekhar, Walter Baade; Knut Lund-
mark is standing between Chandrasekhar and Eddington. Courtesy of Yerkes
Observatory.

(1898–1974) in 1934.[22] It would thus seem that the California sun was already
favorable to wild speculations, since they wrote:

> With all reserve we advance the view that a super-nova represents the
> transition of an ordinary star into a *neutron star*, consisting mainly of
> neutrons. Such a star may possess a very small radius and an extremely
> high density. As neutrons can be packed much more closely than ordi-
> nary nuclei and electrons, the "gravitational packing" energy in a *cold*
> neutron star may become very large, and under certain circumstances,
> may far exceed the ordinary nuclear packing fractions. A neutron star
> would therefore represent the most stable configuration of matter as
> such.

The idea that such exotic stars might exist remained a speculation, however, until a
pulsar was found in the center of the Crab nebula supernova in 1968 (see section 5.4).
    On theoretical grounds, the Russian-born American physicist George GAMOW
(1904–1968) pointed out in 1937 that neutrons can be packed much more closely

---

[22] *Proc. Natl. Acad. Sci.*, **20**, 259 (1934).

than ordinary nuclei and electrons. Specifically, for densities rising into the range 10^9–10^{10} g cm^{-3}, protons and electrons are squeezed into neutrons. It follows that the pressure supplied by the electrons tends to disappear as the density rises. But then, the argument that was applied by Ralph Fowler to the motions of electrons in a white dwarf also applies to the motions of the neutrons. Accordingly, the neutrons also supply pressure, although for them to do so requires densities of the order of 10^{14} g cm^{-3}. At these densities a normal star like the sun would have collapsed into a *neutron star* with a radius of only 10 km.

Two years later, in 1939, Robert OPPENHEIMER (1904–1967) and his Canadian associate George VOLKOFF (1914–2000) at the University of California at Berkeley were the first to calculate reasonably realistic equilibrium configurations for neutron stars. The equation of state of the nuclear matter was obtained by considering a degenerate, relativistic gas obeying Fermi's quantum statistics. The gross features of the models—their masses, radii, and density distributions—were determined from Einstein's theory of gravitation, which necessarily replaces Newtonian mechanics when the densities become so large. Calculating detailed equilibrium models along these lines, they found that a stable neutron star can exist only in a finite range of masses (from $0.1M_\odot$ to $0.7M_\odot$) and a finite range of densities (from 10^{14} to 10^{16} g cm^{-3}). Although much effort has gone into improving the equation of state of nuclear matter, most of the conclusions reached by Oppenheimer and Volkoff have survived, though the masses are now believed to lie in the range from $0.1M_\odot$ to as much as $2M_\odot$ or possibly even $3M_\odot$. Three decades were to elapse, however, before their results found practical application to actual celestial bodies.

4.6 THE PULSATION THEORY OF VARIABLE STARS (II)

In 1880 Ritter suggested that most periodic variable stars were pulsating, but his idea did not find favor and was generally ignored (see eq. [3.11]). Yet a clue to the nature of the variable star δ Cephei was provided in 1895, when the Russian astronomer Aristarch BELOPOLSKY (1854–1934) discovered that the radial velocity of that star also varies with the same period. He therefore suggested that δ Cephei was a spectroscopic binary similar to those discovered by Pickering and Vogel in 1889. In 1897, however, Belopolsky observed periodic radial-velocity variations of another Cepheid variable, η Aquilae, and found that minimum brightness and the zero of velocity did not coincide, as would be expected for a simple binary model in which the light variations are due to eclipses. Later, in 1899, Karl Schwarzschild found that the changes in brightness of the Cepheid variables are accompanied by changes in effective temperature, therefore suggesting that their variability is a temperature effect. The conclusive death blow for the binary-star hypothesis came from Shapley, however, who pointed out in 1914 that the recently discovered giant nature of the Cepheid variables would have made these stars larger than their presumed binary separations. He therefore concluded that "the explanation that appears to promise the simplest solution of most, if not all, of the Cepheid phenomena is founded on the rather vague conception of periodic pulsations in the masses of isolated stars."[23]

---

[23]*Astrophys. J.*, **40**, 448 (1914).

An earlier suggestion by Moulton in 1909 was that most periodic variable stars were changing from a prolate to an oblate ellipsoid of revolution (see eq. [3.14]). The main problem with this theory was that the period of the radial-velocity variation should be twice that of the light variation, rather than of equal length. Shapley's suggestion was made more specific by Eddington who, apparently oblivious of Ritter's work, discussed anew the radial pulsations of a star and their actual maintenance in a set of highly influential papers.[24]

Radial Pulsations

In appendix D we present Eddington's mathematical formulation of the small radial adiabatic pulsations of a centrally condensed star. When $\gamma > 4/3$, there is an infinite discrete set of modes of oscillation. Eddington calculated numerically the fundamental mode of the standard model and, in agreement with Ritter's original result, found that its period is indeed inversely proportional to the square root of the mean density. In 1941 the German-born astrophysicist Martin SCHWARZSCHILD (1912–1997), then at Columbia University in New York City, calculated the first three overtones of Eddington's standard model. As expected, he found that the nth overtone has n nodes along the radius. At the same time, the Belgian astrophysicist Paul LEDOUX (1914–1988), then at the Yerkes Observatory, and the Lithuanian-born mathematician Chaim-Leib PEKERIS (1908–1993) in Cambridge (Mass.) jointly published in the *Astrophysical Journal* a useful approximate expression for the lowest mode of radial oscillation:

$$\sigma_0^2 \approx (3\bar{\gamma} - 4) \frac{|W|}{I}, \qquad (4.19)$$

where W is the potential gravitational energy, I is the moment of inertia with respect to the star's center, and $\bar{\gamma}$ is the pressure average of the adiabatic exponent. As they showed, if the ratio of central to mean density is not too large, eq. (4.19) provides a reasonable approximation for the lowest radial frequency.

Now, the prevalence and regularity of the Cepheid variables imply that the energy dissipated during the oscillations must be replenished in one way or another. As was pointed out by Eddington, the heat continually liberated in a star is an abundant source from which the energy required to maintain the oscillations might be derived. However, this heat can be made available as mechanical work only if the star behaves as a thermodynamic engine, i.e., if the excess heat is added to the stellar matter when it is at a high temperature and subtracted when it is at a low temperature. In section 5.6 we shall see that Eddington's original idea of a *valve mechanism* is of paramount importance in explaining stellar pulsations. His first suggestion was that the maintaining energy for the oscillations came from the variation with temperature and density of the rate of liberation of subatomic energy. This was not confirmed by subsequent stellar model calculations, however. His second suggestion rests on the idea that the pulsations are confined to the outer layers of a star, where a changing opacity could act as a valve. Specifically, if the atmosphere of a pulsating star has a layer in which the opacity increases during contraction, the excess radiation is

---

[24] *Mon. Not. Roy. Astron. Soc.*, **79**, 2 (1918); **79**, 177 (1919); **101**, 182 (1941); **102**, 154 (1942).

then prevented from escaping and so exerts pressure on the overlying layers; this is the mechanism that provides the energy required to continue the oscillations. In 1941, after the large predominance of hydrogen in stars was fully recognized, Eddington suggested that it was the *ionization of hydrogen* in that layer that might increase its heat capacity, heat being thus accumulated in the form of ionization energy at compression and liberated by recombination during expansion. His second suggestion was not confirmed either but, as will be explained in section 5.6, it paved the way for a more adequate solution.

Nonradial Pulsations

Throughout the 1920s and 1930s, emphasis was laid strongly on purely radial oscillations, probably on the grounds that small nonradial motions would be rapidly damped out in a real star. In fact, these nonradial oscillations present a problem that is, in any case, of great mathematical complexity. Pekeris was the first to show, in 1938, that this problem is equivalent to the solution of two simultaneous second-order differential equations. To illustrate the problem, he considered in detail the exact case of an initially homogeneous gaseous sphere (see eq. [D.13]). The problem was further discussed in 1941 by Cowling (figure 4.11), who showed that a polytropic gas sphere exhibits three infinite discrete sets of modes of oscillation: the *acoustic modes* (or *p* modes), which are largely actuated by pressure variations; the *gravity modes* (or *g* modes), which correspond to stable or unstable motions depending on the actual stratification in the equilibrium configuration; and the *Kelvin modes* (or *f* modes), which are analogous to the volume-preserving modes of an incompressible fluid sphere (see eq. [3.13]). However, it was not until 1966 that the more realistic case of a system containing both convectively stable and unstable regions was solved by Ledoux and one of his students in Liège, Paul SMEYERS (b. 1934), who later became a professor at the Katholieke Universiteit te Leuven. As they showed, the gravity modes then split into two discrete spectra: one set of modes corresponding to unstable motions in the regions where convection prevails, and another set corresponding to stable oscillations which are largely confined in the radiative regions.

At this point it is evident that the general discussion of the nonradial oscillations of a star was becoming a rather complex mathematical problem; and it would not have been mentioned here were it not for the fact that, four decades later, these nonradial oscillations were to find a practical application in the new fields of *helioseismology* and *asteroseismology* (see sections 6.3 and 6.5). In the intervening years, this highly theoretical subject was thoroughly discussed by Ledoux and his associates at the Institut d'Astrophysique de Liège which was, for a time, an important center of stellar-pulsation studies. Pekeris did not pursue his astronomical studies, however, but he remained very active in many other fields of research. In 1948 he emigrated to Israel, where he became head of the Department of Applied Mathematics at the Weizmann Institute of Science in Rehovoth.

Figure 4.11 Thomas George Cowling circa 1960. A second-generation astrophysicist, he entered the subject of stellar structure when Eddington and others had already established the discipline on a firm foundation. Cowling has made very important contributions to the subject, but above all he was one of the first to stress the importance of magnetic fields in solar and stellar problems. Courtesy of the Royal Astronomical Society, London.

4.7 THE EARLY STUDIES OF STELLAR ROTATION

William de Wiveleslie ABNEY (1843–1920) in London was the first scientist to express the view that the axial rotation of single stars could be determined from measurements of the widths of spectral lines. His suggestion was based on a consideration of the opposing Doppler shifts of a spectral line from receding and approaching stellar limbs. In 1877 he suggested that the effect of a star's rotation on its spectrum would be to broaden all lines and that "other conditions being known, the mean velocity of rotation might be calculated."[25] Abney's paper was severely criticized by Vogel in Potsdam because some stellar spectra contained both broad and narrow lines. Vogel also noted that spectrum photography would not help resolve the problem since the apparent line widths in stellar spectrograms depend on

---

[25] *Mon. Not. Roy. Astron. Soc.*, **37**, 278 (1877).

exposure time, plate sensitivity, and length of development. Abney's suggestion found little favor among his contemporaries. The reason may be the enormous weight accorded to the opinion of Vogel, who for many years dominated all thought in the field of stellar spectroscopy. What an irony that only a few years after his premature negation of Abney's ideas, Vogel himself introduced the photographic method into stellar spectroscopy and completely reversed his stand in 1898, expressing himself in favor of the hypothesis that stellar rotation does produce a measurable broadening of spectral lines.

The next major advance was made in 1909 by Frank SCHLESINGER (1871–1943), then at the Allegheny Observatory in Pittsburgh, who presented convincing evidence of axial rotation in the brightest star of the *eclipsing binary* δ Librae. He noticed that just before and just after light minimum, the radial velocities he measured departed from their expected values. A positive excess was observed just before mideclipse, while after mideclipse the departure was found to be negative. Schlesinger concluded that this occurrence could be produced if the brightest star rotates around an axis so that, at the time of partial eclipse, the remaining portion of its apparent disk is not symmetrical with respect to its axis of rotation. One year later, he observed a similar phenomenon in the eclipsing system λ Tauri. In 1924, at the University of Michigan, Richard ROSSITER (1886–1977) positively established the effect in β Lyrae and gave a complete curve of the residuals in velocity during the eclipse, while Dean MCLAUGHLIN (1901–1965) found the same effect in Algol. These were the first accurate measurements of the axial rotation of stars.

Another approach to the determination of rotational velocities was provided by *spectroscopic binaries* not known to be variable in light. In 1919 Walter Adams and Alfred JOY (1882–1973) at Mount Wilson studied the binary W Ursae Majoris. They observed that "the unusual character of the spectral lines is due partly to the rapid change in velocity during even [their] shortest exposures but mainly to the rotational effect in each star, which may cause a difference in velocity in the line of sight of as much as 240 km s^{-1} between the two limbs of the star."[26] Then, in 1929, a systematic study of rotational line broadening in spectroscopic binaries was undertaken jointly by Grigori SHAJN (1892–1956) at the Simeis Observatory in the Crimea and Otto STRUVE[27] (1897–1963) at the Yerkes Observatory. (Their collaboration took place by mail.) Shajn and Struve positively established that in spectroscopic binaries of short period, at least, line broadening is essentially a result of rotation.

The next step was to extend these measurements to *single stars*. Indeed, the possibility existed that rapid rotation occurred only in binary stars, perhaps because tidal forces could produce synchronization of axial rotation and orbital revolution. Since the sun has a very slow axial rotation, about 2 km s^{-1} at the equator, most astronomers at the time believed that the rotation of all other single stars was probably also small.

---

[26] *Astrophys. J.*, **49**, 190 (1919).

[27] He was actually the second Otto Struve and the fourth generation of astronomers, since he was the grandson of Otto Struve (1819–1905) and the great-grandson of Wilhelm Struve (1793–1864), who were both closely associated with the Pulkovo Observatory, near St. Petersburg (see section 2.6). See A. H. Batten, *Resolute and Undertaking Characters: The Lives of Wilhelm and Otto Struve* (Dordrecht: Reidel, 1988); D. E. Osterbrock, *Yerkes Observatory: 1892–1950* (Chicago: University of Chicago Press, 1997).

During the years 1930–1934, a systematic study of stellar rotation was undertaken by Struve, in collaboration with his Yerkes colleagues Christian ELVEY (1899–1972) and Christine WESTGATE (b. 1908). The measurements were made by fitting the observed contour of a spectral line to a computed contour obtained by applying different amounts of Doppler broadening to an intrinsically narrow line contour having the same equivalent width as the line observed. Comparison with an observed line profile gave the projected equatorial velocity $v_e \sin i$ along the line of sight. These early measurements indicated that the values of $v_e \sin i$ fell into the range 0–250 km s^{-1} and may occasionally be as large as 400 km s^{-1} or even more. As early as 1930 it was found that the most obvious correlation between $v_e \sin i$ and other physical parameters is with spectral type, with rapid rotation being peculiar to the earliest spectral types. This was originally recognized by Struve and later confirmed by a number of statistical studies of line widths in early-type stars by Westgate. The same conclusions were reached almost simultaneously by Shajn in the Soviet Union. The O-, B-, A-, and early F-type stars frequently have large rotational velocities, while in late F-type stars and later types rapid rotation occurs only in close spectroscopic binaries. A study of rotational line broadening in early-type close binaries was also made by the Dutch astronomer Egbert Adriaan KREIKEN (1896–1964) in Java. From his work it is apparent that the components of these binaries have their rotational velocities significantly diminished with respect to single, main-sequence stars of the same spectral type. The following year, 1936, Pol SWINGS (1906–1983) in Liège properly established that in close binaries of short period axial rotation tends to be either perfectly or approximately synchronized with the orbital motion.

Some Theoretical Results

Over the years 1919–1925, at a time when the axial rotation of single stars had not yet been measured, several astrophysicists made important contributions to the theory of rotating stars. Let us briefly review these early studies.

In 1919 Jeans (figure 4.12) conceived the idea that sequences of uniformly rotating, centrally condensed bodies may not mirror the classical Maclaurin-Jacobi sequences (see section 3.4). Jeans was a brilliant applied mathematician and was particularly skillful at formulating complex problems. It was a simple matter for him, therefore, to construct sequences of uniformly rotating polytropes, with the index n being a measure of central concentration (see eq. [3.8]). He found that for values of n larger than $n_c \approx 0.8$ the Maclaurin-Jacobi pattern does not prevail;[28] that is to say, when $n \geq n_c$ (giving a ratio of central to mean density larger than about 3), the sequences of uniformly rotating polytropes do not develop into triaxial configurations, but instead terminate by the balancing of the gravitational and centrifugal accelerations at the equator. Jeans therefore concluded that if one imagined a slowly contracting sequence of such figures, *equatorial breakup* would eventually occur; in other words, any further contraction would result in matter being shed

---

[28] J. H. Jeans, *Problems of Cosmogony and Stellar Dynamics* (Cambridge: Cambridge University Press, 1919), 165–186. More recent calculations by Richard JAMES (b. 1935) in Manchester indicate that one has $n_c = 0.808$ (*Astrophys. J.*, **140**, 552 [1964]).

Figure 4.12 Sir James Jeans, who applied mathematics to many branches of physics and
astronomy with marked success. He was particularly skillful at formulating
complex astronomical problems and was also a very successful popular writer.
Courtesy of the Royal Astronomical Society, London.

from the equator in a continuous stream, with the centrifugal force outweighing
gravity at the equator. This semiquantitative picture led Struve to suggest in 1931
that the emission-line B stars, which are the most rapidly rotating of all stars, could
well be uniformly rotating B-type stars on the verge of equatorial breakup, with
surrounding gaseous rings.

During the 1920s, as more was discovered about the physical processes taking
place in stars, it became clear that energy is transported mainly by radiation—not
by convection—in the deep interior of the vast majority of stars. This led Milne to
construct, in 1923, the first detailed model for a slowly rotating star in strict radiative
equilibrium (see eq. [4.10]). Although Milne is particularly remembered for his important contributions to the theory of stellar atmospheres,[29] he was also an excellent
applied mathematician. Making proper use of perturbation techniques, he was thus

---

[29]E. A. Milne, *Handbuch der Astrophysik*, **3**, part I, 65 (1930). This key work was reprinted in
Selected Papers on the Transfer of Radiation, edited by D. H. Menzel (New York: Dover Publications,
1966), 77.

Figure 4.13 The Swedish theoretical astronomer Hugo von Zeipel, who is best known for his comprehensive study of rotating stars in radiative equilibrium. Courtesy of the Uppsala University Library.

able to show that uniform rotation has two general effects on the structure of a star. First, because the centrifugal force takes from the pressure part of the burden of supporting the weight of the overlying layers, he found a reduction in the total luminosity of a star when it is compared to its nonrotating counterpart having the same mass. Second, because a uniformly rotating star is slightly oblate, he also found that part of the mass in the equatorial belt is supported by the centrifugal force, whereas this is not the case in the polar regions. He therefore concluded that the pressure and hence the net outward flux of energy must also be smaller at the equator than at the poles. In other words, the nonsphericity of a rotating star induces a dependence of surface temperature on latitude, with the polar regions appearing hotter than the equatorial belt.

The next important step was made in Uppsala by Hugo VON ZEIPEL (1873–1959; figure 4.13), who showed in 1924 that in a chemically homogeneous, uniformly rotating star in strict radiative equilibrium, the outward flux of energy is proportional to the effective gravity, which combines the effects of the gravitational attraction and centrifugal force. (This is known as *von Zeipel's law of gravity darkening*.) However, he also found that the liberation of energy at any point in the star's interior must be given by

$$\epsilon = \text{constant} \times \left(1 - \frac{\Omega^2}{2\pi G\rho} \right), \tag{4.20}$$

where Ω is the constant angular velocity of rotation. Clearly, since the actual energy sources in a star cannot possibly satisfy such a condition, at least one of the assumptions made in deriving eq. (4.20) must be relaxed. This fact was first discussed, independently, by Vogt in Heidelberg and by Eddington in Cambridge.[30] Specifically, they pointed out in 1925 that the breakdown of strict radiative equilibrium in a uniformly rotating star will set up slight rises of temperature and pressure over some areas of any given level surface and slight falls over other areas. A pressure gradient between the equator and the poles will thus ensue, causing a flow of matter primarily in planes passing through the axis of rotation. A quantitative study of the physical properties of this large-scale *meridional circulation* was originally made by Eddington in 1929. Unfortunately, he overestimated the speed of these currents in the sun by about five orders of magnitude. This was a most regrettable error because Eddington was so formidable and influential a person that it misled his successors into believing that the stars were well mixed. A more adequate evaluation of the circulation velocities in the bulk of a star was obtained in the early 1950s, but the first complete discussion of the meridional flow and its deflection by the star's rotation was not made until 1995.[31]

4.8 SOLAR AND STELLAR HYDRODYNAMICS

The sun and most of the stars appear to remain in a state of hydrostatic equilibrium under the action of their own gravitation and centrifugal force of axial rotation. Detailed observation of the sun has demonstrated that this balance is only approximate, since the sun's surface shows traces of internal motion, both around the rotation axis and in meridian planes passing through the axis. As was noted in section 2.4, Carrington was the first to demonstrate in the 1850s that the outer visible layer of the sun does not rotate like a solid body (see eq. [2.14]). Convincing evidence for a mean meridional motion of sunspots was not found until 1942, however, when Jaakko TUOMINEN (1909–1989) in Helsinki positively established the existence of a slow equatorward migration of sunspots at heliocentric latitudes lower than about 20° and a slow poleward migration at higher latitudes. All these problems are the domain of *astrophysical fluid dynamics*—a field that developed quite slowly during the first half of the twentieth century.

The first mathematical research on solar rotation was done independently by Wilsing and by Sampson in the 1890s. For the sake of simplicity, however, both of them assumed the material velocity in the sun to be wholly one of pure rotation. Accordingly, they found that the present-day differential rotation of the sun had to be regarded as transitory, as it was the remnant of some initial distribution of angular velocity, so that in the course of time it would be slowly transformed by the effects of molecular friction into a solid-body rotation. In 1898 Ernest Julius WILCZYNSKI

---

[30]H. Vogt, *Astron. Nachr.*, **223**, 229 (January 1925); A. S. Eddington, *The Observatory*, **48**, 73 (March 1925).

[31]See J. L. Tassoul, *Stellar Rotation* (Cambridge: Cambridge University Press, 2000), 93.

Figure 4.14 The Norwegian hydrodynamicist Vilhelm Bjerknes, who in 1917 was the founder
of an important school of geophysics in Bergen. His own scientific contributions
have had a great influence upon the advance of dynamic meteorology, and he was
the first in 1926 to apply hydrodynamical concepts to the solar activity problem.
Courtesy of the Norwegian Academy of Science and Letters.

(1876–1932) in Potsdam further argued that besides this secular change in angular velocity there could be periodic changes—say with an eleven-year period—that would somehow explain the periodicity of sunspots. No further progress was made until 1941, when Aleksandr LEBEDINSKY (1913–1967) in Leningrad (now St. Petersburg) recognized the importance of anisotropic turbulent friction in the problem of explaining the maintenance of the sun's equatorial acceleration. Specifically, he pointed out that the violent dynamical processes observed on the sun's surface were a clear indication that turbulent friction in the sun is considerably greater than molecular friction. Unlike the latter, however, turbulent friction in the sun is anisotropic. Accordingly, he found it essential in his calculations to make allowance for a difference in momentum transfer between the radial (i.e., in the direction of gravity) and lateral directions. This proved to be the clue to the problem, which led him to show that turbulent friction, if anisotropic, does indeed prevent the equalization of angular velocity on the surface of the sun. Five years later, in 1946, the

Polish-born Norwegian scientist Jeremi WASIUTYŃSKI (b. 1907) went further into the problem by making a detailed study of the mechanical and thermal effects of turbulence in rotating stars. At this juncture the theoretical problem of solar rotation was quietly abandoned for almost twenty years, but it took several decades before the importance of Lebedinsky's work was fully recognized.[32]

No survey of this period would be complete without special mention of the Norwegian school of fluid mechanics.[33] To Vilhelm BJERKNES (1862–1951; figure 4.14), its founder, we owe a systematic study of physical hydrodynamics. Although his interests were mainly developed in the specific context of meteorology, they paved the road to many astrophysical applications. In this respect, let us mention the pioneering work of Svein ROSSELAND (1894–1985) and Gunnar RANDERS (1914–1992), who both contributed to the hydrodynamics of rotating stars.[34] Of particular importance also is the stability analysis of a rotating compressible fluid by Halvor SOLBERG (1895–1974) and Einar HØILAND (1907–1974). To the latter we owe an important stability criterion, which asserts that dynamical instability occurs in a rotating star whenever the specific angular momentum $\Omega \varpi^2$ (where Ω is the angular velocity and ϖ is the distance from the rotation axis) decreases from the poles to the equator on a surface of constant entropy. This criterion, which was derived in 1941, generalizes to compressible fluids the well-known Rayleigh criterion, which states that an incompressible circular vortex is unstable when $\Omega \varpi^2$ decreases with increasing distance from the rotation axis.[35]

Sunspots and the Solar Cycle

To the best of our knowledge, Bjerknes was the first to suggest, in 1926, a theory explaining both the sunspots and their twenty-two-year cycle.[36] In this theory, a sunspot is considered the seat of vortex motions around a vertical axis, somewhat similar to the tropical cyclones on the earth. Thence, it is shown that the vortex motions in a sunspot lead to a centrifugal pumping effect, by means of which the gases in the central regions are lifted from below and adiabatically cooled. This explains why the sunspots are cooler, and hence darker, than the photosphere. It is

---

[32]A. I. Lebedinsky, *Astron. Zh.*, **18**, No. 1, 10 (1941); J. Wasiutyński, *Astroph. Norvegica*, **4**, 1 (1946).

[33]See A. Eliassen, "Vilhelm Bjerknes and his Students," *Annu. Rev. Fluid Mech.*, **14**, 1 (1982).

[34]During the 1920s and early 1930s, Rosseland also contributed much to the theory of spherically symmetric stars. He is best known for the *Rosseland mean mass absorption coefficient*, or simply the *Rosseland mean*, which is a sort of harmonic mean of the opacity over the frequencies.

[35]In the late 1950s independent studies showed that, in the presence of a weak magnetic field, there exist magnetically induced motions that become unstable when the angular velocity Ω decreases with increasing distance ϖ from the rotation axis (E. P. Velikhov, *J. Exptl. Theoret. Phys.*, **36**, 1398 [1959]; S. Chandrasekhar, *Proc. Natl. Acad. Sci.*, **46**, 253 [1960]). An apparent paradox therefore arises as the strength of the magnetic field tends to zero. This paradox was resolved in the early 1970s, when it was recognized that *two* distinct instabilities are actually involved in the problem: (i) a *weak* instability, which generates at best small-scale turbulent motions when $d\Omega/d\varpi < 0$ and $d(\Omega\varpi^2)/d\varpi > 0$, and (ii) a *strong* instability, which reshuffles the angular momentum distribution when the stronger condition $d(\Omega\varpi^2)/d\varpi < 0$ is satisfied (D. J. Acheson and R. Hide, *Rep. Prog. Phys.*, **36**, 159 [1973]). By the late 1990s it was widely accepted that the former instability plays an important role in generating turbulence in accretion disks (S. A. Balbus and J. F. Hawley, *Rev. Modern Phys.*, **70**, 1 [1998]).

[36]*Astrophys. J.*, **64**, 93 (1926).

also assumed that the magnetic field of a sunspot is due to the whirling motion of ionized matter. To explain the numerous collective properties of sunspots, Bjerknes further assumed the existence, in each hemisphere, of two subphotospheric zonal vortices around the sun's rotation axis. These gigantic vortex rings, which have opposite rotations to account for the change of polarity in alternate cycles, are carried round by a large-scale meridional circulation. One vortex ring is supposed to travel from about 35° latitude toward the equator just below the photosphere, which explains the equatorward migration of sunspots. At the same time, the other vortex ring—the one which has given rise to the previous sunspot cycle—sinks into deeper layers to the higher latitudes carried by the same meridional flow. Specifically, wherever a local sinuosity in these vortex rings rises to the surface and cuts the photosphere, a sunspot of the corresponding cycle makes its appearance or may be discovered. Unfortunately, observation shows that only a minority of sunspots exhibit recognizable vortex motions. Moreover, it is quite impossible to explain how a whirling motion could produce magnetic fields as strong as those observed in sunspots. Bjerknes' work on the twenty-two-year cycle has had lasting value, however, and has undoubtedly stimulated much research in the field.

The next step toward the resolution of the sunspot problem came in 1943 with new developments in the theory of ionized matter permeated by large-scale magnetic fields. This was originally achieved by the Swedish physicist Hannes ALFVÉN (1908–1995; figure 4.15), who suggested a new theory according to which the magnetic field of a sunspot is the primary phenomenon.[37] As there is apparently no possibility of explaining the magnetic field of a sunspot as produced by the spot itself, Alfvén thus starts out by postulating that there are isolated regions in the sun's core which are unstable and give rise to disturbances in the form of closed magnetic rings. These whirling rings travel along the lines of force of the sun's general magnetic field in the form of magnetohydrodynamic waves, which we now call *Alfvén waves*. After some time, each of these whirling rings will thus hit the photospheric layers and be reflected. During this phase, two sections of the torus intersect the sun's surface, and Alfvén identifies them with a pair of sunspots or, more precisely, a bipolar sunspot group. Alfvén also assumes that the sun's general magnetic field is a dipole field. Accordingly, since the disturbances are transmitted along the magnetic lines of force with a speed of about one meter per second, the wave front first intersects the solar surface at high latitudes, later proceeding toward the equator. This explains the wandering of the sunspot zone from high to low heliocentric latitudes. In this theory, one can also derive the lower temperature of the sunspots from the low hydrostatic pressure caused by their magnetic fields.

Specific aspects of this theory were further discussed by Alfvén's fellow countryman Claës WALÉN (1911–1983), who made a much closer investigation of the shape of these magnetohydrodynamic waves and of the manner in which they produce sunspots. In doing so Walén brought essential clarification to the theory of Alfvén waves, even though their application to the sunspot problem did not with-

---

[37] *Arkiv Mat. Astron. Fys.*, **29A**, No. 11, 1 (1943); **29A**, No. 12, 1 (1943); see also *Mon. Not. Roy. Astron. Soc.*, **105**, 3 (1945).

Figure 4.15 The Swedish physicist Hannes Alfvén, who was awarded the Nobel Prize for physics in 1970 for "fundamental work in magnetohydrodynamics with fruitful applications in different areas of plasma physics." He was one of the first after Cowling to realize that the universe is full of electrically conducting gases. Courtesy of the Royal Institute of Technology, Stockholm.

stand the passage of time.[38] The Alfvén-Walén pioneering work on sunspots should not be ignored, however, because it shows that the problem of how sunspots are generated is reduced to a consideration of how the sunspot magnetic fields are produced. In that sense, their work heralds the more elaborate discussions that were made in the 1950s and 1960s (see section 5.8).

---

[38] Little known is the fact that Walén in 1946 was the first to investigate the properties of an electrically conducting fluid when its electrical conductivity may be considered as infinite. Assuming the conductivity to be infinite, he demonstrated that the magnetic lines of force move with the fluid like a material substance and behave as though they are permanently attached to the fluid (*Arkiv Mat. Astron. Fys.*, **33A**, No. 18, 1 [1946]).

Chapter Five

The Golden Age: 1940–1970

I admit that most of the [mechanical] arts have been invented only little by little and that it required a rather long sequence of centuries to bring watches, for example, to their present point of perfection. But is not the same true of the sciences? How many of the discoveries that have immortalized their authors had been prepared by the works of preceding centuries, sometimes being already brought to their maturity, to the point where they required just one step more to be accomplished?
—Jean Le Rond D'ALEMBERT

Globular clusters are tight groupings of stars, containing on the average several hundred thousand stars, which differ from the loosely grouped and less populous galactic clusters in their spatial distribution. In the late 1910s the American astronomer Harlow Shapley, a former student of Russell at Princeton and later director of the Harvard College Observatory from 1921 to 1952, examined several globular clusters in detail using the Mount Wilson 60-inch reflector. This led him to recognize, in 1920, that the Milky Way—our galaxy—is much larger than was generally suspected and that the sun is not located near its center. He also found that the H-R diagrams for globular clusters differed drastically from that obtained by Russell for stars close to the sun (see figure 2.12). Specifically, he found that there seemed to be hardly any stars in the upper part of the main sequence above the position of the sun, and that red giants were much brighter than blue giants, whereas in Russell's diagram the red giants were of almost the same brightness as the blue giants. These results clearly suggest that the stars in globular clusters represent a type of star population that is markedly different from that observed in the solar vicinity.

These differences were finally explained almost twenty-five years later by the German astronomer Walter BAADE (1893–1960), who was on the staff of the Mount Wilson Observatory during World War II.[1] Technically classified as an "enemy alien," Baade was restricted to the Mount Wilson–Pasadena area, but was allowed to make observations of galaxies with the 100-inch telescope. Thus, while his

---

[1] The year 1931 had brought Baade an invitation to join the staff at Mount Wilson; he accepted immediately, resigning his position at Hamburg and becoming a permanent resident in the United States until 1958. When his fellow countryman Rudolf MINKOWSKI (1895–1976) was forced by Hitler and the Nazis to resign his professorship at Hamburg in 1935, Baade helped him to emigrate to the United States, thus enabling the two men to resume at Mount Wilson their former collaboration. Upon his retirement in 1958 Baade spent six months in Australia, where he used the 74-inch telescope at Mount Stromlo near Canberra, before returning to Göttingen as Gauss professor.

colleagues were assigned to wartime research, he had ample opportunity to use that large reflector—and this at a time when Los Angeles and neighboring towns were blacked out as a military precaution. Blessed with a particularly dark sky, Baade was able to resolve the nucleus of the Andromeda galaxy and several elliptical galaxies into stars. This led him to formulate in 1944 the revolutionary concept of two kinds of *stellar population* in the Milky Way and other galaxies. Broadly speaking, Baade described as population I all those single stars and galactic star clusters whose H-R diagrams approximately resemble that of the sun's neighbors. Population II includes stars in globular clusters, in the nuclei of spiral galaxies, in elliptical galaxies, and those of very large motion in the Milky Way. Actually, five years earlier, Lyman SPITZER, Jr. (1914–1997), then at Harvard College Observatory, had predicted the existence of two extreme types of stellar system, "Sc spirals" and "globular systems," rich in interstellar gas and dust or free of it, containing high-luminosity stars or not containing such stars. Baade himself initially resisted the interpretation that his two populations were young and old stars, respectively, as was suggested by several theoreticians; he later reversed his stand. By the early 1950s, it was generally accepted that all population II stars are old as compared with population I stars, the latter generally being in gaseous, star-forming regions.[2]

Independent observational work has shown that it is convenient to divide population I into three subpopulations (young, intermediate, and old) and likewise population II into two subpopulations (mild and extreme). Significant differences in average properties between these subpopulations have been observed, in addition to the basic difference in their H-R diagrams. One of them is a difference in their kinematical properties, as determined from a study of their space velocities and spatial distributions. For example, it was found that the most extreme population I objects—the blue giants of spectral types O and B—are confined in a flat system to the galactic plane, whereas the most extreme population II objects—the globular clusters in the galactic halo—move in large orbits about the galactic center and occupy a gigantic spherical volume reaching to extremely large distances from the galactic plane. This result strongly suggests, therefore, that there is a correlation between the kinematical behavior of stars in our galaxy and their age. The existence of such a correlation was strongly reinforced through the discovery of a special type of very loose galactic clusters called *stellar associations*. Their importance was first pointed out in the late 1940s by the Soviet astronomer Viktor AMBARTSUMIAN (1908–1996) of the Byurakan Observatory, on Mount Aragats, north of the Armenian capital Yerevan. He noted that these associations are roughly spherical in shape despite their differential revolution about the galactic center. He therefore concluded that they were formed recently and so must be expanding rapidly in all directions, with velocities of the order of 5–10 km s^{-1}, to mask the tendency toward elongation due to the shearing effect in the Milky Way. If so, then, their ages cannot exceed a few millions or tens of millions of years. This is quite young compared with the age of the sun, which is of the order of 4.5 billion years.

---

[2] See L. Spitzer, Jr., and J. P. Ostriker, eds., *Dreams, Stars, and Electrons: Selected Writings of Lyman Spitzer, Jr.* (Princeton, N.J.: Princeton University Press, 1997), 3; see also D. E. Osterbrock, *Walter Baade: A Life in Astrophysics* (Princeton, N.J.: Princeton University Press, 2001), 133.

Another important difference between the two populations is the difference in their chemical compositions, as determined from the study of their spectral lines (see section 4.1). Indeed, several independent studies made in the early 1950s showed that the abundance of all the elements heavier than helium appears to be much smaller in population II stars as compared with population I stars. Specifically, if Z designates the total mass density of elements heavier than helium expressed in the hydrogen density as unit, they found that $Z = 0.003$ for the oldest population II stars and $Z = 0.040$ for the youngest population I stars.[3] This implies that the most recently formed stars were produced out of a medium of interstellar gas and dust that was already enriched in heavy elements. Since these elements were probably formed in stellar interiors and then spewed out to space in the final explosions that mark the death of many stars, all this seems to indicate that population II stars are the first generation of stars to be formed in a galaxy, and population I stars are a mixture of all later generations.

In retrospect, one cannot but feel that Baade's concept of two stellar populations came at a most opportune time. Indeed, only after about 1940 did it become clear that *thermonuclear fusion reactions* provide the internal power that allows the sun and stars to shine for extended periods of time, and that the *nucleosynthesis* of heavy elements from lighter ones can account for the abundances we observe of the elements heavier than helium. These fundamental discoveries of nuclear astrophysics—together with the development of fast electronic computers—enabled astrophysicists to carry out the numerical calculations necessary to investigate the structure and evolution of the sun and stars. These theoretical results, in turn, could be compared with the H-R diagrams of star clusters for confirmation or improvement. It is to these new developments that we now turn.[4]

5.1 NUCLEAR REACTIONS AND ENERGY PRODUCTION IN STARS

As was noted in section 4.2, the American physical chemist William Harkins was the first to suggest that the fusion of light elements into heavier ones could provide a basis of explanation for the long-term sources of energy generation in the sun and stars. This important suggestion was repeated in a more precise form in 1920, when Jean Perrin and Arthur Eddington independently pointed out that, if four atoms of hydrogen could be fused into one atom of helium, energy would be released in adequate quantity. Unfortunately, at that time the details of how a proton could pass through the repulsive Coulomb barrier and penetrate into a nucleus were quite obscure. Then, in 1928, the Russian physicist George Gamow developed a quantum mechanical theory for the probability of interpenetration of two colliding

---

[3] M. Schwarzschild, *Structure and Evolution of the Stars* (Princeton, N.J.: Princeton University Press, 1958; New York: Dover Publications, 1965), 27.

[4] For reprints and English translations of many key papers pertaining to the period 1930–1975, see K. R. Lang and O. Gingerich, eds., *A Source Book in Astronomy and Astrophysics 1900–1975* (Cambridge, Mass.: Harvard University Press, 1979). See also O. Gingerich, "Reports on the Progress in Stellar Evolution to 1950," in *Stellar Populations*, edited by P. C. van der Kruit and G. Gilmore, I.A.U. Symposium No. 164 (Dordrecht: Kluwer, 1995), 3.

nuclei. A year later, Robert d'Escourt ATKINSON (1898–1982), a young English physicist at Göttingen, and the German physicist Friedrich Georg HOUTERMANS (1903–1966) extended Gamow's theory and calculated the rates of thermonuclear reactions between different elements in the conditions existing in stellar interiors. This was the first attempt at a quantitative theory of nuclear energy generation within stars. They tentatively called their paper "How Can One Cook Helium Nuclei in a Potential Pot?," but when it was published in the *Zeitschrift für Physik*, the editor changed the title to a less imaginative one: "On the Question of the Possibility of the Synthesis of Elements in Stars." Specifically, they found that the most effective nuclear interactions were those between the nuclei of hydrogen and some other light element, which they could not identify because of lack of data at that time.

Atkinson pursued their work alone and, by 1931, was already led to suggest that the observed relative abundances of the heavy elements might be explained by the synthesis of heavy nuclei from hydrogen and helium by successive proton captures. In 1936 he also showed that the most likely nuclear reaction within stars is the collision of two protons to form a nucleus of deuterium (i.e., heavy hydrogen) and a positron (i.e., a positively charged, electronlike particle).[5]

Gamow, who defected from the Soviet Union in 1933, moved to George Washington University in Washington, D.C., where he was soon followed by the Hungarian physicist Edward TELLER (b. 1908) who left Hitler's Germany in that fateful year of 1933. In March 1938, Gamow and Teller together with Merle TUVE (1901–1982) from the Carnegie Institution organized the Fourth Washington Conference on Theoretical Physics to bring astronomers and physicists together to discuss the stellar-energy problem (see figure 5.1). At this conference, their student Charles Louis CRITCHFIELD (1910–1994; figure 5.2) outlined his ideas on a chain of reactions starting with proton-proton collisions and ending with the synthesis of helium nuclei. Hans Albrecht BETHE (b. 1906; figure 5.3), a German émigré to the United States, was so stimulated by the meeting that he decided to collaborate with Critchfield in an attempt to calculate the rate of this process. Making use of formulas derived by Gamow and Teller for the probability of nuclear penetration within stars, Bethe and Critchfield were able to show that the *proton-proton (pp) chain* predicts the correct energy production for the sun (see appendix G).[6]

In the same year, 1938, the German physicist and philosopher Carl Friedrich VON WEIZSÄCKER (b. 1912; figure 5.3) proposed that all the chemical elements were already formed before the formation of stars as we observe them today. If so, then, we are no longer limited to reactions that begin from hydrogen and helium. In the year 1938 a similar idea was independently developed by Bethe, but his full discussion did not appear in print until March 1939. A detailed study of the various thermonuclear reactions that may occur in stellar interiors was thus carried out simultaneously by Bethe in the United States and Weizsäcker in Germany. They found the remarkable fact that the group of reactions involving carbon and nitrogen form a cycle, which repeats again and again as long as the hydrogen lasts. In this

---

[5] See G. Gamow, *A Star Called the Sun* (New York: Viking Press, 1964); I. B. Khriplovich, "The Eventful Life of Fritz Houtermans," *Physics Today*, **45**, No. 7, 29 (1992).

[6] See G. Gamow, *My World Line, An Informal Autobiography* (New York: Viking Press, 1970), 136.

Figure 5.1 Some of the participants in the Fourth Annual Conference on Theoretical Physics
in Washington, D.C. (March 1938). *Left to right*: (*seated*) L. H. Thomas, George
Gamow, Svein Rosseland, Harlow Shapley, J. A. Fleming, ..., R. Gunn; (*standing*) Willis E. Lamb, Jr., C. L. Critchfield, Wolfgang Pauli, Mario Schönberg, S.
Chandrasekhar, R. d'Escourt Atkinson, M. S. Rosenblum, Francis Perrin (Jean
Perrin's son), Edward Teller, ... From K. C. Wali, *Chandra* (Chicago: University
of Chicago Press, 1991).

cycle, known as the *carbon-nitrogen (CN) cycle*, the nuclei of carbon and nitrogen
pass through a series of transformations and reappear at the end of each cycle (see
Appendix G). Weizsäcker argued that the CN cycle provided the main source of
energy for the main-sequence stars. At the same time, Bethe speculated that the
CN cycle is predominant in the upper main-sequence stars, whereas the pp chain
predominates for the main-sequence stars less massive than the sun.

During the next decade or so, the relative importance of these two sets of reactions
for the sun was a matter of considerable debate. The reason for the long vacillation
between the CN cycle and the pp chain is very simple. They show a markedly
different dependence on the temperature T, and whereas the pp-chain rate increases
approximately as T^5, the CN cycle proceeds as T^{16}. Now, it so happens that
the central temperature of the sun is such that the two processes could be equally
efficient for that star. Any small change in the measured nuclear cross sections or
in the calculated central temperature of the sun would thus throw the balance very
strongly one way or the other. By the early 1950s, however, theoretical models
had been used to show that the pp chain provides the main portion of the nuclear
energy in the sun, but that the CN cycle appears to make a small contribution too.
The sun thus happens to be near the dividing line between the two processes: In all

Figure 5.2 Charles Critchfield, who proposed the proton-proton (*pp*) chain as competitive
with the carbon-nitrogen cycle. From G. Gamow, *A Star Called the Sun* (New
York: Viking Press, 1964). Courtesy of the Gamow Estate.

stars more massive than slightly over $1M_\odot$, the CN cycle prevails, whereas all less
massive stars live on the *pp* chain. The sun is thus a very average star.

Now, in the early 1950s many astronomers believed that the cosmic abundance
scale of the chemical elements in the universe could be explained in terms of the
gigantic explosion, the Big Bang, that had caused the recession of the galaxies.
Indeed, by 1948 Ralph ALPHER (b. 1921), Bethe, and Gamow[7] had proposed a theory
in which the early universe, just after the Big Bang, consisted of a very hot gas of
neutrons at about 10^{10} K. Protons were formed by neutron decay, and successive
captures of neutrons led to the formation of the heavier elements. Unfortunately,
it was soon found that the chain broke down at helium, as the nucleus produced
by the capture of a neutron by a helium nucleus was unstable, and hence would
immediately disintegrate. This led to the idea that *nucleosynthesis* in stars explains
the observed abundances of the heavier elements, whereas the observed hydrogen,
helium, and deuterium must have been produced in the early stages of the universe.
Although the existence of markedly different abundances in the young and old stellar

---

[7]The theory was actually developed by Alpher and Gamow, who added Bethe's name to their paper
in order to make a pun on the first three letters of the Greek alphabet—α, β, and γ.

Figure 5.3 Carl von Weizsäcker (*left*) and Hans Bethe (*right*), who simultaneously and independently proposed the carbon-nitrogen (CN) cycle. Bethe was awarded the
Nobel Prize for physics in 1967 "for his contributions to the theory of nuclear reactions, especially his discoveries concerning energy production of stars." This was
the first Nobel Prize awarded for an astronomical achievement, but several other
astronomers—Eddington, Hale, Russell—had been nominated before. From G.
Gamow, *A Star Called the Sun* (New York: Viking Press, 1964). Courtesy of the
Gamow Estate.

populations was convincing evidence of the formation of heavy elements in stars,
the first direct proof was furnished in 1952 by Paul Willard MERRILL (1887–1961) at
Mount Wilson, who discovered the presence of the short-lived element technetium
in certain late-type giants. Since all the isotopes of this element are radioactive,
with half-lives shorter than a few million years, technetium could not have been
present in red giants when they formed billions of years earlier, thus proving that
nuclear synthesis must be occurring in these stars.

The next important progress was made, independently, by Öpik and Salpeter, who
showed that powerful nuclear reactions could synthesize carbon and the heavier elements in an evolved star, once the *pp* chain and CN cycle had exhausted its hydrogen
content. Ernst Julius ÖPIK (1893–1985) was an eminent Estonian astronomer who
had made his home in Tartu. Following the much feared Soviet occupation of Es-
tonia in 1944, however, Öpik and his family left the mother country and eventually
settled down in Northern Ireland, where he became a distinguished staff member of
the Armagh Observatory. Edwin Ernest SALPETER (b. 1924; figure 5.4) had a very
similar fate since, in 1939, he fled Hitler's forces in his native Austria and settled
with his family in Australia. Having obtained his Ph.D. degree in England, Salpeter
began his career in 1949 at Cornell University (Ithaca, N.Y.) as a research associate
and was named a full professor in 1957.

Figure 5.4 Edwin Salpeter, who thoroughly investigated in 1952 the three-alpha reaction for the later stages of stellar evolution. Belatedly, he was awarded the 1997 Crafoord prize with Fred Hoyle in recognition of the outstanding importance of their solution of the carbon problem. From G. Gamow, *A Star Called the Sun* (New York: Viking Press, 1964). Courtesy of the Gamow Estate.

As we shall see in section 5.3, Öpik first suggested in 1938 that both the main-sequence stars and the red-giant stars shine because of thermonuclear reactions, following a sequence well defined by successive increases in central temperature. Later, in 1951, he realized that after the central temperature of a giant star reaches about 4×10^8 K, all the helium will be converted into carbon by triple collisions of helium nuclei. The following year, Salpeter, unaware of Öpik's discussion, presented in much greater detail the arguments for the synthesis of carbon by what today we call the *three-alpha reaction* (see appendix G). Actually, both Öpik and Salpeter qualitatively considered the synthesis of heavier elements like ^{16}O, ^{20}Ne, etc., by the capture of additional helium nuclei at temperatures of about 10^9 K. A genuine quantitative discussion of the relevant nuclear reactions was not made until 1954, however, when the English astrophysicist Fred HOYLE (1915–2001)

Figure 5.5 "B²FH." *Left to right*: Margaret Burbidge, Geoffrey Burbidge, William Fowler,
and Fred Hoyle in July 1971 with the steam train presented to Fowler in honor
of his sixtieth birthday. Hoyle was knighted in 1972. Fowler was awarded the
Nobel Prize for physics in 1983 "for his theoretical and experimental studies of
the nuclear reactions of importance in the formation of the chemical elements in
the universe." Appropriately, in his autobiography he wrote, "I am very fortunate
the Nobel prize can be given for team work. I consider it to be an award to the
Kellogg Radiation Laboratory." It remains a mystery why Hoyle's name was not
bracketed with Fowler's in the 1983 Nobel Prize award. Courtesy of Dr. E. M.
Burbidge.

worked out in detail the synthesis of the chemical elements from carbon to nickel.[8]
Parenthetically, note that Hoyle had also previously argued, in 1946, that under the
conditions of temperature and density necessary for statistical equilibrium between
the nuclei and the free protons and neutrons, the most abundant elements would be
distributed about the element iron, the *iron group of elements* as they eventually
became called. This is in agreement with the observational data on abundances
which show that, indeed, there is a marked peak at the element iron.

By the mid-1950s it was widely accepted that nuclear reactions occurring in stars
act to produce fusion of lighter nuclei into heavier ones, which are then dispersed into
the interstellar medium in the terminal phases of the stellar lifetime. This basic idea
has been developed by several nuclear physicists and astronomers into an elaborate
theory for explaining the observed abundances of the elements. This was essentially a multidisciplinary venture to which a large number of scientists contributed:
Alastair CAMERON (b. 1925), William FOWLER (1911–1995), Fred Hoyle, Edwin

---

[8]For interesting personal comments on that important period of nuclear astrophysics, see F. Hoyle,
Quart. J. Roy. Astron. Soc., **27**, 445 (1986). See also L. Mestel, *Astron. Geophys.*, **42**, No. 5, 23 (2001).

Figure 5.6 At the meeting of the American Astronomical Society at Montréal in 1964, Alastair
           Graham Walter Cameron discusses why there is so much helium in the universe.
           Courtesy of Dr. A.G.W. Cameron.

Salpeter, and their associates on the formation of heavy atoms; Geoffrey BURBIDGE
(b. 1925), Margaret BURBIDGE (b. 1919), Jesse GREENSTEIN (1909–2002), and their
associates on the abundance features of stars. The classic paper on the production
of medium-weight and heavy elements in stellar interiors was published in 1957 in
Reviews of Modern Physics by the Burbidges, Fowler, and Hoyle. In this paper,
which is often referred to as B^2FH (figure 5.5), they explained that chemical elements could be produced by at least eight different types of synthesizing process,
if one believes that only hydrogen is primeval. In addition to the element-building
reactions considered above, they paid special attention to slow and rapid neutron
capture and proton capture, since these mechanisms are the only way in which elements heavier than iron can be produced. At the same time, a detailed study of these
possibilities was carried out independently by Cameron (figure 5.6), who was then
working for Atomic Energy of Canada Limited at Chalk River, Ontario. Appropriately, he emphasized strongly that nuclear fuels might burn explosively, leading
to rapid neutron capture during supernova explosions. With these important works
nuclear astrophysics had become a mature science.

5.2 CALCULATION OF STELLAR STRUCTURE

The history of stellar structure prior to 1938 can be separated into two distinct peri-ods. During the first period, from about 1845 to the 1910s, observers and theoretical astronomers alike had little or no interest in the internal structure of the sun and the stars. This explains why all earlier investigations in this field were made by out-siders to the astronomical community, each of them working in isolation.[9] At that time it was generally believed that gravitational energy that was converted into ther-mal energy by contraction was the only source of maintenance of a star's heat and radiation. Actually, the first quantitative study of the distribution of temperature within a star was made by Lane in 1870 under the assumption that the outward flow of energy was by convective motions. The second period began in the 1910s, when Eddington found it necessary to break away from Lane's pioneering investigation on one fundamental point, namely, the mode of energy transfer within stars. Specif-ically, he showed that the energy was commonly transported radially outward by the process of radiative transfer. Meanwhile, as the gravitational contraction theory was becoming untenable, there was also a gradual acceptance of the idea that the liberation of subatomic energy within a star could sustain its luminosity for extended periods of time. The whole situation was surveyed in 1939 by Chandrasekhar in his important monograph *An Introduction to the Study of Stellar Structure*. At this time the nuclear reactions responsible for stellar energy generation were being discov-ered, and Chandrasekhar's book was the first to include a discussion of these new developments.

Throughout the 1920s and 1930s, no doubt due in part to Eddington's preemi-nence, astrophysics flourished at Cambridge University, which became for a time the mecca of theoretical studies in stellar structure and evolution. During the 1940s, however, this British leadership rapidly passed on to the United States, where elec-tronic computers, developed during World War II mostly for military purposes, made it possible to carry out numerical calculations of staggering proportions. Of partic-ular importance is the pioneering work of Martin SCHWARZSCHILD[10] (1912–1997; figure 5.7), who had obtained his Ph.D. degree at Göttingen in 1935, but was forced to leave Germany as Hitler's persecution of Jews intensified. He then emigrated to

---

[9]For the record, let us recall the professions of these pioneers, at the time when they made their contributions to the nascent science of astrophysics: Mayer had a medical practice in Heilbronn, Ger-many; Waterston was a naval instructor in Bombay, India; Helmholtz was a professor of physiology at Königsberg in East Prussia, Germany (now Kaliningrad, Russia); Lane was working at the Office of Weights and Measures in Washington, D.C., U.S.A.; and Ritter was a structural engineer who taught mechanics in Aachen, Germany.

[10]He was the son of Karl Schwarzschild (1873–1916), a German astronomer and physicist who achieved great things despite a short lifespan (see figure 2.9). Karl's exceptional ability in science was evident in a paper on the theory of celestial orbits he wrote at the age of sixteen. In 1901 he became full professor and director of the observatory at the University of Göttingen, leaving in 1909 to become director of the Astrophysical Observatory at Potsdam. He served actively during World War I, first in Belgium and France, then on the Eastern Front, in Russia. There he contracted pemphigus, a metabolic disease of the skin that was then incurable. He died in May 1916, at Potsdam. By his own request he was buried in Göttingen. For more details about Martin Schwarzschild's life and family, see L. Mestel, *Biographical Memoirs of Fellows of the Royal Society*, **45**, 469 (1999).

Figure 5.7 Martin Schwarzschild in 1982. Courtesy of Princeton University Observatory.

the United States in 1937, eventually becoming a naturalized American citizen in 1942. After relatively short appointments at Harvard and Columbia Universities, he accepted a full professorship at Princeton University in 1947. There, he was one of the first to recognize the potential of the early computers developed at the Institute for Advanced Study by John VON NEUMANN (1903–1957) and to use them for calculating numerical models of stars. From 1953 to 1977, Martin Schwarzschild did most of his work on stellar structure in collaboration with Richard HÄRM (1909–1996), an Estonian émigré to the United States, who was an exceptionally skilled calculator. These early developments are described in the book *Structure and Evolution of the Stars*, published by Martin Schwarzschild in 1958, which served as a starting point for the many workers who became active in this field during the 1960s.

As was shown in sections 3.2 and 4.3, the basic conditions of equilibrium for a spherically symmetric star are

$$\frac{dp}{dr} = -\frac{Gm}{r^2}\rho,\tag{5.1}$$

$$\frac{dm}{dr} = 4\pi r^2 \rho, \tag{5.2}$$

$$\frac{dL(r)}{dr} = 4\pi r^2 \rho \epsilon, \tag{5.3}$$

and, for radiative equilibrium,

$$\frac{dT}{dr} = -\frac{3}{4ac}\frac{\kappa\rho}{T^3}\frac{L(r)}{4\pi r^2}, \tag{5.4}$$

or, for convective equilibrium,

$$\frac{dT}{dr} = \frac{\gamma - 1}{\gamma}\frac{T}{p}\frac{dp}{dr}, \tag{5.5}$$

where all symbols have their standard meanings (see eqs. [4.5], [4.6], [4.8], [4.10], and [4.15]). Modifications must be made to eq. (5.3) during the rapid expansion and contraction phases.

To these equations we must add an equation of state and the equations for the opacity and the energy generation. For a perfect gas, one has

$$p = \frac{\mathcal{R}}{\mu}\rho T + \frac{1}{3}aT^4 \tag{5.6}$$

(see eqs. [4.2]–[4.4]). Now let X, Y, and Z be the fractions of matter, by weight, which consist of hydrogen, helium, and heavier elements, respectively ($X+Y+Z = 1$). Hydrogen and helium are entirely ionized in a star's interior and, for many purposes, the heavier elements may be assumed to be almost entirely ionized. The mean molecular weight is, therefore,

$$\mu = \frac{1}{2X + \frac{3}{4}Y + \frac{1}{2}Z}. \tag{5.7}$$

The detailed forms of the functions ϵ and κ, which depend upon the physical circumstances, are rather complicated and are usually defined numerically or by means of interpolation formulas. White dwarfs and the cores of red giants require constitutive equations distinct from eqs. (5.6) and (5.7); moreover, thermal conduction by degenerate electrons becomes of paramount importance in these objects. During the 1960s, extensive tabulations of opacities were calculated by Arthur Nelson Cox (b. 1927) and his associates at the Los Alamos Scientific Laboratory in New Mexico.

For the calculation of the structure and evolution of a star, four differential equations are generally solved for four unknown functions, for which two boundary conditions are specified at the center and two at the surface (see eqs. [5.1]–[5.5]). For massive stars consisting of a convective core and a radiative envelope, it is often a good approximation to take $p = 0$ and $T = 0$ at the free surface. In main-sequence stars of solar mass or less, however, there should be no convective core because they live on the pp chain which is weakly dependent on temperature; yet a wholly radiative model predicts too high a central temperature and hence too high a luminosity for these stars. This inconsistency was resolved in 1953 by

Donald OSTERBROCK (b. 1924), then at Princeton University, who followed up earlier indications by Biermann, Öpik, and Strömgren that convection in the outer layers of a star may affect the internal structure. Specifically, he integrated a composite model consisting of a radiative core and a convective envelope which does explain the salient features of low-mass main-sequence stars. Following Osterbrock and others, one often uses the simple adiabatic relation $p = KT^{\gamma/(\gamma-1)}$ throughout the convection zone out to the very edge, so that the usual boundary condition that p and T tend simultaneously to zero is satisfied regardless of the value of the constant K.[11]

Normally, one considers that evolution takes place through a sequence of equilibrium states, each one of them being characterized by a progressively different chemical composition. Of course, when constructing a model, one must test the temperature lapse rate at each point to determine whether eq. (5.4) or eq. (5.5) must be used to describe the energy transport (see section 4.3). When convection occurs in the outer layers of a star, however, its description is not as simple as that given by eq. (5.5). In this case radiation may carry an appreciable fraction of the energy, and the situation is further complicated by the ionization of hydrogen and helium within the outer convective layers. An approximate procedure for treating this problem was described by Albrecht Unsöld and Erika VITENSE (b. 1923), later Böhm-Vitense, who was then working at Kiel University. By this method, she calculated in 1953 a model for the sun's atmosphere in which the total energy flux is carried by a combination of radiation and convection. Her working theory is based on the assumption that the convective bubbles travel one "mixing length" upward before they break up and lose whatever identity they may have. Unfortunately, the theory itself has no way of determining the mixing length, so that in some cases it leaves a great uncertainty in the structure and extent of the outer parts of a stellar model. Important theoretical progress was made by the American hydrodynamicist Edward SPIEGEL (b. 1931) in the 1960s, but we are still far from having anything like a complete theory of turbulent convection in stars.[12]

Until the late 1950s most investigators used the *fitting method* to solve the four basic equations of stellar structure. Specifically, a pair of trial integrations, one starting from the center and one from the surface, was carried forward until they met at some interior fitting point. If the four dependent variables did not match at that point, subsequent trial integrations were made until they were continuous across the fitting point. The boundary-value problem was then solved and the model completed. This iterative procedure, which was extensively used by Martin Schwarzschild and Richard Härm during the 1950s, was also pursued for all evolutionary phases of stellar evolution by Chushiro HAYASHI (b. 1920) and his associates at Kyoto University. At about the same time, Louis George HENYEY (1910–1970) and his associates at the University of California at Berkeley developed a powerful method for automatic solution of the equations of stellar structure, suitable for electronic computers and applicable to a wide range of phys-

---

[11] See, e.g., M. H. Wrubel, *Handbuch der Physik*, **51**, 1 (1958).
[12] E. A. Spiegel, in *Stellar Evolution*, edited by R. F. Stein and A.G.W. Cameron (New York: Plenum Press, 1966), 143.

ical conditions during the lifetime of a star, including those encountered during the rapid expansion and contraction phases. The *Henyey method*, as it became known, involves transforming the four basic equations explicitly to difference equations and solving them by modern techniques such as relaxation procedures. The idea of the method is as follows: one starts with an initial guess for all values of the dependent variables, expands the difference equations to first order in a Taylor series to obtain variational equations, and then improves the approximate trial solution by solving the variational equations for corrections in the dependent variables. The basic principle of the method was published in 1959 by Henyey and his associates. Soon afterward, with fast electronic computers becoming more and more widely available, several groups around the world began to develop their own versions of the method. The result was an outpouring of numerical models during the 1960s, which led to a substantial advancement of the field.[13]

5.3 A BRIEF SURVEY OF STELLAR EVOLUTION

Throughout the 1930s, it was tacitly assumed that a star evolving on the main sequence would remain chemically homogeneous, despite the fact that hydrogen was presumably transformed into helium in its central regions. The short turnover time of ascending and descending convective currents was sufficient argument for the effective homogeneity of regions in which turbulent convection prevails. As for those regions in radiative equilibrium, appeal was made to rotationally driven meridional currents that would prevent the development of chemical inhomogeneity (see section 4.7). If so, Strömgren found in the early 1930s that stars with less hydrogen lie above and to the right of the main sequence, therefore suggesting that hydrogen depletion would bring a star from the main sequence to the giant branch (see section 4.4). In 1939, shortly after Weizsäcker and Bethe had worked out the details of the CN cycle, Gamow proposed a more detailed evolutionary scheme using the assumption that stellar energy was supplied by this cycle. Again assuming that a star will remain well mixed throughout its evolution, he showed that stars with less hydrogen are located higher on the main sequence, and thus he suggested that a star would move upward along the main sequence until the whole of its hydrogen was exhausted. He interpreted the mass-luminosity relation as a statistical regularity due to the fact that most of the stars were observed in the lower part of their evolutionary track. Obviously, the other major problem with this theory is the origin and energy sources of the red giants. Indeed, since Gamow assumed that all the stars evolved through a series of homologous equilibrium configurations, his red-giant models had structures similar to those of main-sequence stars. Accordingly, he found that red giants had extremely low central temperatures and densities—so low in fact that the CN cycle could not take place. It was therefore suggested that reactions between

---

[13]The following two textbooks were particularly influential: J. P. Cox and R. T. Giuli, *Principles of Stellar Structure* (New York: Gordon and Breach, 1968); D. D. Clayton, *Principles of Stellar Evolution and Nucleosynthesis* (New York: McGraw-Hill, 1968).

protons and the lightest elements, which would take place at low temperatures, were responsible for the energy production in these stars. Although many of Gamow's ideas on stellar evolution have been superseded, his 1939 paper is nevertheless an interesting pioneering attempt to apply the new results of nuclear physics to slowly evolving stars.

At this juncture it is worth comparing Gamow's 1939 ideas with those proposed ten years later by Vasilii Grigorivich FESENKOV (1889–1972) and developed in the mid-1950s by Alla Genrikova MASEVICH (b. 1918) and her colleagues at the Sternberg Astronomical Institute in Moscow.[14] Their basic assumption is that stars condense onto the main sequence in certain mass ranges and then evolve along it. In particular, Masevich contended that the observed distribution of stellar luminosities for the early-type stars could be obtained, assuming the origin of stars in O associations, only for an evolution down the main sequence with decreasing mass. In one set of models, she determined evolutionary paths for two main-sequence stars of types O7 and B3. The evolution is first down the main sequence, the star remaining well mixed and continually losing mass in proportion to its total luminosity. After the mass has decreased to about $3M_\odot$, total mixing ceases, and the star branches off the main sequence. A serious difficulty associated with these evolutionary models is that they depend on the assumption of complete mixing. Furthermore, no observational evidence is available that would justify a substantial mass ejection from stars located all along the upper part of the main sequence. As we shall see in section 5.7, however, Fesenkov's 1949 idea of mass loss from main-sequence stars retains its interest when discussing the rotational evolution of stars on the lower part of the main sequence. Mass loss from the O- and B-type stars will be discussed in section 6.2.

The first decisive step toward the resolution of the red-giant problem was made by Öpik (figure 5.8) in 1938. Specifically, he proposed that both the giant and main-sequence stars shine via the same processes of atomic transmutation and that these processes follow a sequence depending on successive increases in central temperature. According to Öpik, the key to understanding the red giants lay in the fact that a main-sequence star generally develops a convective core when nuclear burning occurs in its central regions. Provided that there is no interchange of matter between the core and the surface, the star will then settle into a compound structure consisting of a hydrogen-burning core in convective equilibrium surrounded by an inert hydrogen envelope in radiative equilibrium. Once all the core hydrogen is converted into helium, another stage of evolution for the compound model starts: the gravitational contraction of the core, which is gradually transformed into a superdense and superhot nucleus. With the rapid increase in temperature in the shell adjacent to the contracting core, nuclear energy is then released in an intermediate shell. The rapid increase of energy generation with increasing temperature and density causes this hydrogen shell to generate too much energy, and the whole of the radiative envelope expands to form a giant star.

---

[14]For a comprehensive discussion of these models, see G. R. Burbidge and E. M. Burbidge, *Handbuch der Physik*, **51**, 134 (1958).

Figure 5.8   The Estonian-born astronomer Ernst Öpik at work in his office at Armagh Obser-
             vatory in Northern Ireland. He published dozens of research papers on subjects
             ranging from stellar evolution to the orbits of meteors. He was also an excellent
             observer—an unusual trait in a theoretician. Courtesy of Sir Patrick Moore.

Another important step forward was made in 1942 by the Brazilian physicist Mario
SCHÖNBERG[15] (1914–1990) who collaborated with Chandrasekhar at the Yerkes
Observatory. They investigated the equilibrium configurations of stellar models
in which a radiative envelope surrounds either a convective core of higher mean
molecular weight, with all the energy being generated in the core, or an exhausted
isothermal core of higher mean molecular weight, with the energy being generated in
a thin shell located at the core-envelope interface. Their calculations showed that for
each assumed value of the ratio of the core/envelope mean molecular weights μ_c/μ_e
there is a maximum mass of the exhausted isothermal core beyond which no fit can
be made with the radiative envelope to make an equilibrium model. This critical
mass is a function of the ratio μ_c/μ_e, and has its largest value for a homogeneous

---

[15]Mário Schenberg, as he was known to his fellow countrymen, made only two contributions to
astrophysics, one with Gamow and the other with Chandrasekhar (G. Gamow and M. Schönberg, *Phys.
Rev.*, **59**, 539 [1941]; M. Schönberg and S. Chandrasekhar, *Astrophys. J.*, **96**, 161 [1942]). He was very
active in many fields of theoretical physics, however, and made frequent stays in Europe and North
America during the 1940s and early 1950s. In 1953 he became the director of the Physics Department
of the University of São Paulo. Unfortunately, he was forced into retirement in 1961, when a military
regime was installed in the country. He went twice to jail, in 1965 and 1968, in defense of his political
and democratic principles. Following a general amnesty, he came back to the University of São Paulo
in 1980 as a Professor Emeritus.

star ($\mu_c/\mu_e = 1$), but diminishes to a lower limit of about 10% of the mass of the star when $\mu_c/\mu_e = 2$. This is known as the *Schönberg-Chandrasekhar limit*. As they showed, the existence of such a limit implies that the lifetime of a star on the main sequence is limited to the time it takes to burn 10% of its hydrogen into helium. After the star has reached this limit, the core must therefore contract under its own gravity. This, in turn, leads to adjustments of the outer layers, which begin to expand, causing the star to move in the H-R diagram from the main sequence to the giant region, in agreement with Öpik's original suggestion.

By this time it had become clear that a nonhomogeneous structure was capable, in principle, of accounting for the extended envelopes and the associated low surface temperatures of the red giants. This conclusion was reinforced by a critical reappraisal by Peter Alan SWEET (b. 1921) in Glasgow of the problem of rotationally driven meridional currents, which had been assumed to maintain chemical homogeneity in a star (see section 4.7). Actually, when Eddington first calculated the circulation velocities in 1929, he concluded that they required about 10^7 years to mix the center and surface in the sun. Calculations made by Sweet in 1950 showed this conclusion to be a great overestimate, and the time to mix the sun was found to be of the order of 10^{12} years. In 1953 the problem was further discussed by the English astrophysicist Leon MESTEL (b. 1927), who showed that the condition for mixing is even more stringent than that derived by Sweet. Mestel's semiquantitative analysis strongly suggested that no continuous mixing between core and envelope can take place since the nonspherical distribution of mean molecular weight set up by the rotationally driven currents themselves tends to choke back the motion.[16] By the mid-1950s it was widely accepted that rotation does not cause effective mixing in the vast majority of stars.

The next significant advance in our knowledge of stellar evolution came in 1952 from the work of Allan SANDAGE (b. 1926) of Mount Wilson and Palomar Observatories, and Martin Schwarzschild, who considered the quasiequilibrium states through which an unmixed model passes after reaching the Schönberg-Chandrasekhar limit. Their starting point is a chemically inhomogeneous model with an isothermal core which has reached the limiting mass, a shell source burning on the CN cycle, and a radiative envelope. Then they consider a series of evolving models in which it is assumed that the hydrogen-exhausted core is contracting so that an additional source of energy is present in the core. They found that as the cores contract, liberating gravitational energy, the envelopes greatly expand. It was therefore concluded that stars initially on the main sequence will, after developing a chemical discontinuity between the core and the shell, evolve rapidly to the right in the H-R diagram, amply covering the giant region. Figure 5.9 illustrates speculative evolutionary tracks in the H-R diagram for stars of different masses. The main conclusion of this work appears to be that a single discontinuity of chemical composition can explain the properties of the red giants. As was pointed out by the English astrophysicist Roger TAYLER (1929–1997), however, such an assumption is essentially approximate since, as massive stars evolve, the convective core retreats so that a zone of continuously

---

[16]See P. A. Sweet, *Mon. Not. Roy. Astron. Soc.*, **110**, 548 (1950); L. Mestel, *ibid.*, **113**, 716 (1953). See also J. L. Tassoul, *Stellar Rotation* (Cambridge: Cambridge University Press, 2000), 93 and 145.

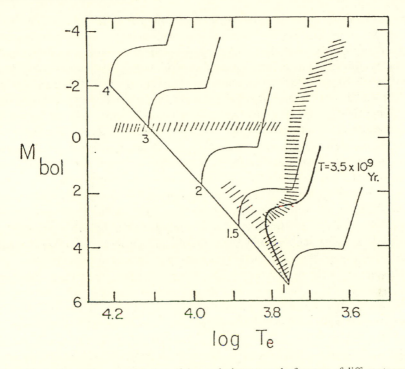

Figure 5.9 A schematic representation of the evolutionary tracks for stars of different masses. The schematic Hertzsprung-Russell diagram for the globular clusters M3 and M92 is shown by hatched markings for comparison. The heavy line is the theoretical appearance of the diagram 3.5 billion years after the formation of the stars. From A. R. Sandage and M. Schwarzschild, *Astrophys. J.*, **116**, 463 (1952).

varying composition is set up between the core and the envelope. Tayler in 1954 therefore derived evolutionary model sequences in which account is taken of this intermediate zone, as the hydrogen in the interior is gradually exhausted and the convective core shrinks in size. He found that, although his individual tracks are not identical to those depicted in figure 5.9, his H-R diagram for stars of different masses but of the same age gives qualitative agreement with the lower part of the observed H-R diagram for globular clusters.[17]

Further work along these lines was carried out in 1955 by Fred Hoyle and Martin Schwarzschild, who attempted to construct evolving models that would explain all the details of the observed features of the H-R diagrams of globular clusters. The evolutionary tracks were computed for stellar masses of $1.1M_\odot$ and $1.2M_\odot$, which correspond approximately to the masses at the top of the truncated main

[17]This early work on chemically inhomogeneous models was soon followed by numerous investigations of the evolution of stars from the main sequence to the giant stage (see, e.g., R. S. Kushwaha, *Astrophys. J.*, **125**, 242 [1957]; J. M. Blackler, *Mon. Not. Roy. Astron. Soc.*, **118**, 37 [1958]; C. B. Haselgrove and F. Hoyle, *ibid.*, **119**, 112 [1959]). For a review of these and other pioneering calculations made during the 1950s, see D. H. Menzel, P. L. Bhatnagar, and H. K. Sen, *Stellar Interiors* (New York: John Wiley & Sons, 1963).

sequence in the observed H-R diagrams. They found that their models can explain the evolution of stars as far as the giant branch in globular clusters provided electron degeneracy in the core is taken into account, the chemical compositions of the models used being characteristic of extreme population II stars. They also found that provision for a surface convection zone is essential, since with it their models could make the transition from evolution at almost constant luminosity on the way over to the giant branch to rapidly increasing luminosity and slowly varying surface temperature in the giant branch itself. Obviously, the next problem is to consider how a giant star evolves after the contraction of its degenerate core causes the central region to become more hot and dense. This was originally discussed by Martin Schwarzschild and Richard Härm in 1962, who showed that helium eventually begins to burn abruptly in the degenerate core of a $1.3 M_\odot$ red giant to form carbon via the 3α reaction. Subsequent calculations have shown that stars more massive than about $2.25 M_\odot$ evolve from the main sequence to central helium burning without developing a degenerate core, and thus without going through a central *helium flash*. In all cases, the ignition of the 3α reaction is explosive in the sense that the rate of nuclear-energy production in the core greatly exceeds the rate at which energy can be transported radially outward. The core therefore expands, causing the envelope to shrink and the surface temperature to increase. In other words, the net effect of helium burning is to stop a star from moving farther upward and to the right in the H-R diagram. The star then moves back toward the main sequence, parallel to the horizontal band that connects the red giant branch with the main sequence in the H-R diagram of globular clusters (see figure 5.10). At the same time, with increasing central temperatures, additional nuclear reactions involving helium and carbon can produce oxygen, neon, magnesium, and eventually up to iron. (Heavier elements are formed successively by the slow capture of neutrons by nuclei near the iron peak.) By the mid-1960s, then, a coherent picture of the evolution of stars had emerged, even though the details of the most advanced phases were still very uncertain at the time. The American astrophysicist Icko IBEN, Jr. (b. 1931; figure 5.11) was undoubtedly the most steadfast contributor to this progress.[18]

By the mid-1950s both observational and theoretical evidence had been given that the structure of a star is determined by its original chemical composition, its mass, and its age. Studies of stellar evolution had also reached the general conclusion that a star spends most of its lifetime on the main sequence, that the age of some globular clusters is of the order of 4 billion years, and that only massive stars burn their hydrogen fast enough to enter the giant branch during the lifetime of our galaxy. Another interesting advance in the field came from the comparative study of galactic and globular clusters. Indeed, since the stars in each cluster have most probably condensed from a common cloud of interstellar matter, we may reasonably assume that they have, at least to a first approximation, the same composition and the same age. Hence, the variation on the H-R diagram of a single cluster would be due to the stars being of different mass. Figure 5.12 illustrates a composite H-R diagram for

---

[18] For a detailed description of stellar evolution within and off the main sequence, see I. Iben, Jr., *Annu. Rev. Astron. Astrophys.*, **5**, 571 (1967); **12**, 215 (1974); I. Iben, Jr., and A. Renzini, *ibid.*, **21**, 271 (1983). See also R. Kippenhahn and A. Weigert, *Stellar Structure and Evolution* (Berlin: Springer-Verlag, 1990).

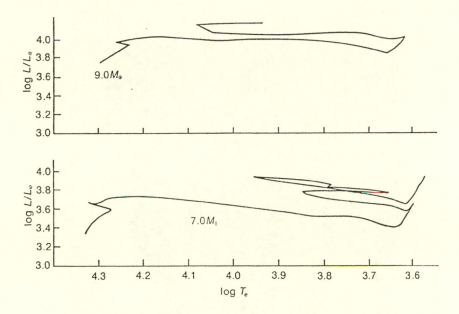

Figure 5.10 Post-main-sequence evolutionary tracks on the theoretical H-R diagram for stars
having $M = 9M_\odot$ (after Iben) and $M = 7M_\odot$ (after Hofmeister et al.). From I.
Iben, Jr., *Astrophys. J.*, **140**, 1631 (1964); E. Hofmeister, R. Kippenhahn, and A.
Weigert, in *Stellar Evolution*, edited by R. F. Stein and A.G.W. Cameron (New
York: Plenum Press, 1966), 263.

several galactic clusters and one globular cluster (M3) superimposed by fitting the
lower part of the main sequence of each cluster on one line. The interpretation of
this diagram was originally made by Sandage in 1957. The more massive the star,
the shorter is its lifetime on the main sequence. As the age of a cluster increases, the
point where the observed H-R diagram turns off from the main sequence therefore
moves down the remaining main sequence to the fainter, less massive stars. Accordingly, the clusters can be dated on figure 5.12 by the main-sequence lifetime of
the stars that have just left the main sequence. These *turn-off points* were converted
into cluster ages in the right-hand scale of figure 5.12. The cluster M67 is an old
galactic cluster, showing broad deviation from the initial main sequence, whereas
the cluster NGC 2362 is a very young one and shows little deviation.

Further significant progress in our understanding of stellar evolution has come
from the study of the youngest clusters. The first indication that the main sequence
of very young clusters might differ considerably from the so-called *zero-age main
sequence* was obtained in 1953 by the Soviet astronomer Pavel PARENAGO (1906–
1960) from his detailed study of the stars in the Orion nebula association. He found
that the hottest, and therefore the most massive, stars of this association lie on
the main sequence, whereas the less massive cool stars fall distinctly above the
zero-age main sequence. Three years later, the American astronomer Merle WALKER
(b. 1926) began an extensive study of very young clusters that confirmed Parenago's

Figure 5.11 The American astrophysicist Icko Iben, Jr., explaining stellar evolution at the
blackboard. The picture was taken in the fall of 1964, the year he became a faculty member at the Massachusetts Institute of Technology (Cambridge, Mass.).
Later, in 1972, he joined the faculty of the University of Illinois in Urbana-
Champaign. For more than three decades, Iben has investigated numerically the
structure and evolution of stars beyond the main-sequence stage. Courtesy of
Dr. I. Iben, Jr.

original finding. Using the evolutionary tracks obtained by Henyey and his associates,[19] he explained that the less massive, cooler members of the cluster NGC
2264 must still be in the process of contracting gravitationally from the prestellar
medium. He concluded that this cluster, which is embedded in a gas cloud, is only
3 million years old, and that its less massive stars have not yet become hot enough
to ignite thermonuclear reactions.

In 1956, when Walker published his analysis of the cluster NGC 2264, it was
generally assumed that pre-main-sequence stars had to be in radiative equilibrium
and that they would slowly brighten as they contracted toward the main sequence.
A major reversal occurred in the ideas when Hayashi (figure 5.13) convincingly
showed that convection would enable a contracting star to release its energy much
faster than by radiation, so that initially a star would appear in the upper right-

---

[19]L. G. Henyey, R. LeLevier, and R. D. Levée, *Publ. Astron. Soc. Pacific*, **67**, 154 (1955).

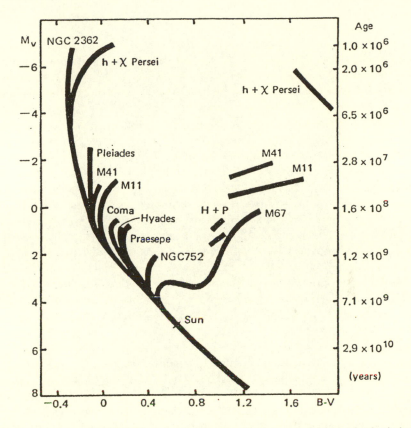

Figure 5.12 A composite H-R diagram, showing the absolute magnitudes and color indexes
of stars in a variety of clusters. The age of the cluster, defined by the position
along the main sequence at which its stars turn off to the right, is marked on the
vertical scale at the right. From A. R. Sandage, *Astrophys. J.*, **125**, 435 (1957).

hand corner of the H-R diagram.[20] According to Hayashi's calculations, convection
dominates the structure of a newborn star so that it moves rapidly downward on
the diagram while maintaining its surface temperature nearly constant. As the
star contracts and heats, a radiative zone develops near the center and gradually
includes more and more mass. The star's evolutionary path on the H-R diagram
then turns sharply to the left until it reaches the main sequence. (This is roughly the
inverse behavior of a main-sequence star on its way over to the giant branch.) As
Hayashi showed, if the contraction toward the main sequence proceeds in this way,
the massive stars in a young cluster should have reached the main sequence while
the less massive ones are still in their pre-main-sequence contraction phase. This
is apparently confirmed by Walker's observations of the very young cluster NGC
2264. In fact, many of the stars belonging to that cluster are eruptive variables of

---

[20]C. Hayashi, *Publ. Astron. Soc. Japan*, **13**, 450 (1961).

Figure 5.13 The Japanese astrophysicist Chushiro Hayashi, who is best known for his work on the pre-main-sequence contraction phase. In 1995 the Inamori Foundation awarded him the Kyoto Prize for basic science, stating that he "has developed powerful, original theories and contributed immeasurably to the development of his field." The picture was taken in May 1968 at the NASA Goddard Space Flight Center, Greenbelt, Md. Courtesy of Dr. C. Hayashi.

the so-called *T Tauri* type, that is, dwarf stars that show rapid and erratic variations in brightness, probably associated with the fact that they are still contracting in the fringes of obscuring clouds of interstellar dust.

5.4 POSTGIANT EVOLUTION AND STELLAR REMNANTS

For the sake of clarity, let us first restore the chronology of the early events in the life of a star. By the early 1960s it was generally accepted that stars are continually forming from condensations in the interstellar gas and dust.[21] At that time, however,

---

[21] Photographs of the Orion nebula show the presence of what appear to be small knots in the interstellar medium. These are known as *Herbig-Haro objects*, after their joint discoverers—George Howard HERBIG (b. 1920) at Lick Observatory and Guillermo HARO (1913–1988) at Tonantzintla Observatory in Mexico. It was found, in 1954, that one of these objects apparently contained two faint stars that had not been visible eight years earlier. This observational work strongly suggested, therefore, that they were newly formed stars that are still growing as they pull interstellar gas and dust into themselves (see, however, section 6.1).

the entire process of star formation was not yet completely understood, but it seemed inevitable that the primitive stars were heated by gravitational contraction until they began to shine. Thereafter, as was described in section 5.3, they go on shrinking until the central temperatures are hot enough to permit the atomic synthesis of hydrogen into helium. When these nuclear reactions take place fast enough to supply the energy being radiated away, gravitational contraction stops and the stars have arrived on the main sequence. The characteristic time of this pre-main-sequence phase, known as the *Hayashi contraction*, varies from 10^5 years for what will eventually become a B-type star on the main sequence to some 10^9 years for the smallest M-type stars; the determining factor is the total mass of the forming star.

Detailed numerical calculations show that the main-sequence lifetime of a star is about

$$t_{\mathrm{ms}} \approx 10^{10} \, \frac{M/M_\odot}{L/L_\odot} \text{ years,} \qquad (5.8)$$

where M and L are the mass and luminosity, respectively. To a good approximation, the mass-luminosity relation along the main sequence is about $L \propto M^4$, so it follows that massive stars evolve more rapidly than those of small mass. Since the present age of the sun is 4.5 billion years, it thus has several billion years left before it rapidly leaves the main-sequence area and moves into the red-giant region. However, main-sequence lifetimes as short as 10^6 years and as long as 10^{12} years are to be expected. Calculations also show that early-type main-sequence stars have convective cores which decrease in size as one goes to later-type stars. Around spectral type F, at $T_{\mathrm{eff}} \approx 7{,}500$ K, an outer convective envelope begins to develop; the convective envelope deepens as one moves downward along the main sequence.

As we know from the pioneering work of Martin Schwarzschild and coworkers, when the hydrogen content in the center of a main-sequence star is too low for nuclear reactions to supply enough energy to support the star against gravity, contraction again becomes an important energy source as the inner parts of the star contract while its outer parts expand. The star then rapidly becomes brighter and cooler and moves upward and to the right on the H-R diagram, toward the red-giant region. We then have a (degenerate or nondegenerate) core of inert helium, a shell-like area where hydrogen in the envelope is burning, and the hydrogen envelope itself. Eventually conditions in the center will become extreme enough for the nuclear burning of helium to form carbon via the 3α reaction. The giant star will then settle into a second period of nuclear burning, although the core burning of helium during the giant stage will not last nearly as long as the core burning of hydrogen during the main-sequence stage. In fact, the current theory of stellar evolution shows that both the main sequence and the giant region are heavily populated because stars spend a large part of their lives in these regions, with only a short amount of time being spent in between.

The advent of helium core burning in a star is generally held to be the end of evolution into the red-giant region. This event, which was briefly discussed in section 5.3, reverses the direction of motion in the H-R diagram and sends the star back toward the main sequence along a nearly horizontal track. During this stage, the evolution of the star takes it through one or more unstable states, as we observe that some variable stars occur in this part of the H-R diagram (see section 5.6). Not unexpectedly, the postgiant evolutionary phases are much more uncertain than the phases that have been discussed so far. This is no doubt due in part to the fact that the whole question of mass loss during postgiant evolution was, and still is, very much unsettled. The question naturally arises, therefore: What happens to a star as it evolves beyond the helium-burning stage? Generally speaking, the very last phases of the life of a star are poorly understood, and the exact manner in which these stages are attained is even less understood. Nevertheless, by the end of the 1960s it was widely accepted that three qualitatively different kinds of terminal stage may be discerned: the ubiquitous white dwarfs, the neutron stars (observed as pulsars), and the highly speculative black holes.[22]

White Dwarfs and Planetary Nebulae

Observationally, *white dwarfs* are stars that lie well below the main sequence, mostly to the left in the H-R diagram (see figure 2.12). They were first recognized as a distinct class of stars, having stellar masses and planetary dimensions, in the 1910s. The puzzle of the white dwarfs was resolved by Ralph Fowler and others, who showed that the pressure of a degenerate electron gas is sufficient to support an object of stellar mass against its own gravity at precisely the radii of white dwarfs. The theory was worked out in detail by Chandrasekhar in 1935, who found that there is a critical stellar mass above which stable white dwarfs cannot exist (see eq. [4.18]). This was a major discovery, for the existence of this critical mass—now called the *Chandrasekhar limit*, about $1.4 M_\odot$—has profound consequences for the terminal stages of stellar evolution: Stars that end their evolution with masses in excess of this limit are doomed to ultimate collapse, with mass loss occurring in explosive events, as observed in supernovae. In some cases, the violence of the explosion does not wholly disrupt the star, however, so that a stellar remnant of even greater density than a white dwarf may be left behind: either a *neutron star* or a *black hole*.

Knowledge of the thermal structure of white dwarfs did not begin until 1940, when Robert MARSHAK (1916–1992), then a student with Bethe at Cornell University, calculated the thermal conductivity of an electron gas under conditions of strong degeneracy. He found this conductivity to be so high that a radiating white dwarf has the bulk of its mass as a nearly isothermal core, while its thin nondegenerate surface layers act as a blanket, thus keeping down the luminosity. Typically, for white-dwarf luminosities of $10^{-2} L_\odot$, the temperature of the isothermal core is about 10^7 K. Now, by the time a star has reached the white-dwarf stage, it probably will have used all nuclear-energy sources, since at these temperatures it can support only the burning of hydrogen, which has been exhausted. There is thus little doubt that white dwarfs are in the ultimate stage of stellar evolution. The thermal energy of the degenerate electrons is not available, however, since the electrons are already crowded in all available low-energy states. If so, then what is the source of the energy they radiate? This question went unanswered until 1952, when Mestel

---

[22]For comprehensive reviews and reprints of key papers, see H. Gursky and R. Ruffini, eds., *Neutron Stars, Black Holes and Binary X-Ray Sources* (Dordrecht: Reidel, 1975).

showed that the only source available is the thermal energy of the nondegenerate nuclei throughout the star. Calculations indicate that this source of heat is amply adequate to support luminosities between $10^{-3} L_{\odot}$ and $10^{-2} L_{\odot}$ for about 10^9 years. A white dwarf thus loses heat quite slowly, becoming gradually cooler and fainter and ultimately ending its life as a cold inert body. Except for details, Mestel's theory of white-dwarf evolution has successfully resisted the passage of time.[23]

Now, the evidence accumulated throughout all these years indicates that the last stages of activity before an evolved star settles down to die as a white dwarf are represented by the *planetary nebulae*. These objects—so named because their disk-like shape was reminiscent of planets when seen through a small telescope—are expanding shells of rarefied gas around an extremely hot star, as was shown by the Dutch astrophysicist Herman ZANSTRA (1894–1972) and a few others in the early 1930s. One of the most intriguing problems about planetary nebulae is their progenitors and how they form. The hypothesis that red supergiants are immediate parents of planetary nebulae was proposed in 1956 by the Soviet astrophysicist Iosif SHKLOVSKII (1916–1985; figure 5.14), who pointed out that the observed expansion velocities of planetary nebulae, of the order of 20 km s^{-1}, are comparable with the escape velocities from the most extended red supergiants.[24] By the late 1960s the place of a planetary nebula in the evolutionary scheme of a star was still unclear, except perhaps for the final evolution of the central star to a white dwarf and of the gaseous shell into interstellar space. A much more definite picture emerged during the early 1970s, however, when detailed evolutionary calculations showed that the most probable progenitors of planetary nebulae are luminous red supergiants. These last results strongly suggest, therefore, that a planetary nebula comprises a short transitory stage, lasting a few times 10^4 years, in the evolution of a medium-sized star in the process of losing its outer layers as it changes from a supergiant star to a white dwarf.

Supernovae

We saw in section 2.7 that major explosions in stars—*supernovae*—are relatively rare events.[25] These spectacular outbursts of the most massive stars occur in a single galaxy only once in several centuries. We are faced, however, with a rather complex

[23] See especially the following pioneering papers: L. Mestel, *Mon. Not. Roy. Astron. Soc.*, **112**, 583 (1952); **112**, 598 (1952); L. Mestel and M. A. Ruderman, *ibid.*, **136**, 27 (1967).

[24] According to a competitive theory developed during the 1950s at the Byurakan Observatory, planetary nebulae are not formed from any existing stars in the Milky Way but rather from prestellar objects which evolve into proper stars, producing expanding envelopes as a by-product. This view has its roots in Ambartsumian's idea that explosions from a very dense state are associated with the formation of stars, star clusters, galaxies, and even the universe itself. As we know, the theory according to which the nuclei of planetary nebulae are newborn stars was not corroborated by observational data; it was finally abandoned by its proponents in favor of a theory based on Shklovskii's idea. See G. A. Gurzadyan, *Planetary Nebulae* (New York: Gordon and Breach, 1969), 278–304; G. A. Gurzadyan, *The Physics and Dynamics of Planetary Nebulae* (Berlin: Springer-Verlag, 1997), 430; D. Lynden-Bell and V. Gurzadian, *Astron. Geophys.*, **38**, No. 2, 37 (1997).

[25] For a detailed historical account of the observational data pertaining to stellar explosions, see D. Leverington, *A History of Astronomy* (Berlin: Springer-Verlag, 1995), 171.

Figure 5.14 The Soviet astrophysicist Iosif Shklovskii, who was known throughout the world for his brilliant and imaginative theoretical work on a broad range of problems. An independent thinker in politics as well as in science, he was seldom permitted to travel to Western countries during the Cold War. From H. Friedman, *Physics Today*, **38**, No. 5, 104 (1985). Courtesy of Gertrude Friedman.

situation that was originally recognized by Rudolf Minkowski in 1941: there are at least two types of supernova. The type I supernovae rise to a typical absolute magnitude of -16, and they have broad emission bands in their spectra; they are also deficient in hydrogen. Type II supernovae attain about absolute magnitude -14 at maximum brightness, and they have spectra resembling those of ordinary novae with hydrogen emission lines. The type I supernovae decrease in magnitude after maximum initially much faster than type II supernovae. Moreover, the ejected shells of type I supernovae have lower velocities of expansion than those of type II. Type I supernovae are related to the old population II stars, and type II supernovae to the young population I stars. These differences strongly suggest that type I supernovae occur in stars that are older and less massive than the stars responsible for type II supernovae.

The first quantitative theory that accounts for the catastrophic events leading to supernova flare-ups was proposed by Gamow and Schönberg in 1941. They suggested that the collapse of a supernova might be the result of a sudden drop of pressure in its interior caused by the emission of neutrinos, that is, massless particles carrying energy at the speed of light and interacting only very weakly with matter. Specifically, they showed that at a temperature of about 10^8 K a nucleus can alternately capture and release an electron while emitting a neutrino and an antineutrino. After one cycle the nucleus has the same composition, but it has lost the energy associated with the two neutrinos. Gamow accordingly gave this particular neutrino-producing reaction the name of *Urca process*, after the Casino da Urca in Rio de Janeiro, where also, no matter how one plays the game, one always seems to lose. Although it was later shown that neutrinos may be emitted during nuclear reactions other than the Urca process, the work of Gamow and Schönberg is nevertheless important because it focused attention on the fact that sizable neutrino losses could lead to a tremendous speeding up of the later stages of stellar evolution after helium burning.

By the early 1970s it was generally accepted that stars with initial mass up to $4M_\odot$ produce white dwarfs, even though the mass of the white dwarf itself cannot be more than $1.4M_\odot$. The reason lies in the fact that the white dwarf is only the central remnant of the original star. In fact, much of the initial mass has been blown away earlier in the star's life; some of its mass was shed by the star during its red-giant phase, in the form of stellar winds, and the remainder was blown off during the planetary-nebula stage. As was originally pointed out by Fred Hoyle and William Fowler in 1960, however, a different fate awaits a star of much larger initial mass. Specifically, they argued that the explosion of type II supernovae follows the implosion of nondegenerate core material in very massive stars, whereas the explosion of type I supernovae results from the ignition of degenerate nuclear fuel in the core of less massive stars. The first working hydrodynamical models of these phenomena were made in the United States by David ARNETT (b. 1940), Stirling COLGATE (b. 1925), William ROSE (b. 1935), and a few others.[26]

For a star in the intermediate mass range from about $4M_\odot$ to $8M_\odot$, several independent studies made during the late 1960s and early 1970s have confirmed their views, namely, that the collapse of such a star generates so much heat that the central temperature of the star becomes high enough for carbon nuclei to react with each other, at about 6×10^8 K. With the onset of carbon burning in the degenerate core, a sequence of nuclear reactions then occurs in which progressively heavier elements are built up out of carbon. At the same time, a runaway effect develops in which the temperature dramatically increases, detonating the degenerate carbon core and causing the entire star to explode (see, however, section 6.6).

---

[26]F. Hoyle and W. A. Fowler, *Astrophys. J.*, **132**, 565 (1960); W. A. Fowler and F. Hoyle, *Astrophys. J. Suppl.*, **9**, 201 (1964); S. A. Colgate and M. H. Johnson, *Phys. Rev. Letters*, **5**, 235 (1960); S. A. Colgate and R. H. White, *Astrophys. J.*, **143**, 626 (1966); W. D. Arnett, *Nature*, **219**, 1344 (1968); *Astrophys. Space Science*, **5**, 180 (1969); W. K. Rose, *Astrophys. J.*, **155**, 491 (1969). For a comprehensive review of the state of the art by the mid-1970s, see S. A. Colgate, in *Neutron Stars, Black Holes and Binary X-Ray Sources*, edited by H. Gursky and R. Ruffini (Dordrecht: Reidel, 1975), 13. See also W. D. Arnett, in *Essays in Nuclear Astrophysics*, edited by C. A. Barnes, D. D. Clayton, and D. N. Schramm (Cambridge: Cambridge University Press, 1982), 427.

If the initial mass of the star exceeds about $8M_\odot$, numerical calculations have shown that such a violent detonation does not occur, so that the successive stages of burning continue until finally the element iron is reached. At this point the process stops because nuclear reactions involving iron do not yield energy; instead they absorb energy. Since we can get no energy from building up heavier nuclei, the central pressure therefore drops sharply, and the star inexorably contracts until this contraction turns into a catastrophic implosion of the core. Supposedly, the collapsing material has become so hot and dense that it bounces and generates a shock wave, which blows off the star's outer layers. In the late 1960s, following the suggestion made by Colgate and White from the Lawrence Radiation Laboratory (Livermore, Calif.), these layers were believed to be dramatically ejected into outer space by the extreme heating caused by the deposition of neutrino energy in the outer layers. As we shall see in section 6.6, however, the correctness of their views was not established until the final decades of the century.

In any case, by the early 1970s the following picture was slowly emerging from the calculations. The explosion that follows the disintegration of the iron-group nuclei into alpha particles and neutrons in a very massive star is as violent as the explosion that follows the detonation of carbon in less massive stars. In both cases, the explosion and the ejection of matter are observed as a *supernova*. All or a large fraction of the material within the star disperses in this explosion, therefore enriching the interstellar medium with heavy elements. Theoretical studies have shown that in some cases the entire star is shattered in the explosion; in other cases, however, neutrino losses may affect the evolution to such an extent that a highly compressed remnant of the star's core remains behind. Actually, the collapse of a star of intermediate mass leads to a core mass above the Chandrasekhar limit, thus carrying it beyond the white-dwarf stage. Then, as was shown in section 4.5, the matter compresses to a new equilibrium state with densities of about 10^{14} g cm^{-3} or more, compared to the densities of 10^8 g cm^{-3} for white dwarfs. At this point, the material has become virtually incompressible as a degenerate neutron gas. The star's core is then said to be a *neutron star*, with a radius of some 10 km, and a mass of the order of $1M_\odot$ to $3M_\odot$.

Neutron Stars and Pulsars

As explained in section 4.5, the concept of the neutron star was originally considered in the 1930s, but these objects remained hypothetical until the serendipitous discovery of *pulsars*, that is, variable radio sources having very short and extremely stable periods, originally thought to be pulsations. In the summer of 1967, a team of British scientists at the Mullard Radio Astronomy Observatory headed by Antony HEWISH (b. 1924) were making a series of observations with their radio telescope. Jocelyn BELL (b. 1943), later Bell Burnell, was assigned the task of investigating fluctuations in the strength of radio waves from distant galaxies. Unexpectedly, she observed signals consisting of regular bursts of radio noise at perfectly spaced intervals of about one second, each burst lasting no more than one-hundredth of a second.[27] Speculation immediately suggested the possibility that intelligent beings—the so-called little green men—might be beaming a message to the earth, or that the signals were merely a form of interference generated by mechanical devices on earth. Both these possibilities were soon eliminated, however, mainly because many other similar sources were soon found, and their extraterrestrial origin proved by the fact that they all moved with the stars.

Not unexpectedly, the announcement in February–April 1968 of the discovery of four pulsars, with periods ranging from 0.25 to 1.33 seconds, led to a flurry of activity among theoreticians. The Austrian-born American astrophysicist Thomas GOLD (b. 1920; figure 5.15) of Cornell University was the first to propose, in May 1968, the currently accepted explanation of pulsars as rapidly rotating neutron stars with a strong dipole magnetic field whose axis is inclined to the rotation axis.[28] Although there were as yet not really enough clues to identify the mechanism of radio emission, Gold argued that this emission derives its energy from the rotational energy of the neutron star, and that it is a result of relativistic velocities set up in a corotating magnetosphere, leading to radiation in the pattern of a rotating beacon. If so, then the extraordinary stability of the pulse period was due to the rotation of the neutron star, whereas the extreme shortness of the radio bursts was a consequence of the smallness of the star itself, which is no more than 10 km in radius. Gold's farsightedness is particularly apparent in the concluding remarks of his paper:

> If this basic picture is the correct one it may be possible to find a slight, but steady, slowing down of the observed repetition frequencies. Also, one would then suspect that more sources exist with higher rather than lower repetition frequency, because the rotation rates of neutron stars are capable of going up to more than $100/s$, and the observed periods would seem to represent the slow end of the distribution.

Both predictions were soon to be confirmed by radio and optical observations.

The intimate connection between pulsars and supernova remnants became more evident in October 1968, when a team of Australian radio astronomers found a pulsar with an extremely short period—0.089 second—in the Gum nebula, which was created by the blast shock wave from a supernova in the constellation Vela.[29] Two months later, a pulsar with an even shorter period—0.033 second—was discovered in the Crab nebula, therefore suggesting that the supernova remnant, probably a collapsed neutron star, was indeed the source of the radio pulses.[30] A very faint star near the center of the nebula was positively identified optically as the source of the radio pulses when, in February 1969, several groups of American observers detected

---

[27]A. Hewish, S. J. Bell, J.D.H. Pilkington, P. F. Scott, and R. A. Collins, *Nature*, **217**, 709 (February 1968). A few weeks later, three other pulsars were reported in the following paper: J.D.H. Pilkington, A. Hewish, S. J. Bell, and T. W. Cole, *Nature*, **218**, 126 (April 1968). See also Jocelyn Bell Burnell's after-dinner speech in the *Eighth Texas Symposium on Relativistic Astrophysics*, edited by M. D. Papagiannis (New York: New York Academy of Sciences, 1977), 685.

[28]T. Gold, *Nature*, **218**, 731 (May 1968).

[29]M. I. Large, A. E. Vaughan, and B. Y. Mills, *Nature*, **220**, 340 (October 1968).

[30]D. H. Staelin and E. C. Reifenstein, *Science*, **162**, 1481 (December 1968); J. M. Comella, H. D. Craft, R.V.E. Lovelace, J. M. Sutton, and G. L. Tyler, *Nature*, **221**, 453 (February 1969).

164

CHAPTER 5

Figure 5.15 The Cornell astrophysicist Thomas Gold in 1963. A versatile phenomenologist who has contributed to many branches of astronomy and geophysics, he was the first in 1968 to propose the lighthouse model for the pulsars. Courtesy of Dr. T. Gold.

fast light pulsations having the same frequency as the radio pulses.[31] In fact, this radiation came from the very peculiar star that Baade and Minkowski had identified in 1942 as the stellar remnant of the Crab nebula supernova explosion. (Astronomers had observed it for years without noticing the optical blinking because the rapidity of the pulses concealed the true nature of the star.) Gold's oblique-rotator model received further dramatic support when it was shown in May 1969 that the period of the Crab pulsar was slowly increasing with time at the rate of 10^{-5} second per year.[32]

Added support for the identification of pulsars as neutron stars was given by Gold in January 1969, when he showed that the loss of rotational energy involved in the slowing down of the Crab pulsar was exactly that required to light up the Crab nebula itself. Actually, this and related theoretical contributions revived interest in the work

[31]W. J. Cocke, M. J. Disney, and D. J. Taylor, *Nature*, **221**, 525 (February 1969); R. E. Nather, B. Warner, and M. Macfarlane, *ibid.*, **221**, 527 (February 1969); J. S. Miller and E. J. Wampler, *ibid.*, **221**, 1037 (March 1969).

[32]D. W. Richards and J. M. Comella, *Nature*, **222**, 551 (May 1969).

of Shklovskii, who had argued in 1953 that both the radio and optical emissions of this nebula come from the *synchrotron process*, that is, electromagnetic radiation emitted by high-speed electrons spiraling about a strong magnetic field. Thus, as the spinning neutron star is losing its rotational energy, the presence of a very high magnetic field allows this energy to reappear in the form of the kinetic energy of fast-moving electrons. Subsequent work has shown that pulsars can accelerate electrons to the relativistic energies required to explain both the synchrotron radiation of the Crab nebula and the beamed electromagnetic waves that are observed as pulsars.[33] Appropriately, by December 1970 the Crab pulsar had already been observed over the whole electromagnetic spectrum, from radio waves to high-energy γ rays, with most of the radiated energy in the X-ray region. However, it was not until several years later, in the mid-1970s, that optical and γ-ray pulses were observed from the so-called Vela pulsar in the Gum nebula.

Black Holes and X-Ray Sources

As was noted, by the early 1970s it was generally accepted that a star of about $1 M_\odot$ eventually uses up its nuclear fuel and contracts to become a white dwarf, whereas a star with less than $3 M_\odot$ of material at the end of its evolution contracts beyond the white-dwarf state and becomes a much denser neutron star. For a contracting or imploding stellar remnant of more than $3 M_\odot$, however, the force of gravity overwhelms all other forces. This problem was originally discussed in 1939 by the American physicists Robert Oppenheimer and Hartland SNYDER (1913–1962), who pointed out that at these high densities Einstein's general theory of relativity was needed to describe the dynamics of a collapsing star. Actually, their work has its roots in earlier work by the German polymath Karl Schwarzschild, who had solved Einstein's field equations in 1916 to show that the space-time outside a spherical mass has a singular behavior at what is now called the *Schwarzschild radius*, $R_S = 2GM/c^2$. (Here M is the total mass, G the constant of gravitation, and c the speed of light.) Making use of this result, Oppenheimer and Snyder were able to demonstrate that, to an external observer, matter collapses in infinite time within the radius R_S and that this radius defines a trapped surface from which light cannot escape to infinity. Theoretically at least, very massive stars can collapse completely to such a final state, the *black hole*, that the density approaches infinity.[34]

Since no form of electromagnetic radiation can escape from the surface of a black hole, a massive collapsed star would thus move through space as did the original star, but unseen by any external observer. Its gravitational field would still be present, however, so that the collapsed star would attract and capture any material it encountered. This led Shklovskii and his colleagues in Moscow to argue

---

[33]The following pioneering papers are particularly worth noting: N. S. Kardashev, *Astron. Zh.*, **41**, 807 (1964) (*Soviet Astron.*, **8**, 643 [1965]); F. Pacini, *Nature*, **216**, 567 (1967); T. Gold, *ibid.*, **221**, 25 (1969); P. Goldreich and W. H. Julian, *Astrophys. J.*, **157**, 869 (1969); J. P. Ostriker and J. E. Gunn, *ibid.*, **157**, 1395 (1969).

[34]The concept of the black hole dates back to 1784, when Michell suggested that, if stars were massive enough, light would be prevented by gravity from leaving the surface. A similar suggestion was made in 1796 by Laplace, who published a proof of the result that a sufficiently massive and condensed body would appear invisible.

in the 1960s that an invisible black hole could be detected if it were close enough to a
large visible companion. As was originally suggested by Igor Dmitrievich NOVIKOV
(b. 1935) and Yakov Borisovich ZELDOVICH (1914–1987; figure 5.16), gas would
then be drawn from the visible companion onto the invisible black hole, where the
gas might become hot enough to emit X rays.[35] They also noted that the presence
of a black hole might be deduced from an analysis of the orbital elements of the
visible companion.

Although extraterrestrial X-ray sources were already observed in the 1960s, it
was not until the early 1970s that X-ray observations made by the *Uhuru*[36] satellite
provided a tentative indication that black holes do actually exist. Cygnus X-1 is so
named because it was the first X-ray source discovered in the constellation Cygnus.
Observations show that the X-ray intensity of this object varies irregularly and as
rapidly as within 0.1 second, sometimes showing variations over times as short as
0.001 second. This indicates that the X-ray emitting region must be less than 0.001
light-second in size (less than 300 km). Cygnus X-1 was identified in 1972 as the
invisible companion of a blue supergiant whose spectrum shows that it is a member
of a spectroscopic binary with a period of 5.6 days.[37] The conservative estimate
of about $12M_\odot$ for the supergiant leads to a mass larger than $3M_\odot$ for the X-ray
emitting body. Since the mass of a neutron star cannot be greater than approximately
two or three solar masses, this meant that the optically invisible X-ray source Cygnus
X-1 was the best candidate yet for a black hole.

5.5 EVOLUTION OF CLOSE BINARY STARS

By the early 1950s it was generally recognized that main-sequence stars eventually
deplete all their central hydrogen reserves and evolve rapidly to the right in the H-R
diagram. As was shown in section 5.3, it is the development of chemical inhomogeneities resulting from hydrogen burning that causes the core of a main-sequence
star to contract while the surrounding layers expand, thus moving the star as a whole
from the main sequence toward the giant region. Detailed calculations show that
the rate at which this expansion occurs depends crucially on the star's mass (see
eq. [5.8]). In single stars, expansion could go on unhampered by external influences until the excess of potential energy released in their central regions has been
expended. In close binary stars, on the other hand, the proximity of a companion
surrounds the expanding star with an invisible barrier which the volume of the star
cannot exceed. If and when a given star has reached this limit, its further radial
expansion is arrested, so that a continuing tendency to expand will necessarily bring
about a mass transfer to its companion. As we shall see, the existence of these invisible barriers—the so-called Roche lobes—is probably the most important factor
determining the evolution of close binary stars.

---

[35] See their addendum 2 (p. 827) in *Nuovo Cimento Suppl.*, **4**, 810 (1966).

[36] Uhuru means "freedom" in Swahili; the satellite was launched in December 1970 from a platform
off the coast of Kenya, on the seventh anniversary of that country's independence.

[37] L.L.E. Braes and G. K. Miley, *Nature*, **232**, 246 (1971); R. M. Hjellming and C. M. Wade, *Astrophys.
J. Letters*, **168**, L21 (1971); B. L. Webster and P. Murdin, *Nature*, **235**, 37 (1972); C. T. Bolton, *ibid.*,
240, 124 (1972).

Figure 5.16 Chandrasekhar with Igor Novikov and Yakov Zeldovich, at the General Relativity Conference in Warsaw (1962). Belatedly, Chandrasekhar was awarded
the Nobel Prize for physics in 1983 "for his theoretical studies of the physical
processes of importance to the structure and evolution of the stars." From K. C.
Wali, *Chandra* (Chicago: University of Chicago Press, 1991).

Kuiper in 1941 was the first to realize that the components of binary stars are
so greatly condensed toward the center that their gravitational attraction can be
approximated by that of a pair of central mass points. Such a simple model was
originally developed by the French mathematician Édouard Albert Roche in 1873
(see section 3.4). Let M_I and M_{II} denote the masses of the two components and
let D be their mutual separation. We choose a rotating frame of reference with the
origin at the center of gravity of the mass M_I. The x axis points toward the center
of gravity of the mass M_{II}, and the z axis is perpendicular to the orbital plane. The
effective gravity—that is, the gravitational attraction corrected for the centrifugal
force—at any point P can be described as the gradient of a potential Ψ, where

$$\Psi = G\frac{M_I}{r_I} + G\frac{M_{II}}{r_{II}} + \frac{1}{2}\Omega^2\left[\left(x - \frac{M_{II}}{M_I + M_{II}}D\right)^2 + y^2\right], \qquad (5.9)$$

in which r_I and r_{II} are the distances from the centers of gravity of the two masses. Let
us further assume that the rotational angular velocity occurring in eq. (5.9) is equal
to the Keplerian orbital angular velocity, so that the axial rotations are perfectly
synchronized with the orbital revolution. We thus let

$$\Omega^2 = \frac{G(M_I + M_{II})}{D^3}. \qquad (5.10)$$

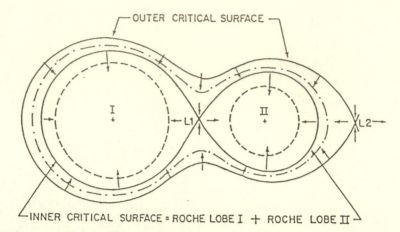

Figure 5.17 The inner and outer critical surfaces of the binary Roche model plotted in the equatorial plane. The arrows indicate the direction of the effective gravity. For a detached binary system, the two stellar surfaces (dashed curves) both lie beneath the inner critical surface; for a contact binary system having a common gaseous envelope, the stellar surface (dash-dotted curve) lies between the inner and outer critical surfaces. From F. H. Shu, S. H. Lubow, and L. Anderson, *Astrophys. J.*, **209**, 536 (1976).

If we adopt D as the unit of length and GM_I/D as the unit of potential, we can then write (except for an additive constant)

$$\Psi = \frac{1}{r_I} + q\left(\frac{1}{r_{II}} - x\right) + \frac{1}{2}(1+q)\left(x^2 + y^2\right), \tag{5.11}$$

where $q = M_{II}/M_I$ is the mass ratio.

Figure 5.17 represents a section of the equipotentials $\Psi = constant$ cut by the orbital plane $z = 0$. Quite generally, level surfaces corresponding to high values of Ψ form separate lobes enclosing each one of the two centers of gravity and differing little from spheres. With diminishing values of Ψ, the two lobes become increasingly elongated in the direction of their common center of gravity until, for a certain critical value $\Psi = \Psi_{in}$ characteristic of each mass ratio, both lobes come into contact to form a dumbbell-like configuration. This will henceforth be called the *inner critical surface*, and its two lobes will be called the *Roche lobes*. Note that the Roche lobes unite at a point where the effective gravity vanishes (i.e., at the so-called Lagrangian point L_1). For even smaller values of Ψ, the connecting part of the dumbbell will open up so that single level surfaces enclose both components, thus providing us with a convenient representation of a contact binary having a common gaseous envelope. Below a critical value $\Psi = \Psi_{out}$ ($< \Psi_{in}$) characteristic of each mass ratio, however, gravitational confinement of a binary star against the expansive tendency of its internal pressure is no longer possible. An inspection of figure 5.17 shows that this *outer critical surface* also contains a point where the effective gravity vanishes (i.e., the so-called Lagrangian point L_2). Mass loss from a binary system, if any, occurs at that point.

Binary-Star Evolution

The evolutionary significance of the Roche lobes has its roots in the work of Kuiper, who suggested in 1941 that gaseous streams may develop in close binaries if one of their components fills its Roche lobe completely, or if the system has a common gaseous envelope filling up its outer critical surface. Subsequent work by Joy, Struve, and their associates has shown that gaseous streams are particularly frequent in close binaries where a main-sequence star is accompanied by a larger but cooler and less massive companion. Following current practice, these binary stars will henceforth be called *Algol-type binaries*, after the well-known prototype. Further important progress was made in 1951, when Parenago and Masevich discovered that the subgiant components of Algol-type binaries do not obey the usual mass-luminosity relation for single main-sequence stars. Specifically, they found that their luminosities and radii are significantly larger than those of main-sequence stars having the same masses. In 1954 Struve also found that the luminosity excesses of these subgiants are correlated with the mass ratios of the two components. These results at once presented a serious problem, since the subgiants were always observed as stars of advanced evolution whereas their more massive companions were still on the main sequence; yet, as was already known at the time, it is always the more massive star that should have evolved faster. This somewhat embarrassing situation was called the *Algol paradox*.

Another important step forward was made in the mid-1950s by the Czech-born astronomer Zdeněk KOPAL (1914–1993; figure 5.18) in Manchester and, independently, by his American colleague John Avery CRAWFORD (b. 1921) in Berkeley. They found that in most Algol-type binaries the fractional dimensions of subgiant components coincide with those of their Roche lobes within the limits of observational errors. This led Kopal to divide the close binary stars into three groups: *detached systems*, in which both components are perceptibly smaller than their Roche lobes; *semidetached systems*, with one star—the contact component—filling its Roche lobe completely and the other being distinctly smaller than its own; and *contact binaries*, in which both stars fill up their Roche lobes and may actually overflow them. As was noted by Kopal, the most striking feature of the semidetached systems is that it is always the less massive component which appears to fill its Roche lobe, while its more massive companion remains well interior to it. This class of systems was thus identified with the Algol-type binaries. Although many subgiants are the contact components of semidetached systems, Kopal nevertheless found a few binaries in which the subgiant component is smaller than its Roche lobe. This lent further support to the idea that the clustering of subgiant components to their Roche lobes was the consequence of an inexorable radial expansion that had been halted by these invisible barriers. If so, then a ready explanation for the presence of undersize subgiants in some Algol-type binaries is that these stars are still in the process of radial expansion.

In order to reconcile the observed features of *semidetached systems* with the then current theoretical ideas about post-main-sequence evolution, Crawford made the bold hypothesis of mass transfer between the two components. Specifically, he assumed that the observed subgiants had originally been more massive but had lost so much material to their mates by Roche-lobe overflow that the original mass

Figure 5.18 Zdeněk Kopal, a classical astronomer whose career lasted nearly sixty years. He
is best known for the pioneering work he did during the 1950s on the structure
and evolution of close binary stars. He also developed a lifelong interest in
the analysis of the light curves of eclipsing binaries. His 1941 iterative method
brought him on a collision course with Russell from Princeton, who firmly
adhered to the view that nothing remained to be added to his own work of 1912 on
eclipsing binaries. The Czech-born Kopal, who was active in Cambridge (Mass.)
from 1938 to 1951, eventually became an American citizen—an allegiance he
has retained ever since. From the early 1950s his principal academic allegiance
was to the University of Manchester, where he held the chair of astronomy.
The portrait was taken on his sixtieth birthday in April 1974. From J. Dyson,
Astrophys. Space Science, **213**, 171 (1994).

ratios were eventually reversed, as we observe. Crawford's evolutionary scheme
was subsequently popularized by Hoyle, who described it as a "strange dog-eats-dog
evolution."[38] This was a very speculative approach indeed, since by the mid-1950s
the theory of stellar evolution beyond the main-sequence stage was still in its infancy.

Application of this idea was originally made in 1960 by the Canadian astronomer
Donald MORTON (b. 1933), who was then working in Princeton and so was particularly well informed about the recent work of Martin Schwarzschild and coworkers

---

[38] J. A. Crawford, *Astrophys. J.*, **121**, 71 (1955); F. Hoyle, *Frontiers of Astronomy* (London: Mercury
Books, 1955), 198.

on post-main-sequence evolution. By making use of detailed numerical models, Morton was able to show that a large fraction of the mass of the originally heavier component of a binary star was transferred to its companion on a thermal time scale, which is of the order of 10^5 years for a $10M_\odot$ star. This process, which is difficult to observe because of the short time scale involved, was found to take place as soon as the initially heavier star fills up its Roche lobe. Morton also showed that it is during this evolutionary phase that the mass ratio is reversed, therefore supporting Crawford's hypothesis. Not unexpectedly, further important progress was made starting in the late 1960s, when it became clear that the Henyey method developed to study the evolution of single stars could be modified easily to compute evolutionary sequences of interacting close binaries. This led to a sudden flood of papers about mass exchange and angular momentum transfer in semidetached systems. And so was born a new subfield of theoretical astrophysics: binary-star evolution.[39]

Contact Binaries

We next turn to the *contact binaries*, which have both components filling or overfilling their Roche lobes. Practically all known contact systems are eclipsing binaries. The light curves of these extremely close systems have a sinusoidal appearance, which is due to the severe tidal distortion of the components. They also have eclipse minima of almost equal depth, thus implying very similar effective temperatures for both components. In fact, this remarkable property of the contact binaries seems to be continuous over a wide range of spectral types, from stars as early as O type to stars as late as K type. They range in orbital period from about 6 days to about 6 hours. The *W Ursae Majoris systems* are a very distinct subgroup, characterized by periods of less than a day. These late-type stars have long been of special interest since, as was shown by Shapley in 1948, they are about twenty times more numerous per unit volume than all other eclipsing binaries. This is an important result because it strongly suggests that they are not a passing stage in the evolution of a binary star, but must be presently evolving on a nuclear time scale.

The similarity of effective temperatures for both components would not be surprising if contact binaries consisted of two identical stars. However, for some as yet unknown reason, these binaries always consist of dissimilar components with unequal masses. Furthermore, the less massive component is generally overluminous for its mass and its companion underluminous in terms of the conventional mass-luminosity relation for single main-sequence stars. To the best of our knowledge, Struve was the first to recognize in 1948 that their anomalous mass-luminosity relation might be causally related to the existence of a common envelope that redistributes and radiates away the luminosities emanating from the two independent cores. This important suggestion was further discussed by Yoji OSAKI (b. 1938) of the University of Tokyo and, independently, by Leon LUCY (b. 1938), a British

---

[39]For detailed reviews of these early works, see R. Kippenhahn and A. Weigert, *Zeit. Astrophys.*, **65**, 251 (1967); M. Plavec, *Adv. Astron. Astrophys.*, **6**, 202 (1968); B. Paczyński, *Annu. Rev. Astron. Astrophys.*, **9**, 183 (1971). For an overall discussion, see C.W.H. de Loore and C. Doom, *Structure and Evolution of Single and Binary Stars* (Dordrecht: Kluwer, 1992).

émigré who was then working at Columbia University.[40] They found that, as a consequence of their Roche-lobe geometry and similar effective gravities and temperatures, the components of contact binaries have luminosity ratios roughly equal to the first power of their mass ratio rather than the fourth power or so observed for any two single main-sequence stars.

Of particular importance is the work of Lucy, who showed how the existence of unevolved W Ursae Majoris systems with unequal components could be reconciled with stellar evolution theory. By assumption, a common convective envelope initially surrounded both components, thus leading to a large-scale lateral transfer of energy from the more massive component to the less massive one, roughly equalizing the effective temperatures over the whole surface. Although the exact nature of the mechanism that had initially equalized these temperatures was not fully understood, Lucy's 1968 equilibrium models attracted particular attention at the time. The agreement with observation was not perfect, however, and on theoretical grounds his models were somewhat too restrictive. Yet Lucy's work was undoubtedly a major breakthrough in the theory of contact binaries in the sense that it stimulated a great deal of activity, and hence controversy, during the 1970s.

Classical and Dwarf Novae

Another significant advance in our knowledge of close binaries and their evolution was the discovery by Walker in 1954 that the *classical nova* DQ Herculis is actually part of an eclipsing binary with the remarkably short period of 4 h 39 min. Later, in 1956, Joy was able to show that the *dwarf nova* SS Cygni is a spectroscopic binary with the short period of 6 h 38 min. The question then arose as to whether all novae and novalike stars are close binary systems, and whether the sudden eruptions were somehow triggered by the binary nature of these objects. The binary-star model was strongly advocated by Struve and his Chinese-born American colleague Su-Shu HUANG (1915–1977) in the mid-1950s, but it was their fellow countryman Robert Paul KRAFT (b. 1927) who played a leading role in showing in the early 1960s that membership in a short-period binary system is a prerequisite for a star to become a nova. As Kraft showed, one component is usually a blue white dwarf and the other is usually a red star, of about the same mass, which is overflowing its Roche lobe. Further work by several other astronomers has confirmed this conclusion.

Following these early works, it was widely accepted that classical nova outbursts are associated with the white-dwarf components of close binary stars. The companion stars are large, cool red stars which are filling their Roche lobe and losing matter, through the Lagrangian point L_1, which ultimately reaches the surface of the small hot white dwarf. Because of its angular momentum, however, this mass outflow cannot be immediately accreted, so that a disk is formed around the white dwarf by matter transferred from the late-type star to its hot companion. As this accretion continues, a layer of hydrogen-rich material is then built up on the surface of the white dwarf. Several theoretical studies have shown that the bottom of this

---

[40]Y. Osaki, *Publ. Astron. Soc. Japan*, **17**, 97 (1965); L. B. Lucy, *Zeit. Astrophys.*, **65**, 89 (1967); L. B. Lucy, *Astrophys. J.*, **151**, 1123 (1968); **153**, 877 (1968).

layer is gradually compressed and heated until it reaches the ignition temperatures for the hydrogen-burning reactions. A thermonuclear runaway then occurs, and the hydrogen-rich envelope of the white dwarf is thrown off, producing a nova outburst.[41] By the mid-1970s, however, it was already clear that the source of the eruptions of dwarf novae was not the same as that for classical novae, and that the accretion disk around the white-dwarf component could be responsible for most of the increased luminosity during dwarf-nova eruptions. At about the same time, preliminary attempts were also made to explain the explosion of type I supernovae as originating in interacting binary systems, but there was apparently no general consensus on exactly what was happening during the interaction. As we shall see in section 6.6, interest in this theory was revived during the 1980s.

5.6 THE PULSATION THEORY OF VARIABLE STARS (III)

As was noted in section 2.7, the *period-luminosity relation* for the Cepheid variables resulted from the discovery in 1912 by Henrietta Leavitt of Harvard College Observatory that a group of variable stars in the Small Magellanic Cloud show a correlation between their periods and apparent magnitudes. Hertzsprung recognized at once the importance of Leavitt's relation and attempted to find its zero point, for then the observed period of a Cepheid variable would immediately indicate its mean absolute magnitude and hence its distance. Shapley (figure 5.19) was the first to make extensive use of this method in deriving the distances of the globular clusters and hence an estimate of the size of our galaxy. The calibration was difficult to perform, however, because few Cepheid variables in our galaxy are near enough for their trigonometric parallaxes to be measured. He thus had to depend upon the less accurate method of statistically estimating the distances of isolated variable stars, with the help of their radial velocities and proper motions. He also assumed that nearby objects have the same properties as those in remote systems and that consequently the absolute magnitudes of any two Cepheid variables are equal if they have the same period. Shapley's relation, which is indicated by a dashed line in figure 5.20, also includes the RR Lyrae stars.

The Revised Period-Luminosity Relation

Shapley's period-luminosity relation was used successfully at Mount Wilson by Hubble and his coworkers in determining the distances of the nearer galaxies. Their numerical estimates were widely accepted until 1952, when Baade positively established that the absolute magnitudes of the Cepheid variables had to be considerably revised upward. Specifically, if one assumes that the RR Lyrae stars and the Cepheid variables in the Andromeda galaxy M31 obey Shapley's relation, then the apparent photographic magnitude, m_{pg}, of the RR Lyrae stars in this system should be 22.4

---

[41] S. Starrfield, J. W. Truran, W. M. Sparks, and G. S. Kutter, *Astrophys. J.*, **176**, 169 (1972); S. Starrfield, W. M. Sparks, and J. W. Truran, *Astrophys. J. Suppl.*, **28**, 247 (1974). For an overall discussion of these and other models, see E. L. Robinson, *Annu. Rev. Astron. Astrophys.*, **14**, 119 (1976).

Figure 5.19 At the International Astronomical Union meeting in Moscow in 1958, Harlow
Shapley (*right*) is seen with another giant, Ejnar Hertzsprung. About forty
years earlier, they had already attempted to calibrate Leavitt's period-luminosity
relation. Courtesy of Owen Gingerich.

in an exposure of 30 minutes. They should, therefore, have been easily accessible
to the 200-inch Hale reflector at Palomar Mountain, which had become operational
around 1950. Yet Baade could reach only the brightest population II stars with
such an exposure. Since it was already well established that they were about 1.5
magnitudes brighter than the RR Lyrae stars, Baade rightfully concluded that the
latter were to be found in M31 at $m_{pg} = 23.9$ and not at $m_{pg} = 22.4$ as predicted on
the basis of Shapley's relation. And since the mean absolute magnitude of the RR
Lyrae stars was rather certainly fixed at $M_{pg} = 0.0$, the inescapable conclusion was
that the absolute magnitude of the Cepheid variables in M31 was underestimated by
1.5 magnitudes, corresponding to an error in the resulting distance by a factor of 2.

The consequences of Baade's 1952 discovery were enormous: The globular
clusters in M31 and in our galaxy turned out to have closely similar luminosities, with the latter somewhat smaller than M31. Moreover, Hubble's characteristic time scale—the so-called age of the universe—had to be increased from
about 1.8 billion years to about 3.6 billion years. (This was a definite improvement but, as we shall see in section 6.6, the value was modified again in subse-

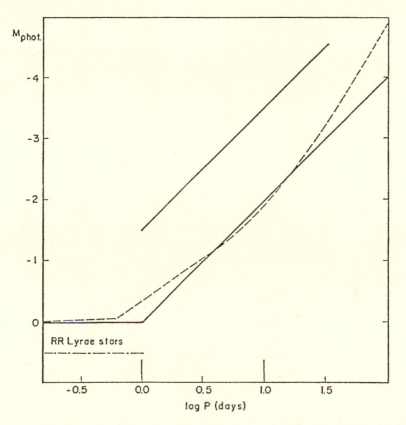

Figure 5.20 The period-luminosity relations (schematic) for RR Lyrae stars and Cepheid
variables. The dashed line indicates Shapley's relation, the solid line Baade's
(the top line applies to type I cepheids, the bottom one to type II cepheids). The
dash-dotted line is the correction that was later suggested for the RR Lyrae stars.
From O. Struve and V. Zebergs, *Astronomy of the 20th Century* (New York:
Macmillan, 1962).

quent years.) Obviously, this spectacular error of 1.5 magnitudes is attributable to
the fact that the Cepheid variables are not a homogeneous system of stars, all of approximately the same physical characteristics, but consist of two different groups. In
fact, these two types of variable star have different absolute magnitudes for the same
period, but Shapley and others had been calculating absolute magnitudes and distances as if the two types were alike. Following Baade, we shall thus separate them
into types I and II, corresponding to his general division of all stellar objects into
populations I and II. The classical, or type I, cepheids (such as δ Cephei) are those
for which the period-luminosity curve must be raised; that is, they are intrinsically
brighter by 1.5 magnitudes than the type II cepheids (such as W Virginis) of corresponding periods. The revised period-luminosity relation appears in figure 5.20.

The classical cepheids are thus the most luminous; they are of extreme population I and are found in spiral galaxies. Type II cepheids, also known as W Virginis stars after their prototype, are found in globular clusters and other metal-poor population II systems. Note that for the RR Lyrae stars the mean absolute magnitudes are independent of the period, and their value, formerly taken as zero, has now become about $+0.6$ on the average.

The Pulsation Mechanism

By 1952 it was widely accepted that certain types of variable star—such as the long-period variables, the classical cepheids, the W Virginis stars, and the RR Lyrae stars—owe their changes to periodic radial expansions and contractions. As was noted in sections 3.3 and 4.6, the mathematical theory of the small radial adiabatic pulsations of a star was originally discussed by Ritter in 1880 and further developed by Eddington in 1918. The theory was studied in great detail during the 1920s and 1930s as it provided a good description of many properties of these stars, such as their periods and locations in the H-R diagram. But it gave no information about the actual cause of the radial pulsations since, in the absence of viscous dissipation, the star would be a conservative system. Eddington was the first to point out that the driving mechanism must add heat to the stellar matter during compression, therefore enhancing the subsequent expansion, and that heat must be lost during expansion, therefore diminishing the resistance to the subsequent contraction. Later, in 1941, he also made the important suggestion that the thin outer layer of a star in which hydrogen changes from neutral to ionized ($H \rightleftharpoons H^+$) might be the seat of the driving mechanism of a pulsating star.

It was not until 1953, however, that a clear possibility of understanding the cause of pulsation was found by Sergei Alexandrovich ZHEVAKIN (1916–2001) at the Radio Physics Institute in Gorky (now Nijni-Novgorod). He suggested then and explored in more detail in later papers that the region of second ionization of helium ($He^+ \rightleftharpoons He^{++}$) might be a suitable location for a valve mechanism of the type discussed by Eddington. Unaware of Zhevakin's publications, the American astrophysicists John Paul COX (1926–1984) and Charles Allen WHITNEY (b. 1929) made a similar suggestion in 1958. Soon afterward, in 1962, the effectiveness of He^+ ionization as a driving mechanism was first conclusively demonstrated by the linear nonadiabatic calculations of the American astrophysicist Norman BAKER (b. 1931) and his German colleague Rudolf KIPPENHAHN (b. 1926). Subsequent calculations by Cox and his associates in the United States have amply confirmed their conclusions, and leave little reasonable doubt that He^+ ionization is the major source of driving for most kinds of variable star. Of particular importance is the work of Robert Frederick CHRISTY (b. 1916) at the California Institute of Technology in Pasadena, who performed in 1962 the first nonlinear nonadiabatic calculations of stellar pulsation. It was also suggested by Christy that hydrogen ionization could in some cases make a significant contribution to the valve mechanism. Numerous detailed calculations, while confirming the correctness of this suggestion, have shown that, nevertheless, He^+ ionization is the major source of driving for the Cepheid variables and RR Lyrae stars. It was found, however, that hydrogen ionization might well be the dominant source of driving for many of the long-period variables (see figure 2.15).[42]

Complex numerical calculations were made during the 1960s and 1970s to follow the evolution of theoretical stellar models of radially pulsating stars. These calculations—notably those made in 1965 by the German astrophysicist Emmi HOFMEISTER (b. 1933), later Meyer-Hofmeister—have been remarkably successful in reproducing the global properties of many variable stars, especially the existence of a narrow area occupied by the RR Lyrae stars and both types of cepheids in the H-R diagram and their period-luminosity relation (see figure 6.10). By the mid-1960s they could already reproduce quite accurately individual properties of these stars, such as the symmetric skewness of their light curves and the phase lag between their mechanical pulsation and brightness variation (see figure 5.21). As a matter of fact, so successful were these theoretical developments that, by the mid-1970s, attention was increasingly paid to nonradial stellar pulsations. These and related matters will be discussed further in section 6.5.

5.7 STELLAR ROTATION AND MAGNETIC FIELDS

Struve in 1930 was the first to point out that rapid rotation of single stars is confined to the earliest spectral types, whereas rotation ceases to be found in the middle F spectral types and later (see section 4.7). This conclusion was strengthened by a number of statistical studies of line widths by Struve and his associates in the years 1930–1934. They also found that supergiants of early and late spectral types, and normal giants of type F and later, never show conspicuous rotations. At this juncture the problem was quietly abandoned for fifteen years. Interest in measurements of axial rotation in stars was revived in 1949 by Arne SLETTEBAK (1925–1999), who was based initially at the Yerkes Observatory and later at the Perkins Observatory of the Ohio State University.

The main results pertaining to stellar rotation were assembled by Slettebak in 1970 and are summarized in figure 5.22. In this figure the mean projected equatorial velocities for single, normal, main-sequence stars are compared with the mean observed $v_e \sin i$ values for giant and supergiant stars, emission-line B stars (type Be), metallic-line (Am) stars and peculiar A-type (Ap) stars, and population II objects. The narrow cross-hatched curve represents the distribution of rotational velocities for main-sequence stars, as derived by Helmut Arthur ABT (b. 1925) of the Kitt Peak National Observatory (Tucson, Arizona). The open circles indicate mean rotational velocities obtained by the Soviet astronomers Alexander BOYARCHUK (b. 1931) and

---

[42]For comprehensive reviews of stellar radial pulsations, see S. A. Zhevakin, *Annu. Rev. Astron. Astrophys.*, **1**, 367 (1963); R. F. Christy, *ibid.*, **4**, 353 (1966); J. P. Cox, *Theory of Stellar Pulsation* (Princeton, N.J.: Princeton University Press, 1980). For a general discussion of chaotic pulsations in stellar models, see J. R. Buchler, in *Nonlinear Astrophysical Fluid Dynamics*, edited by J. R. Buchler and S. T. Gottesman (New York: New York Academy of Sciences, 1990), 17.

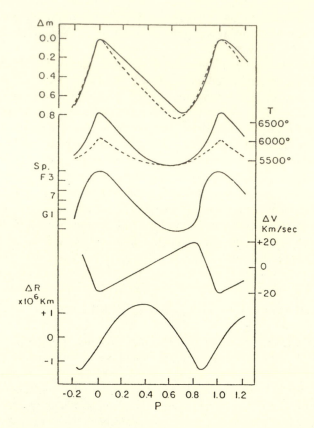

Figure 5.21 Periodic variations of δ Cephei. The continuous curve at the top is the observed
light curve (Δm); the dashed curve shows the change in magnitude if the star's
radius stayed constant. The second continuous curve (T) shows the *color* temperature, the temperature of a black body that matches the observed relative
energy distribution; the dashed curve gives the *effective* temperature, the temperature at which a black body of the same surface area would radiate the same
total energy. The other curves show variation in spectral type (Sp.), radial velocity (ΔV), and radius (ΔR), respectively. The abscissa is the phase expressed
in fraction of the period (after W. Becker, *Sterne u. Sternsysteme* [Darmstadt:
Steinkopff, 1950]). From O. Struve, B. T. Lynds, and H. Pillans, *Elementary
Astronomy* (New York: Oxford University Press, 1959). By permission, ©1959
by Oxford University Press, Inc.

Ivan KOPYLOV (b. 1928) for stars belonging to the luminosity classes III and IV;
they are connected by a broad cross-hatched band, suggesting the uncertainties in
the means for these giant stars.[43]

---

[43] As explained in section 2.5, the Harvard stellar classification system, which was universally adopted
in 1922, is basically a one-dimensional temperature sequence. However, since stars of the same temperature may differ in their absolute magnitudes, it was soon realized that at least another parameter

The distribution of rotational velocities along the main sequence is quite remark-able: Rotation increases from very low values in the F-type stars to some maximum in the B-type stars. Notice, however, that the early-type giants rotate more slowly than the main-sequence stars of corresponding spectral types, whereas for the late A and F types the giants rotate more rapidly than their main-sequence counterparts. As was shown by Sandage and Slettebak in 1955, this behavior can be interpreted as an evolutionary effect. Specifically, the observed rotational velocities of these stars are consistent with the hypothesis that a few hundred million years ago they were main-sequence stars of earlier spectral type, perhaps B or early A. Since they are now larger in size than they were when on the main sequence, conservation of angular momentum has reduced their rotational velocities. But then, this drop in rotation is compensated by the steeper drop in rotation along the main sequence, so that these evolving giant stars still have larger equatorial velocities than their main-sequence counterparts.

Supergiants and population II stars are shown schematically near the bottom of figure 5.22. As was noted, the supergiants of all spectral types never show conspicuous rotation. The apparent rotational velocities of population II stars are also small, a result originally due to Greenstein in 1960. Going to the other extreme, we note that the Be stars are shown separately in figure 5.22, with arrows indicating that their mean rotational velocities are in reality larger than shown. The fact that these stars are the most rapidly rotating of all stars was first shown in 1932 by Struve (figure 5.23) and Swings, who was then visiting the Yerkes Observatory. A subclass of the Be stars was defined by Merrill in 1949 and are known as "shell stars." They are characterized by extremely broad absorption lines, which, when interpreted as due to axial rotation, make them as a class the most rapidly rotating Be stars, with the largest $v_e \sin i$'s being in the neighborhood of 400 km s^{-1}.

The mean rotational velocities for the Am and Ap stars are shown in the box near the bottom of figure 5.22, centered about 40 km s^{-1}. Slettebak in 1954 was the first to show that the mean rotational velocities of these objects are three to four times smaller than the means for normal main-sequence stars of corresponding spectral types. By the late 1960s there was as yet no consensus about the actual mechanism that causes the abnormal abundances in the spectra of the Am and Ap stars, but it was generally accepted that they are intrinsically slow rotators and that very few of them have equatorial velocities in excess of 100 km s^{-1}. This led Abt and others to suggest that a low rotational velocity is a necessary but not sufficient condition for the occurrence of chemical abundance anomalies. If so, then what is the mechanism responsible for the abnormally low rotation rates of the Am and Ap stars when compared to normal stars of corresponding temperatures and luminosities?

---

was needed to classify stars. A new system was originally suggested in 1938 by the Yerkes astronomer William Wilson MORGAN (1906–1994), who added another dimension to the Harvard classification system by introducing five luminosity classes, labeled from I to V, to cover the range of luminosities from supergiants (I) to main-sequence stars (V). This empirical classification, usually referred to as the MKK system, rests on an *Atlas of Stellar Spectra* published in 1943 by Morgan and his American colleagues Philip KEENAN (1908–2000) and Edith KELLMAN (b. 1911). For a detailed historical account of this and other spectral classifications, see J. B. Hearnshaw, *The Analysis of Starlight: One Hundred and Fifty Years of Astronomical Spectroscopy* (Cambridge: Cambridge University Press, 1986), 255.

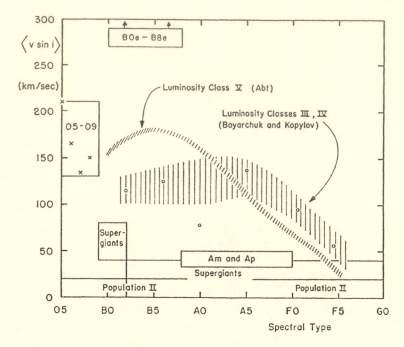

Figure 5.22 Mean projected equatorial velocities for a number of different classes of stars
as compared with normal main-sequence stars. From A. Slettebak, in *Stellar
Rotation*, edited by A. Slettebak (Dordrecht: Reidel, 1970), 5. By permission
of Kluwer Academic Publishers.

As was shown by Abt and his associates in the 1960s, the Am stars are usually or
always members of spectroscopic binaries, while almost none of the normal A-type
stars are found in binaries with orbital periods less than 100 days. Abt therefore
attributed the slow rotation of the Am stars to the deceleration induced by *tidal
interaction* in close binary systems. In contrast to the Am stars, however, the slow
rotational velocities of the Ap stars do not appear to be due to such an interaction.
An important clue toward the resolution of this problem was provided in 1947, when
Horace BABCOCK (b. 1912) at Mount Wilson measured a strong magnetic field in 78
Virginis, a Cr-Eu star, using the Zeeman effect (see section 2.2). Two years later,
he made the bold suggestion that an oblique rotating magnetic dipole could account
for the spectroscopic and magnetic variations of this and other Ap stars. The so-
called *oblique-rotator model* was independently developed in 1950 by the Australian
astronomer Douglas STIBBS (b. 1919), then at Mount Stromlo Observatory, while
Armin DEUTSCH (1918–1969) and George PRESTON (b. 1930) at Mount Wilson and
Palomar became its most ardent advocates. The discovery of strong magnetic fields
in the Ap stars, ranging up to several kilogauss but typically a few hundred gauss,
gave a whole new dimension to the study of these stars, since their slow rotation
rates could now be attributed to some kind of *magnetic braking*. However, there
has been considerable controversy as to whether most of their angular momentum
is lost before or during the main-sequence phase.

Figure 5.23 The Russian-born American astronomer Otto Struve at his desk in the Yerkes
            Observatory director's office (1946). He is best known for his work in stellar
            spectroscopy and for establishing the presence of hydrogen and other elements
            in interstellar space. Struve wrote more than nine hundred papers, many of them
            popular ones, and several books. Courtesy of Yerkes Observatory.

Schatzman's Braking Mechanism

An inspection of figure 5.22 shows that appreciable rotational velocities are common among the *normal* O-, B-, and A-type stars along the main sequence, but that
they virtually disappear somewhere near spectral type F5. The cause of the slow
rotation for main-sequence stars cooler than about F5 has been a matter of considerable debate. Struve in 1955 pointed out that most of the angular momentum
in the solar system resides in the planets. Hence he suggested the possibility of
undetected planetary systems being a normal feature for the cooler main-sequence
stars. Although this explanation has retained its attractiveness well into the 1960s,
a more generally accepted interpretation is the idea of angular momentum loss
through magnetized winds and/or sporadic mass ejections from stars with deep surface convection zones, with the angular momentum being transferred outward via
a large-scale magnetic field. Such a process was originally proposed in the late
1950s by the French astrophysicist Evry SCHATZMAN (b. 1920).[44] In this model
magnetic braking occurs only in stars with deep convective envelopes which have
solarlike flares giving rise to corpuscular emission and magnetically enhanced loss

---

[44]See especially E. Schatzman, *Annales d'astrophysique*, **25**, 18 (1962). For a detailed mathematical
analysis of this and related problems, see L. Mestel, *Stellar Magnetism* (Oxford: Clarendon Press, 1999),
240.

of angular momentum. Since stellar structure theory shows that convection virtually disappears in the envelopes of main-sequence stars hotter than spectral type F5, it thus follows that angular momentum loss must preferably occur in stars later than F5. In other words, the spectral type that separates rapid rotators from slow ones also separates the stars that are in radiative equilibrium near their surface from those that have subphotospheric convection zones. The strongest observational support for such a rotation-activity connection came from the work of Olin Chaddock WIL-SON (1909–1994) of Mount Wilson and Palomar Observatories, who attempted to correlate the rotations of main-sequence stars with their chromospheric properties and hence with the extent of their surface convection zones. He found in 1966 that, as one moves down the main sequence, there is a very sudden appearance of Ca II emission in the middle F's, whereas it is never observed among stars earlier than the spectral type F5. Obviously, the close agreement between the onset of large rotational velocities and the termination of active chromospheres is very suggestive of Schatzman's braking mechanism.

To put the relation between age and axial rotation on a firm quantitative basis, several authors during the 1960s obtained projected rotational velocities for stars belonging to open clusters and associations. Detailed statistical analyses have shown that, within each spectral type earlier than F0, the *average* rotational velocities of the, generally younger, cluster stars are almost the same as those of the field stars. This result strongly suggests that the upper main-sequence stars rotate while preserving their total angular momentum. A quite different picture emerges, however, when one compares the mean rotational velocities of field and cluster stars later than spectral type F5. This was originally discussed in 1967 by Kraft, who found that field stars with Ca II emission rotate, on the average, faster than those without it. Since Wilson had given strong evidence in 1963 that stars with Ca II emission are younger, on the average, than most solar-type stars, Kraft therefore concluded that rotation declines with advancing age in these stars. This picture was actually confirmed by Kraft's pioneering estimates of rotational velocity among solar-type stars in the Hyades and Pleiades. Soon afterward, in 1972, from a detailed analysis of these data, Andrew SKUMANICH (b. 1929) at the High Altitude Observatory (Boulder, Colo.) found that Ca II emission and rotational velocities of solar-type stars both decay as the inverse square root of their age. This is known as *Skumanich's law*.

Rotating Stellar Models

With the advent of high-speed computers in the 1960s, significant advances were also made in the study of the internal structure of rotating stars. The main difficulty of the problem lies in the fact that neither the internal stratification of a rotating star nor the shape of its free surface is known in advance. (This is in sharp contrast to the case of a nonrotating stellar model, for which a spherical stratification can be assumed ab initio!) Another difficulty arises because the actual distribution of angular momentum within a rotating star is still largely unknown, so that the rotation law always has to be specified in an ad hoc manner. Progress has been made using various expansion techniques, by variational methods or, in the late 1960s, by full numerical solutions of all the relevant structure equations. Notably, Jeremiah

OSTRIKER (b. 1937) and his associates at Princeton University have developed a powerful numerical method, the so-called *self-consistent-field method*, which was especially designed to calculate differentially rotating models that greatly depart from spherical symmetry.[45] In agreement with Milne's results reported in section 4.7, the overall effect of uniform rotation is to reduce the total luminosity of a main-sequence star of given mass, as indeed one might expect from the reduction of pressure required to balance gravity in its hydrogen-burning core. Not unexpectedly, the influence of rotation on the luminosity of an upper main-sequence star is much enhanced if the central regions spin faster than the surface, a point verified by the calculations of Peter BODENHEIMER (b. 1937) and his Canadian colleague Maurice CLEMENT (b. 1938). By the mid-1960s interest in the effect of nonuniform rotation was growing, also no doubt due to the work of the Princeton physicist Robert DICKE (1916–1997), who advocated the idea that the deep interior of the sun might be more rapidly rotating than its outer layers.[46]

5.8 THE MATURING OF SOLAR PHYSICS

In section 2.3 we pointed out that Harkness and Young had discovered a bright green emission line in the faint continuous spectrum of the solar corona during the 1869 eclipse. Since then, several other bright coronal lines have been observed, which for many decades defied analysis and were supposed to come from an otherwise unknown substance called, in analogy to helium, "coronium." Confidence in coronium's existence declined after 1910, however, as advances in chemistry and physics gradually filled in the gaps in the periodic table of elements. It vanished in 1927 when Ira BOWEN (1898–1973), of the California Institute of Technology, succeeded in explaining the emission lines from gaseous nebulae. For example, he showed that the intense lines near 5,000 Å, which had been ascribed to the unknown element "nebulium," were actually due to doubly ionized oxygen atoms making what are called "forbidden transitions." Spectral lines of this type occur because of the very low gas density; they are known as *forbidden lines* since they cannot occur in normal circumstances. Similar explanations failed with the coronal lines, however, as none of them could be identified with analogous transitions in the ions of abundant solar elements. By that time it was of course assumed that the solar atmosphere's temperature was decreasing with increasing height above the photosphere so that the corona was comparatively cool. Then, in 1934, Wal-

---

[45] J. P. Ostriker and J.W.-K. Mark, *Astrophys. J.*, **151**, 1075 (1968); J.W.-K. Mark, *ibid.*, **154**, 627 (1968); S. Jackson, *ibid.*, **161**, 579 (1970); P. Bodenheimer, *ibid.*, **167**, 153 (1971); P. Bodenheimer and J. P. Ostriker, *ibid.*, **180**, 159 (1973). For a distinct approach, see M. J. Clement, *Astrophys. J.*, **230**, 230 (1979). Stability criteria can be found in the following papers: P. Goldreich and G. Schubert, *Astrophys. J.*, **150**, 571 (1967); K. Fricke, *Zeit. Astrophys.*, **68**, 317 (1968); J. P. Ostriker and J. L. Tassoul, *Astrophys. J.*, **155**, 987 (1969); H. Shibahashi, *Publ. Astron. Soc. Japan*, **32**, 341 (1980).

[46] For comprehensive reviews of the stellar-rotation studies made during the 1950s and 1960s, see P. A. Strittmatter, *Annu. Rev. Astron. Astrophys.*, **7**, 665 (1969); R. P. Kraft, in *Spectroscopic Astrophysics*, edited by G. H. Herbig (Berkeley: University of California Press, 1970), 385; K. J. Fricke and R. Kippenhahn, *Annu. Rev. Astron. Astrophys.*, **10**, 45 (1972); J. L. Tassoul, *Theory of Rotating Stars* (Princeton, N.J.: Princeton University Press, 1978).

ter GROTRIAN (1890–1954) in Potsdam hit upon the idea that the corona might be much hotter than the 6,000 K photosphere. Specifically, he found that the corona's temperature must be an astonishing 350,000 K in order to explain its continuous spectrum. Three years later, he read a report by the Swedish physicist Bengt EDLÉN (1906–1993) describing the emission spectra of iron atoms stripped of nine or ten of their electrons by high-voltage sparks. Since these spectra could explain two of the coronal lines, he communicated his identification to Edlén, who replied to him by stressing the difficulty of securing more pertinent data. To ensure his priority, Grotrian made public his line identification in 1939. This led Edlén to take a closer look at the coronal-line problem. By 1941 he had found that the coronal lines are produced by forbidden transitions in highly ionized iron, calcium, and nickel atoms in a corona with a temperature of at least 2,000,000 K. The bright green emission line at 5303 Å, for instance, is emitted by iron atoms which have lost thirteen electrons, or one-half the number normally contained in the iron atom. In the early 1960s Herbert FRIEDMAN (1916–2000) and his associates at the United States Naval Research Laboratory recorded the X-ray spectrum of highly ionized iron in the solar corona, thereby confirming Edlén's original finding.

Although the discovery of the high temperature of the coronal gas provides a satisfactory explanation of its emission lines, the hydrodynamical mechanism by which the corona is heated remains a matter of considerable debate. Early theories advocated that particles of interstellar matter would be accelerated in the sun's gravitational field to velocities high enough to produce the observed high temperature. However, the amount of heat supplied by the accreted particles does not seem to be sufficient, and the level of maximum temperature would be located at much too great a height above the solar surface. Furthermore, the close relation of the shape and detailed structure of the corona to the underlying chromosphere rather points to an energy source in the sun itself. By the late 1940s it was generally believed that the heating mechanism may be traced back to turbulent motions in the photosphere, which show as granulation, and that the spicules provide the necessary coupling between the photosphere and the supported regions of the chromosphere and the corona. Theories based on shock wave dissipation were independently developed by Ludwig Biermann in Göttingen and by Martin Schwarzschild in Princeton. Acoustic waves generated by photospheric turbulence would propagate up into the higher atmosphere, steepen in shocks, and dissipate their energy to heat the chromosphere and the corona. Subsequent calculations indicate that such a shock heating can account for at least the initial chromospheric temperature rise; unfortunately, they also suggest that the corona itself may not be heated in this way. Moreover, there is no agreement as to the nature of the waves and the mechanism of their damping. Following Alfvén's pioneering work in magnetohydrodynamics, special attention was also paid to the influence of a magnetic field on the structure and heating of the corona in magnetically active regions. By the late 1960s various forms of heating had been extensively discussed in the literature, but no mechanism was generally accepted or even particularly well understood.[47]

---

[47] See M. Kuperus, *Space Science Reviews*, **9**, 713 (1969).

The Solar Wind

About 1950, solar physicists generally believed that a continuous flux of charged particles streamed outward from the sun, causing weak but persistent auroras and constant minor variations in the earth's magnetic field. Indirect evidence from the sun's emission of charged particles had been mounting for over one hundred years. At the dawn of the twentieth century, the penetration of these particles into the earth's magnetic field near the poles was invoked by the Norwegian scientists Kristian BIRKELAND (1867–1917) and Carl STØRMER (1874–1957) to explain the auroras. Further evidence for the sun's emission of charged particles came from the geomagnetic storms that are associated with the disruption of radio, telephone, and telegraph communications. Because these storms usually came a couple of days after a solar flare, the English geophysicist Sydney CHAPMAN (1888–1970) surmised that corpuscular emission from the sun was again a most plausible explanation. In 1929 he and Vincenzo FERRARO (1907–1974), an Italian émigré working in England, developed a quantitative model of the phenomenon and showed that a cloud of ionized gas ejected from the sun, traveling at 1,000 or 2,000 km s^{-1}, would reach the earth in a day or two and ruffle the earth's magnetic field as it passed. Their theoretical model of the sudden commencement and subsequent phases of a geomagnetic storm so closely resembled the disturbances observed on earth that it was widely accepted.

Another very convincing argument for the corpuscular radiation from the sun was put forward by Biermann (figure 5.24) in 1951. He argued that the gas tails of comets always pointed away from the sun because they were affected by the same radial outflow of charged particles that caused auroras and geomagnetic storms. He showed that, contrary to common belief, solar radiation pressure could not exert enough force to drive cometary tails in the opposite direction, but that a stream of charged particles could do so if their velocities were between 500 and 1,000 km s^{-1} and their density between 100 and 1,000 particles per cubic centimeter at the distance of the earth. Biermann's result also made it clear that this outflowing gas could not come merely in bursts or isolated beams, since the cometary tails showed that the corpuscular radiation was blowing continuously in all directions outward from the sun.

The next important contribution to our understanding of the sun's emission of charged particles was made by Chapman, who was then working in Boulder at the High Altitude Observatory. In 1957 he calculated that a solar corona consisting mainly of electrons and protons would, on account of the efficiency of electrons in conducting heat, extend out beyond the earth's orbit. This was in agreement with the curious observational fact that the earth's upper atmosphere gets hotter, rather than cooler, with increasing altitude. Chapman's theory of the solar corona was basically static, whereas Biermann's explanation of cometary tails was clearly kinematic. This apparent contradiction was resolved by the young solar physicist Eugene Newman PARKER (b. 1927) of the University of Chicago, who conjectured that Biermann's corpuscular radiation might be nothing other than the expansion of Chapman's solar corona. In 1958, to evaluate this possibility, he thus replaced Chapman's hydrostatic equilibrium by a stationary dynamic equilibrium and solved

Figure 5.24 The German astrophysicist Ludwig Biermann in his mature years. During his long career, he has contributed to several branches of solar and stellar physics; he was also the first to notice the solar wind's actions in the tails of comets. Courtesy of the Max-Planck-Institut für Astrophysik, Garching.

for the simple case of a spherically symmetric, nonrotating, nonmagnetic sun. The calculation showed that, due to the high temperature, coronal matter is steadily expanding with a velocity that ranges from a value of a few kilometers per second in the lower corona to supersonic values at large distances from the sun. This result was at first considered surprising in view of the traditional idea that the sun's corona is a static atmosphere. Yet these results were consistent with Biermann's estimates of the velocity and density of the solar corpuscular radiation. Parker soon named this outflowing gas the "solar wind" in order to emphasize its hydrodynamic character. The first spacecraft to detect the *solar wind* was the Soviet Union's *Lunik 2*, en route to the moon, in September 1959. The American *Explorer 10*, launched in March 1961, was the first spacecraft to measure the solar wind in detail.[48]

[48]L. Biermann, *Zeit. Astrophys.*, **29**, 274 (1951); S. Chapman, *Smithson. Contr. Astrophys.*, **2**, 1 (1957); E. N. Parker, *Astrophys. J.*, **128**, 664 (1958). For a detailed historical account of the solar wind, see K. Hufbauer, *Exploring the Sun* (Baltimore: Johns Hopkins University Press, 1991), 213.

Solar Flares

By that time it was generally accepted that large *solar flares* sometimes blasted very energetic particles that had a significant influence on the earth and its immediate environment. Actually, the existence of a flare-induced disturbance of the earth's magnetic field was suspected as far back as 1859, when a severe geomagnetic storm followed the spectacular white-light flare observed by Carrington and Hodgson (see section 2.4). Most flares are not so conspicuous in visible light, but usually appear in plages as short-lived brightenings in the hydrogen H_α line at 6,563 Å. Typically, their brightness increases for several minutes, followed by a slow decay which lasts from thirty minutes to one hour (up to three hours for the largest flares). Since the advent of spacecraft observations in the 1960s, we know that flares also radiate at many frequencies, from γ rays and X rays to radio waves; in addition, they emit high-energy particles called *solar cosmic rays*. Nearly all flares occur in the vicinities of active regions with sunspots, and the more magnetically complex the sunspot group, the higher the frequency of flare occurrence. Although flares occur quite randomly on the solar disk, at the peak of the eleven-year sunspot cycle the average occurrence of small flares is hourly and of truly large ones monthly.

Several estimates made during the 1950s indicate that the total energy expenditure in a large solar flare is of the order of 10^{32} erg, or 10^{29} erg s^{-1} since the lifetime of a flare is about 10^3 seconds. The typical area of a large flare may be taken as 10^{19} cm^2, indicating a mean flux of 10^{10} erg cm^{-2} s^{-1} through the flare area. (From eq. [2.11] the flux of emergent radiant energy from the photosphere is 6.5×10^{10} erg cm^{-2} s^{-1}.) Simple calculations show that the energy released in flares is not available from thermal sources, mass motions, or surface nuclear burning. We are thus forced to conclude that the relevant source of energy is in a magnetic field **B**, with an associated energy density $B^2/8\pi$ erg cm^{-3}. After integrating over a flare volume of about 10^{27} cm^3, one finds that some 10^{32} erg of energy are realizable from a magnetic field of the order of 10^3 G. The most fundamental question is, therefore, how so much electromagnetic energy can be released so quickly. Since Cowling's 1945 pioneering work on the diffusion of magnetic lines of force in the sun, it has been known that the time scale involved in converting the magnetic energy of a static sunspot field into heat by Ohmic dissipation is of the order of three hundred years, rather than the few minutes in which a substantial fraction of a flare's total energy release occurs. Although other mechanisms have been suggested to speed up the annihilation of magnetic fields, few ideas have proved more influential than the one originally proposed by Sweet in 1958. The basic principle involves the merging of a pair of bipolar regions with the formation of a neutral sheet in between. Two oppositely directed magnetic fields are thus pushed against each other, thereby interdiffusing in times small compared to the decay time for Ohmic diffusion over distances characterizing the sunspot configuration. Admittedly, Sweet's simple configuration is no longer thought to be completely relevant, but it is historically important because it has stimulated the development of other, more sophisticated, solar flare models.[49]

---

[49]P. A. Sweet, in *Electromagnetic Phenomena in Cosmical Physics*, edited by B. Lehnert, I.A.U. Symposium No. 6 (Cambridge: Cambridge University Press, 1958), 123; T. Gold and F. Hoyle, *Mon.*

The Solar Cycle

We now arrive at what is one of the most interesting but least understood problem of solar physics—the existence of highly localized activity centers and their interrelation through the solar rotation and magnetic field in the *solar cycle*. Since Hale measured large magnetic fields in sunspots for the first time in 1908, it has become widely accepted that magnetic fields hold the key to the phenomenon called solar activity. As illustrated in figure 2.8, most sunspot groups are bipolar and have opposite magnetic orientations in the northern and southern hemispheres. This type of association persists throughout each eleven-year sunspot cycle. Another remarkable feature of sunspot magnetism is the complete reversal of this pattern from one cycle to the next. Thus the length of the true sunspot cycle is not eleven but twenty-two years.

The presence of magnetic fields in sunspots—with fields up to 4,000 G—raises the question whether the sun itself possesses a general magnetic field. The polar patterns in the corona during a total eclipse, especially when the eclipse occurs during a sunspot minimum, clearly suggest the presence of such a field. A measurement of the sun's general magnetic field was not made until the 1950s, however, because very sensitive instruments are required for its detection. Harold BABCOCK (1882–1968) and his son, Horace Babcock, at Mount Wilson and Palomar Observatories finally succeeded in detecting the existence of weak polar fields of about 1 G within 35° of the poles. It is now known that these polar fields reverse polarity around the maximum of sunspot activity and hence with the same period as the eleven-year half of the solar cycle.

Another important feature of solar activity is the equatorial migration of the region where sunspots are formed (see figure 2.8). It is generally believed that the sunspots, and the bipolar magnetic regions, are evidence of a general *toroidal* field of several hundred gauss at some depth beneath the photosphere (that is, a subsurface azimuthal field whose lines of force are circles around the solar axis). Locally this field may be concentrated into small elements which, driven to the surface by convective motions or magnetic buoyancy, may form bipolar sunspot groups. If so, then, the migration of the region of sunspots traces the migration of the subsurface toroidal field during the twenty-two-year magnetic cycle. The question of the origin of solar activity in this rather simplified picture is therefore reduced to the origin and present maintenance of the sun's magnetic field. Hence, we must inquire why the sun should have reversible polar fields and a much larger toroidal field that migrates from high to low heliocentric latitudes. Study of the solar-cycle dynamics during the 1950s and 1960s has for the most part been carried out in the context of dynamo models, in which the magnetic fields are maintained against Ohmic dissipation by

Not. Roy. Astron. Soc., **120**, 89 (1960); N. E. Petschek, in *The Physics of Solar Flares*, edited by W. N. Hess, AAS-NASA Symposium, NASA SP-50 (1964), 425; S. I. Syrovatskii, *Astron. Zh.*, **43**, 340 (1966) (*Soviet Astron.*, **10**, 270 [1966]); P. A. Sturrock, in *Structure and Development of Solar Active Regions*, edited by K. O. Kiepenheuer, I.A.U. Symposium No. 35 (Dordrecht: Reidel, 1968), 471. For clear introductory texts to this and related matters, see E. R. Priest, in *Solar Flare Magnetohydrodynamics*, edited by E. R. Priest (New York: Gordon and Breach, 1981), 1; E. Tandberg-Hansen and A. G. Emslie, *The Physics of Solar Flares* (Cambridge: Cambridge University Press, 1988).

the combined effect of differential rotation and turbulent cyclonic motions in the subphotospheric convection zone.

The principle of *dynamo action* was first suggested by the Irish-born Cambridge physicist Joseph LARMOR (1857–1942) in 1919. It requires that the conducting fluid moves in such a way as to induce electromotive forces capable of maintaining the magnetic field against Ohmic decay. This is by no means a perpetual-motion engine, however, because energy must be supplied by the forces driving the fluid flow and this energy must be converted into magnetic energy. Unfortunately, all early efforts to build a solar dynamo have been hindered by Cowling's discovery, in 1934, that a steady magnetic field symmetric about an axis cannot be maintained by motions likewise axisymmetric. The problem was resolved in 1955 by Parker, who explained how it was possible to maintain an axisymmetric *mean* solar field by dynamo action.

Parker's regeneration mechanism may be understood as follows. In addition to the *toroidal* field, there is a *poloidal* component to the sun's magnetic field (that is, polar fields having their lines of force in planes containing the solar axis). It is easy to produce a toroidal field from a poloidal field by differential rotation since this motion necessarily shears any poloidal field and hence draws out its lines of force into an azimuthal field. However, in order to regenerate the initial poloidal field from the azimuthal component, we know from Cowling's theorem that the motions should not be axially symmetric. As was noted by Parker, this is certainly true of the turbulent convection below the solar surface. Specifically, the deflecting force of rotation acting on convection causes it to become cyclonic, with a rising cell of fluid spinning and twisting the toroidal field lines into loops in the meridian planes; a large number of such loops then coalesce to regenerate the poloidal field.

By the late 1960s there were already several dynamo models that could reasonably explain the maintenance and migration of the solar magnetic fields. Broadly speaking, all of these migratory dynamos involve the production of a toroidal field by nonuniform rotation in the solar convection zone and the global effect of small-scale nonaxisymmetric motions that regenerate the poloidal field. It is illustrative to display oscillatory magnetic fields in the form of a movie, and twelve frames of such a movie are shown in figure 5.25. Note that the toroidal field, which is shown on the left-hand side of each frame, migrates from higher heliocentric latitudes toward the equator. Of particular interest to the nonspecialist is the work of Horace Babcock, who proposed in 1961 a detailed phenomenological model for the solar dynamo, suggested no doubt by his extensive observations of the sun's magnetic field.[50]

Sunspots

Although research on the equatorial migration of sunspots made great strides during the 1950s and 1960s, the sunspot phenomenon itself, with its reduced temperature

---

[50]E. N. Parker, *Astrophys. J.*, **122**, 293 (1955); H. W. Babcock, *ibid.*, **133**, 572 (1961); R. B. Leighton, *ibid.*, **136**, 1 (1969); M. Steenbeck and F. Krause, *Astron. Nachr.*, **291**, 49 (1969). For general presentations of dynamo action in the sun, see T. G. Cowling, *Annu. Rev. Astron. Astrophys.*, **19**, 115 (1981); M. Stix, *The Sun* (Berlin: Springer-Verlag, 1989); P. Foukal, *Solar Astrophysics* (New York: John Wiley & Sons, 1990).

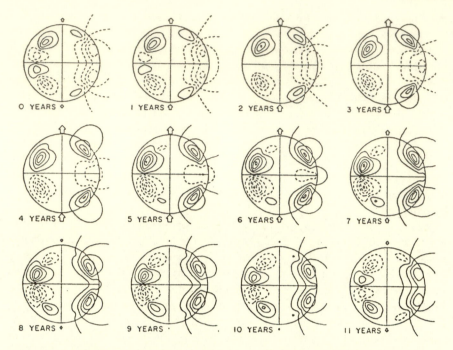

Figure 5.25 A numerical simulation of an oscillatory solar dynamo. The meridional cross sections show contours of constant toroidal field strength on the left, and poloidal lines of force on the right. Solid curves indicate toroidal fields pointing out of the figure, and clockwise poloidal field lines; dashed curves indicate the opposite. Arrows indicate the strength and direction of the polar field. The time is adjusted so that eleven years cover one half cycle. From M. Stix, in *Basic Mechanisms of Solar Activity*, edited by V. Bumba and J. Kleczek, I.A.U. Symposium No. 71 (Dordrecht: Reidel, 1976), 367.

and intense magnetic field, is still a matter of considerable debate.[51] As we know, sunspots represent a disruption in the uniform flow of heat through the sun's outer convective zone. Over the years, various authors have pointed out many of the physical effects that must contribute to the sunspot phenomenon. Biermann in 1941 initiated the modern theory of sunspots with the suggestion that the nearly vertical magnetic field in each sunspot inhibits the upward convective transport of heat from below, thereby providing a direct explanation for a reduced temperature at the visible surface. It was soon recognized, however, that the missing energy flux would be channeled around the sunspot, presumably appearing at the sun's surface as a bright ring around the spot. Since no such bright ring was observed, Parker in 1974 made the bold suggestion that a sunspot is a region of enhanced, rather than inhibited, energy transport. In this approach, then, the magnetic field of a sunspot

---

[51]For an exhaustive discussion of the sunspot phenomenon, see J. H. Thomas and N. O. Weiss, in *Sunspots: Theory and Observations*, edited by J. H. Thomas and N. O. Weiss (Dordrecht: Kluwer, 1992), 3.

generates enhanced Alfvén waves, which propagate rapidly out of the region along the magnetic field. By the 1990s there were thus two possible explanations of why sunspots are dark: *either* the partial suppression by the sunspot magnetic fields of convective energy transport from the underlying layers, *or* the removal of energy from the sunspot by enhanced hydromagnetic wave radiation. Although the first explanation was currently favored, it was not until 1999 that a group of solar scientists from Colorado and New Mexico reported high-photometric-precision observations of bright rings around sunspots.[52] According to these authors, isolated sunspots are commonly seen to be surrounded by a ring of enhanced radiation, the origin of which is probably not bright vertical magnetic elements surrounding the sunspots, but the reemergence of heat blocked by magnetic inhibition of convective energy transport in the spot itself. If so, then, this would lend further support to Biermann's suggestion.

---

[52]M. P. Rast, P. A. Fox, H. Lin, B. W. Lites, R. W. Meisner, and O. R. White, *Nature*, **401**, 678 (1999).

Chapter Six

The Era of Specialization: 1970–

In the future, everyone will be famous for 15 minutes.
  —Andy WARHOL

From the first astronomical use of the telescope by Galileo and others in the 1610s until the accidental discovery of radio waves coming from outer space in 1931, everything astronomers knew about the sun and stars was based entirely on observations made in visible light. During the 1930s, tentative steps were taken in the nascent science of radio astronomy, and experimental rockets were launched in the United States and Germany. Due to World War II, however, real progress was delayed until the late 1940s. At the end of the war, astronomers realized that many recent advances in radar and rocketry, which had a profound effect on the conduct of the war, could be directly applied to radio astronomy and space research. With the advent of the space age, when astronomers became able to place their instruments into satellites orbiting far above the absorbing layers of the earth's atmosphere, many other regions of the electromagnetic spectrum became accessible, in addition to the optical and radio wavelengths.[1] By the 1960s astronomers had already studied various celestial bodies in the infrared, ultraviolet, and X-ray wave bands. Somewhat later, in the 1970s, discrete sources of γ rays were also observed, two of the most intense ones being the Vela pulsar, at the center of the Gum nebula, and the Crab pulsar (see section 5.4). The next two decades saw major developments in infrared astronomy, which led to great changes in our views on the formation of stars and planets from the interstellar medium. Of paramount importance also are the ground-based observations of the so-called five-minute oscillations in the solar photosphere, which made possible the study of the sun's invisible interior. Thus the new field of helioseismology was born at the end of the twentieth century.

Progress in our understanding of the internal constitution of the sun and stars has always depended on scientists who were keenly aware of recent advances in the physical sciences. It is not surprising, therefore, that the pace and direction of advance in theoretical astrophysics strikingly exemplify the alternation of continuity and revolution that characterizes the development of physics since Newton's time. Successive conceptions of the sun and the stars have been discussed at length in the three preceding chapters. As was pointed out in chapter 3, from the mid-1840s to the 1910s, active contributors to solar and stellar physics were few and

---

[1] See D. Leverington, *A History of Astronomy* (Berlin: Springer-Verlag, 1995), 341.

their primary interest was elsewhere. Nonetheless, their solitary work led to the development of solar thermodynamics in the 1850s and, about two decades later, to the first quantitative models for the sun and the stars. The next period opened with the formulation of quantum physics at the dawn of the twentieth century, which paved the way for the quantitative analysis of stellar spectra in the 1920s. Between the two world wars, stellar physics progressed at a much more rapid pace than before. The most steadfast contributors to this progress were a small group of Cambridge astrophysicists led by Eddington, who were at the forefront in theorizing about the internal structure and evolution of the stars. At that time, however, the origin of the energy radiated by the sun and the stars had not yet been fully elucidated, and the theoretical study of solar activity and rotation was still in its infancy. With the 1940s came the maturing of two new sciences, nuclear physics and magnetohydrodynamics, which opened the way for a midcentury flowering of theoretical researches on the sun and the stars. During the 1950s and 1960s, then, most of the ideas that had been slowly developed in the preceding decades suddenly came to fruition as substantial headway was made in both the observational and theoretical fields.

From the viewpoint of a theoretical astrophysicist, the 1950s and 1960s may fitly be designated as the *golden age* of stellar studies. Research on stellar interiors was then viewed as a branch of modern physics, with great emphasis being laid on the atomic and nuclear processes. Actually, the spherically symmetric models discussed in sections 5.3 and 5.4 were so successful in accounting for the major observed properties of stars that the two most challenging problems of stellar hydrodynamics—turbulence and convection—received comparatively much less attention.[2] Moreover, as the source of relatively simple unidimensional problems was rapidly drying up in the late 1960s, theoreticians were increasingly forced to turn their attention to genuine tridimensional problems for which spherical symmetry was no longer an adequate approximation. As we shall see in the present chapter, these problems involve motions of such complexity that the most powerful computers cannot yet infer those motions as a purely deductive process from the basic physical laws underlying stellar hydrodynamics. Theoreticians soon realized, therefore, that progress can be made only through efficient cooperation between theory and observation, for instance in fixing the values of free parameters in which are lumped together all the uncertainties of the theory. This is another way of saying that, from the 1970s onward, stellar hydrodynamics has not kept pace with the accumulation of many new and unexpected discoveries made in observational astronomy at the end of the twentieth century.

6.1 SINGLE, DOUBLE, AND MULTIPLE STARS

Because the very bright giant stars of spectral type O and B have probably existed for only 10 million years since their formation, and because they are found in the spiral arms of galaxies where interstellar gas and dust intermingle with stars, astronomers

---

[2]For a lucid account of turbulence, convection, and related matters pertaining to stellar structure, see V. M. Canuto, *Astrophys. J.*, **524**, 311 (1999).

have long inferred that stars are continually forming in dense concentrations of interstellar matter. Since the early 1970s, there has been a growing realization that all present-day star formation takes place in large *molecular clouds*, but that only a small fraction of the mass of a cloud actively takes part in the star-formation process. Small dark clouds of interstellar dust and gas, known as *Barnard objects* or *Bok globules*, are objects of particular interest since they have often been thought to be the precursors of ordinary stars.[3] During the 1980s, infrared, optical, and radio observations gave evidence that low-mass stars do indeed form in these dark globules. In fact, recent studies of two of the nearest molecular cloud complexes, Taurus and Ophiuchus, revealed that two modes of stellar formation exist: a clustered mode characterized by the formation of a rich cluster of stars from large dense clumps of molecular gas and dust, and an isolated mode in which individual stars form in small, well-separated, dense cores distributed throughout a molecular cloud.

As was noted in section 5.4, *Herbig-Haro objects* have for a long time been recognized as indicators that stars might actually be seen in their very early stages. By the early 1980s several of these objects had been shown to be highly collimated jets of partially ionized gas moving away from very young stars. Pioneering investigations of infrared objects in the late 1960s gave further support to the view that stars are formed in collapsing clouds. The first studied infrared source is the *Becklin-Neugebauer object* in Orion, which is believed to be produced by a circumstellar dust cloud that reradiates the ultraviolet radiation absorbed from a newborn central star. It is apparently associated with the much more extended *Kleinmann-Low object*, which seems to be a compact dust nebula, probably with a newborn star or cluster of stars near its center.[4] By the late 1990s astronomers had actually observed the star-formation process at virtually all wavelengths of the electromagnetic spectrum, from X-ray to radio wavelengths. Of particular importance are the images provided by the Hubble Space Telescope, which demonstrate that collimated outflows and accretion disks embedded in extended circumstellar environments are an integral part of that process. Extensive observations of stellar objects in their pre-main-sequence contraction phase (such as the *T Tauri* stars) have also revealed that a very large fraction of these objects are members of gravitationally bound double (or multiple) systems. Current estimates indicate that the pre-main-sequence binary fraction is up to a factor of two higher than that among main-sequence stars in the solar vicinity. This is clear indication that the usual product of a star-formation event is a binary or multiple system.

Since the early 1970s, there has been a dramatic increase in theoretical research related to the evolution of stars from molecular cloud to main sequence. This increase has been brought about in large part by the development of high-speed computers and powerful numerical methods. Yet, as we shall see, numerous unresolved issues remain. Among the many vexing questions regarding the early phases of stellar evolution, that concerning the angular momentum of stars is of paramount

---

[3]They are named after their American discoverer, Edward Emerson BARNARD (1857–1923), and the Dutch-born American astronomer Bart Jan BOK (1906–1983), who suggested in the 1940s that they might be the sites of star birth.

[4]E. E. Becklin and G. Neugebauer, *Astrophys. J.*, **147**, 799 (1967); D. E. Kleinmann and F. J. Low, *Astrophys. J. Letters*, **149**, L1 (1967).

importance. Indeed, it has long been realized that a typical collapsing protostellar cloud will almost certainly contain far more angular momentum than can be put into a single star. This is the so-called *angular momentum problem*. The questions naturally arise, How is angular momentum transported out of the material that will form the star? To what extent does the formation of disks, binaries, and multiple systems solve the angular momentum problem? and What determines whether a single star or a multiple system is formed? Broadly speaking, three major episodes of angular momentum transfer take place during the pre-main-sequence evolution of a solar-type star. The first occurs during the precollapse phase, when the material is concentrated in dense regions of a molecular cloud. The second occurs during the near-free-fall collapse of a protostar, when angular momentum is approximately conserved until a disklike structure has formed. The third episode, which will be discussed further in section 6.4, occurs during the quasistatic contraction phase to the main sequence.[5]

Collapse and Fragmentation

Prior to the mid-1970s, it was widely accepted that star formation was the result of the collapse and fragmentation of an interstellar cloud. The problem of the fragmentation of an isothermal cloud into stars was originally discussed by Fred Hoyle in 1953. Specifically, he pointed out that, as the gas density increases in the collapsing cloud, somewhat smaller masses become unstable to fragmentation. These smaller condensations each collapse in turn into somewhat smaller aggregations, and so on. As he noted, fragmentation should cease if the gas becomes sufficiently opaque to radiation, so that further contraction is adiabatic rather than isothermal. Later, in 1956, Mestel and Spitzer (figures 6.1 and 6.2) pointed out that this process can be greatly modified by the presence of a large-scale magnetic field. They found that, as long as the field is frozen into the collapsing cloud, the magnetic pressure sets a lower limit to the stellar masses that can be formed in the cloud. If the field is taken as 10^{-6} G in regions where the density is 2×10^{-23} g cm^{-3}, this lower limit is of the order of $500 M_{\odot}$! As they showed, however, there exists a simple mechanism by which a collapsing cloud may manage to condense into stars of much smaller mass. It is based on the fact that magnetic lines of force have no direct effect on the neutral particles but influence only the electrons and the positive ions. As they noted, matter in an ordinary hydrogen cloud is largely neutral, both atomic and molecular, with a small admixture of partially ionized heavier elements with their free electrons, and of dust grains. Since the electrons and positive ions are constrained to follow the magnetic lines of force, it is therefore possible for the charged particles and the magnetic field to drift together out of the gas, leaving the neutral atoms behind. Thus the neutral gas can contract inward, so that the cloud is able to break up into stars of small mass. This process is often referred to as *ambipolar diffusion*.

---

[5]For penetrating reviews of these and related problems, see P. Bodenheimer, *Annu. Rev. Astron. Astrophys.*, **33**, 199 (1995); P. Bodenheimer in *Evolution of Binary and Multiple Star Systems*, edited by Ph. Podsiadlowski, S. Rappaport, A. R. King, F. D'Antona, and L. Burderi, ASP Conference Series, vol. 229 (San Francisco: Astron. Soc. Pacific, 2001), 67.

Figure 6.1 Leon Mestel, a third-generation astrophysicist who, after taking his Ph.D. at
          Cambridge with Hoyle, started his career in 1951–1954 as a research fellow in
          Leeds with Cowling. Mestel has made wide-ranging and important contributions
          to stellar astronomy, but he is best known for his work on cosmic magnetism.
          The portrait was taken in 1977. © Godfrey Argent Studio, London. Courtesy of
          Dr. L. Mestel.

Hoyle's concept of hierarchical fragmentation in a molecular cloud did not withstand the passage of time, however, because observations could not detect velocities
characteristic of collapse in dense concentrations of interstellar matter. Nevertheless, ambipolar diffusion is now believed to be the process by which the classic
magnetic-flux problem is resolved for forming stars. The whole problem was actually reconsidered in 1976 by the Cypriot-born American astrophysicist Telemachos
MOUSCHOVIAS (b. 1945), who showed that even relatively weak magnetic fields
could provide a very effective support against gravity in a molecular cloud. A new
theoretical picture emerged, therefore, in which the formation of dense cores and
their subsequent collapse is initiated by ambipolar diffusion in an otherwise magnetically supported cloud. In more quantitative terms, the characteristic time of this
process, τ_{AD}, is proportional to the ratio n_i/n_e of the number density of the ions and
neutrals. Since it is normally the case that n_i/n_e decreases with increasing density,

Figure 6.2 Chairman of Princeton's astrophysical sciences department from 1947 until 1979, Lyman Spitzer, Jr., made important contributions to interstellar astronomy and to plasma physics. He was also a pioneer in the first attempts to generate power by controlled thermonuclear fusion. His proposal to do this was approved, and the toroidal machine, which he called a "stellarator," was built in the 1950s. In 1985 the Royal Swedish Academy of Sciences awarded him the Crafoord prize for his fundamental pioneering studies of the interstellar medium. Courtesy of Princeton University Observatory.

it follows at once that only in dense regions will the time τ_{AD} be small enough for ambipolar diffusion to be operative. As was noted by Mouschovias, the presence of a magnetic field may account for the observed inefficiency of star formation in molecular clouds, the field itself being expelled from cloud cores wherever large enough densities are attained—thus accounting for the observation that most stars exhibit a relatively weak magnetic field.[6]

At this juncture it is appropriate to briefly review a working "cartoon picture" of how sunlike stars form and evolve, as it was originally proposed by Frank SHU (b. 1943) and his associates at the University of California at Berkeley. Figure 6.3 illustrates their proposed outline of the various stages involved in the birth of

---

[6]T. Ch. Mouschovias, *Astrophys. J.*, **207**, 141 (1976).

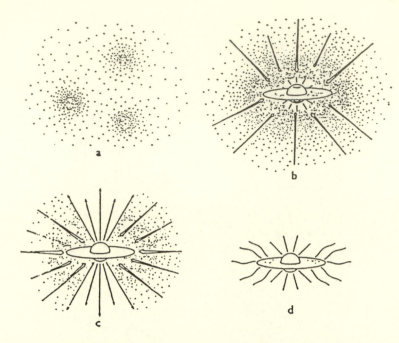

Figure 6.3 The four stages of star formation: (a) Cores form within molecular cloud envelopes as magnetic and turbulent support is lost through ambipolar diffusion;
(b) protostar with a surrounding nebular disk forms at the center of a cloud core
collapsing from inside out; (c) a stellar wind breaks out along the rotational axis
of the system, creating a bipolar flow; (d) the infall terminates, revealing a newly
formed star with a circumstellar disk. From F. H. Shu, F. C. Adams, and S. Lizano,
Annu. Rev. Astron. Astrophys., **25**, 23 (1987). By permission. © 1987 by Annual
Reviews.

stars from molecular clouds. The first phase is the formation of slowly rotating
molecular clouds. The second one begins when a condensing core gravitationally
collapses from "inside out"; this phase is characterized by a central protostar and
a circumstellar disk, both of them deeply embedded within an infalling envelope
of dust and gas.[7] The third phase starts when a powerful wind breaks out along
the rotational axis of the system, reversing the infall and sweeping up the material
over the poles into two outwardly expanding shells of gas and dust. This stage
corresponds to the bipolar outflow phase. As time proceeds, however, more and
more of the rotating inflowing matter will fall preferentially onto the disk rather
than onto the star. At this point, stage d in figure 6.3, the system becomes visible as
a star with a circumstellar disk to all outside observers. Admittedly, there are many
puzzles still associated with this overall evolutionary scheme of a low-mass star.

---

[7]Carl von Weizsäcker, who had originally proposed the carbon-nitrogen (CN) cycle in 1938, was
also the first to show in 1947 that, as a result of turbulent viscosity, a contracting cloud core is slowed
down while its angular momentum is communicated to the surrounding medium, which moves outward,
leaving a protostar and forming a circumstellar disk (*Zeit. Astrophys.*, **24**, 181 [1947]).

It is a most useful one, however, because it accommodates most of what we know theoretically and observationally about the star-formation process.[8]

Some Hydrodynamical Calculations

Another important line of enquiry was initiated in 1969 by Richard LARSON (b. 1941), then working at the California Institute of Technology and later at Yale University (New Haven, Conn.), who made the first reliable hydrodynamical calculations of the gravitational collapse of a spherically symmetric gas cloud, neglecting rotation and magnetic fields. In this ideal case, the collapse is at first isothermal since the cloud is still transparent to its own radiation. As the collapse progresses, however, the density increases more rapidly in the central part than in the outer layers. When the central density rises above 10^{-13} g cm^{-3}, the inner core becomes more opaque, the temperature rises, and the pressure forces are able to halt the collapse. When the central temperature rises above 2,000 K, hydrogen molecules begin to dissociate. This reduces the adiabatic exponent below the critical value 4/3 and causes a dynamical instability (see eq. [4.19]). A second phase of collapse—now almost adiabatic—commences. The bulk of the material eventually comes into hydrostatic equilibrium, but only after the dissociation and ionization of the dominant chemical constituents have been achieved. This phase has a total duration of about 10^6 years. Thence, the evolution of the protostar proceeds through a slow contraction phase that can be viewed as a sequence of quasiequilibrium states. Despite the fact that a spherically symmetric, one-dimensional (1D) collapse is a very particular case of the general tridimensional problem, it is worth noting here because it is a useful guideline for more complex model calculations with less stringent geometrical constraints.

Larson in 1972 was the first to discuss the early phases of the axially symmetric (2D) collapse of a rotating protostar. Broadly speaking, the collapse proceeds initially in the same nonhomologous fashion as for a nonrotating model, and the density distribution again becomes centrally peaked. After a free-fall time, rotational effects begin to dominate, however, so that the system approaches a state of equilibrium with rotational support in the equatorial plane and pressure support along the rotation axis. Larson, who further discussed the problem in 1978, found that at this stage initial nonaxisymmetric perturbations can grow, and fragmentation into a wide binary or a multiple system can result. This was the beginning of an outpouring of numerical studies of 3D protostellar collapse that brought essential clarification to the fragmentation process.[9]

---

[8]F. H. Shu, F. C. Adams, and S. Lizano, *Annu. Rev. Astron. Astrophys.*, **25**, 23 (1987). See also T. P. Ray, *Scientific American*, **283**, No. 2, 42 (2000).

[9]The following pioneering papers are particularly worth noting: R. B. Larson, *Mon. Not. Roy. Astron. Soc.*, **184**, 69 (1978); P. Bodenheimer, J. E. Tohline, and D. C. Black, *Astrophys. J.*, **242**, 209 (1980); A. P. Boss, *ibid.*, **237**, 866 (1980); M. Różyczka, W. Tscharnuter, and H. W. Yorke, *Astron. Astrophys.*, **81**, 347 (1980); S. M. Miyama, C. Hayashi, and S. Narita, *Astrophys. J.*, **279**, 621 (1984); R. H. Durisen, R. A. Gingold, J. E. Tohline, and A. P. Boss, *ibid.*, **305**, 281 (1986). Alternative modes of formation of double and multiple stars have been suggested, for instance, in the collapse of filamentary configurations: P. Bastien, *Astron. Astrophys.*, **119**, 109 (1983); I. Bonnell and P. Bastien, *Astrophys. J.*, **374**, 610 (1991); J. J. Monaghan, *ibid.*, **420**, 692 (1994).

Binary-Star Formation

The general result of these hydrodynamical 3D numerical calculations is that fragmentation does occur, with a wide range of possible outcomes, but that it can explain only the formation of wide binaries, with separations of the order of 100–1,000 AU. Since the orbital angular momenta of wide binaries are of the same order of magnitude as those of cloud cores, the angular momentum problem is therefore solved, at least in principle. But then we are faced with the theoretical difficulty of forming close binaries with orbital angular momenta that are two orders of magnitude smaller. This brings us back to the three mechanisms proposed in section 3.4: simple capture, formation by separate nuclei, and fission.

Current estimates indicate that *capture*, which requires a third body to absorb the excess energy liberated during the process, may occur in regions of very high stellar density (comparable to the cores of globular clusters) but that such an event is far too rare in the low stellar densities of the galactic field. The idea of independent condensations or *separate nuclei* is much out of favor, too, because it would be necessary to put aside the current idea that gravitational collapse occurs through a series of hydrostatic equilibria in which magnetic support gradually seeps away. Another possibility for short-period binaries is *hierarchical fragmentation*, involving subfragmentation during the collapse. An important argument in favor of this approach is that it naturally produces hierarchical multiple systems, as observed.

Another way of forming a close binary is to split a rotating mass into two pieces. This *fission* mechanism was strongly advocated by Jeans in the 1910s and 1920s. Later, in 1953, Raymond LYTTLETON (1911–1995) in Cambridge gave technical arguments why fission could not occur. The matter was reconsidered by Jeremiah Ostriker in 1970, who pointed out the shakiness of Lyttleton's arguments and concluded that "although fission remains unproven, there are now no strong theoretical arguments against the process, and there is considerable observational support for its existence."[10] In the 1970s Norman LEBOVITZ (b. 1935) at the University of Chicago proposed a new fission scenario based on the slow contraction of compressible, homogeneous ellipsoids with internal motions—the so-called Riemann ellipsoids. Lebovitz's basic conjecture is that these ellipsoids, after becoming unstable to infinitesimal disturbances associated with third-order harmonics, will develop a constriction that becomes progressively narrower as the evolution proceeds, resulting in a pair of detached fragments. In view of the inability of a linear stability analysis to predict the outcome of an instability as it progresses to the nonlinear regime, several groups of numerical analysts have followed the evolution of barlike disturbances in rapidly rotating configurations. Since their nonlinear calculations failed to produce binaries, it was therefore concluded that fission is not possible. As was noted by Lebovitz, however, this conclusion is entirely based on the preconception that fission should be the immediate and direct outcome of a barlike instability.

---

[10]J. P. Ostriker, in *Stellar Rotation*, edited by A. Slettebak (New York: Gordon and Breach, 1970), 147.

As he pointed out, its failure to produce binaries reflects only on the barlike route to fission, and not on other initial conditions that could lead to fission.[11]

Tidal Interaction

Further challenging problems arise from the study of double stars whose components are close enough to raise tides on each other's surface. Indeed, *tidal interaction* in a detached close binary will continually change the spin and orbital parameters of the system (such as the orbital eccentricity e, mean orbital angular velocity Ω_0, inclination i, and rotational angular velocity Ω of each component). Unless there are sizable stellar winds emanating from the binary components, the total angular momentum will be conserved during these exchange processes. However, as a result of tidal dissipation of energy in the outer layers of the components, the total kinetic energy of a close binary system will decrease monotonically. Ultimately, this will lead to either a collision or an asymptotic approach toward a state of minimum kinetic energy. Such an equilibrium state is characterized by circularity ($e = 0$), coplanarity ($i = 0$), and synchronism ($\Omega = \Omega_0$); that is to say, the orbital motion is circular, the rotation axes are perpendicular to the orbital plane, and the rotations are perfectly synchronized with the orbital revolution.

By the mid-1970s two distinct evolutionary processes were known that may produce secular changes in the orbital elements of a close binary star. The *tidal-torque mechanism* was originally outlined in 1879 by George Howard DARWIN[12] (1845–1912) with reference to a planet-satellite system. However, it was not until the mid-1960s that the French astrophysicist Jean-Paul ZAHN (b. 1935) revived Darwin's mechanism and applied it to binary-star components having an extended convective envelope. In this approach, then, each component possesses tides lagging in phase behind the external field of force on account of turbulent viscosity in its surface layers. This misalignment of the tidal bulges with respect to the line joining the two centers of mass will therefore introduce a net torque between the components. This torque will cause, in turn, a secular change in spin angular momentum of the individual components, the effects of which will be reflected in secular changes in the orbital elements of the binary. The other process, the so-called *resonance mechanism*, was suggested in 1941 by Cowling, who pointed out that some of the gravity modes of oscillation in a centrally condensed star may enter into resonance with the periodic tidal potential in a close binary. As was originally noted by Zahn in 1975, in a binary component possessing a radiative envelope the resonances of these low-frequency oscillations are heavily damped by radiative diffusion, which operates in a relatively thin layer below the star's surface. Owing to that dissipative process, these oscillatory motions do not have the same symmetry as the forcing

---

[11] For comprehensive discussions of these matters, see N. R. Lebovitz in *Highlights of Astronomy*, vol. 8, edited by D. McNally (Dordrecht: Kluwer, 1989), 129; J. E. Pringle in *The Physics of Star Formation and Early Stellar Evolution*, edited by C. J. Lada and N. D. Kylafis (Dordrecht: Kluwer, 1991), 437; P. Bodenheimer, T. Ruzmaikina, and R. D. Mathieu, in *Protostars & Planets III*, edited by E. H. Levy and J. I. Lunine (Tucson: University of Arizona Press, 1993), 367.

[12] He was the second son of the English naturalist Charles Robert Darwin (1809–1882).

potential. Hence, a net torque is applied to the binary component, which tends to synchronize its axial rotation with the orbital motion.

Until the early 1980s, the observed degree of synchronism and orbital circularity in the short-period close binaries appeared to be in reasonable agreement with Zahn's views on tidal interaction. That is to say, his theoretical predictions based on the above dissipative mechanisms were in agreement with the upper period limits at which these binaries begin to deviate from synchronism or circularity. Unfortunately, more recent observations have revealed that these properties extend up to orbital periods substantially larger than previously held. This is why one of us, Jean-Louis TASSOUL (b. 1938), proposed in 1987 another braking mechanism, which may be much more efficient than the two classical ones but had up to then escaped notice. To understand this *hydrodynamical mechanism* it is important to realize that the standard processes hypothesize that the large-scale motion in an asynchronous binary component is wholly one of rotation. When one relaxes this quite restrictive assumption, however, it is a simple matter to show that mechanically driven meridional currents come into existence. Specifically, it is the forced lack of axial symmetry in an asynchronously rotating body that produces these large-scale transient currents, thereby ensuring that the viscous stresses do indeed vanish at the free surface of the triaxial body. These transient motions, which are quite distinct from the thermally driven Eddington-Vogt currents discussed in section 4.7, cease to exist as soon as synchronization has been achieved. This new mechanism, which is operative in early-type and late-type binaries alike, brings support to the two others, for they are not mutually exclusive, each one of them being operative during different phases of stellar evolution and for different values of the parameters. Admittedly, the hydrodynamical mechanism is a fairly new concept in astronomy; hence, unlike more familiar approaches based on celestial mechanics or resonant interactions with natural modes of oscillation, it has not yet become a part of the astronomical tradition. In this, as in many other debatable issues, it is the accumulation of new observational data that will eventually resolve the controversy.[13]

6.2 EARLY-TYPE STARS

By the late 1920s one of the major problems of the Harvard classification was that the broad emission lines in the spectra of the *Wolf-Rayet stars* remained unidentified (see section 2.5). The recognition that these bright, hot stars lose mass continuously dates essentially from the work of Carlyle BEALS (1899–1979) at the Dominion Astrophysical Observatory (Victoria, B.C.). His early work on Wolf-Rayet stars concerned the interpretation of the so-called *P Cygni line profiles* in their spectra, which typically contain an emission line with a blueshifted absorption component. This feature was first discovered visually in the spectrum of P Cygni by James KEELER (1857–1900) using a spectroscope on the large Lick refractor in 1889. However, it was not until 1929 that Beals interpreted the phenomenon as being due to

---

[13] For an overall review of tidal interaction in close binary systems, see J. L. Tassoul, *Stellar Rotation* (Cambridge: Cambridge University Press, 2000), 207.

mass loss in an extended transparent envelope surrounding the star. The broadening is due to the Doppler effect and the blueshifted absorption feature results from starlight being scattered or absorbed in the envelope.[14] Another major advance was made in 1932 by Bengt Edlén when he identified many of the Wolf-Rayet emission lines with the same features of highly ionized carbon, nitrogen, and oxygen that were observed in the laboratory. The following year, Beals recognized that roughly half of the Wolf-Rayet stars show strong nitrogen lines, whereas the other half show strong carbon lines. They were called WN stars and WC stars, respectively. From the high degree of ionization in the spectra of these stars, Beals deduced very high surface temperatures in the range 50,000–100,000 K, much hotter than previously ascribed to any other stellar types.

Several decades later, in 1967, Donald Morton obtained the first rocket ultraviolet spectra of normal O and B supergiants and showed that many of their line profiles contain an undisplaced emission line with a blueshifted absorption component. This result opened what might be called the modern era of studies of stellar winds and mass loss from early-type stars. We now know that the Wolf-Rayet stars have the largest mass-loss rates, typically 10^{-5}–$10^{-4} M_\odot$ yr^{-1}, with wind terminal velocities of several thousand kilometers per second. The O- and B-type stars, with masses ranging from about $8 M_\odot$ to $100 M_\odot$, are hot ($T > 20,000$ K) and very luminous ($L \approx 10^5$–$10^6 L_\odot$). They have much lower mass-loss rates, ranging from about $10^{-9} M_\odot$ yr^{-1} to $10^{-5} M_\odot$ yr^{-1}, depending on the star's luminosity. The terminal velocity of their winds, of a few hundred to a few thousand kilometers per second, actually exceeds their surface escape velocity. Also known are the so-called *luminous blue variables*, which are very luminous, unstable hot supergiants. (They include such well-known stars as η Carinae and P Cygni.) They have a relatively intense matter outflow, up to several times $10^{-5} M_\odot$ yr^{-1}, and a relatively low terminal velocity, up to a few hundred kilometers per second. These stars are therefore contrasted with the Wolf-Rayet stars with lower expansion velocities, and with the O- and B-type stars with denser and usually slower winds.

During the 1990s, extensive monitoring campaigns showed that the O and B supergiants have winds that are strongly variable. The most prominent features of this variability are the so-called *discrete absorption components*, which migrate through the line profiles from red to blue on a time scale of hours to days. Their cyclical nature strongly suggests that stellar rotation is an important piece in the unsolved puzzle of the wind-variability mechanism. Spectroscopic observations have also revealed that all Wolf-Rayet winds appear to exhibit small-scale temporal fluctuations in their emission lines. The current idea is that we are seeing the manifestation of shock waves superimposed on their radiation-driven winds.[15]

---

[14] A quite similar model had been already suggested in 1904 by the German-born astronomer Jacob Halm, who was then working in Scotland (*Proc. Roy. Soc. Edinburgh*, **25**, 513 [1904]). The time was not yet ripe for Halm's suggestion, however, since it failed to win the attention of his fellow astronomers.

[15] See the numerous review papers in H. Lamers and R. Sapar, eds., *Thermal and Ionization Aspects of Flows from Hot Stars: Observations and Theory*, ASP Conference Series, vol. 204 (San Francisco: Astron. Soc. Pacific, 2000).

A New Mass-Loss Mechanism

A mechanism that explains the mass loss observed for luminous, hot stars was originally discussed in 1967 by Leon Lucy and Philip SOLOMON (b. 1939), who were then working at Columbia University. As they noted, one might first consider Parker's 1958 solar-wind mechanism; that is, one supposes that mass loss results from a hot stellar corona that cannot be contained by the star's gravitational field (see section 5.8). This, however, would require a temperature in the accelerating flow in excess of 10 million degrees, so that ions such as C IV, N V, and Si IV would then be destroyed by collisional ionization. Since these ions do indeed exist in the outflow of early-type supergiants, it follows that mass loss by hot stars is not a consequence of a hot corona. This led them to suggest that the observed mass loss might result from the pressure exerted on the gas by absorption of radiation in selected spectral lines in the ultraviolet region. Detailed hydrodynamical calculations have shown that the force per unit mass due to radiation pressure can indeed exceed the star's surface gravity, so that a continuous outflow of mass occurs. Further calculations have demonstrated that mass-loss rates of $10^{-5} M_\odot \, \mathrm{yr}^{-1}$ could be obtained by considering all the available strong lines. That the driving of winds in the O and B supergiants is based on the interplay between gravity and the star's radiation field has received further support from the relation between the observed terminal velocity of the wind and the escape velocity in these stars. This model also explains the relation between their mass-loss rates and luminosity. Application of these ideas to the Wolf-Rayet stars has not been so successful, however, because several of the approximations made are hopelessly invalid for their dense and highly stratified winds. Nevertheless, recent theoretical work strongly suggests that it is the severe stratification of ionization in the Wolf-Rayet stars, with their inner hot wind gradually changing to a cooler outer wind, that dramatically increases the effectiveness of photon trapping to that in the weak winds of the O and B supergiants. A realistic hydrodynamical modeling of the winds of Wolf-Rayet stars still lies in the future.[16]

The Wolf-Rayet Stars

The Wolf-Rayet stars fall into two groups: first, those associated with O and B supergiants and second, those that are the central stars of planetary nebulae. Those of the first group are young population I objects, those of the second consist of old objects. Actually, there appears to be a clear distinction between these two extremes of age: the young Wolf-Rayet stars have masses of the order of $10 M_\odot$, whereas the masses of those in planetary nebulae are typically below $1 M_\odot$. Up to the mid-1960s the position of the young Wolf-Rayet stars in the scheme of stellar evolution was not very well understood. In particular, it was not clear whether they are stars that are evolving onto the main sequence, as was advocated by the Canadian astronomer Anne UNDERHILL (b. 1920), or whether these stars are in the post-main-sequence stage, as most practitioners in the field were inclined to believe. In 1967 Bohdan

---

[16]L. B. Lucy and P. M. Solomon, *Astrophys. J.*, **159**, 879 (1970); J. I. Castor, D. C. Abbott, and R. I. Klein, *ibid.*, **195**, 157 (1975); D. C. Abbott, *ibid.*, **225**, 893 (1978); C. D. Garmany, and P. S. Conti, *ibid.*, **284**, 705 (1984); L. B. Lucy and D. C. Abbott, *ibid.*, **405**, 783 (1993).

PACZYŃSKI (b. 1940), who was then working in Warsaw, suggested that a binary mass exchange was responsible for producing a young Wolf-Rayet star, an object that is helium rich and highly overluminous for its mass. In this scenario, the binary interaction removes the outer hydrogen-rich layers of an evolving early-type star, leaving the bare helium-rich core with occasionally a little hydrogen left on the surface. As explained in section 5.5, this mass loss occurs during the approach to the red-giant phase of the initially more massive star, which has evolved more rapidly than its companion because of its greater mass. The stripped material is transferred by Roche-lobe overflow to the other star. The binary nature is thereby assumed to be a basic characteristic of *all* young Wolf-Rayet stars. Although many of these stars are spectroscopic binaries, with O or B companions, a few exceptionally well-studied systems seem to have no certain indications of a double nature. Paczyński's theory of Wolf-Rayet stars was therefore abandoned, although the effect of Roche-lobe overflow in binary systems may in some cases play a role in providing the material stripping.

The next decisive step toward the resolution of the problem posed by the Wolf-Rayet stars was made by Peter CONTI (b. 1934) at the University of Colorado in Boulder, who first proposed in 1976 that massive O-type stars could evolve to Wolf-Rayet stars through the actions of their strong stellar winds. Detailed evolutionary calculations have confirmed Conti's scenario so that, by the late 1990s, it was generally accepted that young Wolf-Rayet stars have "bare cores" resulting mainly from stellar winds peeling off single stars initially more massive than about $25 M_\odot$ to $40 M_\odot$. In some cases, as was noted above, massive O-type stars in close binaries may also lose their outer layers from Roche-lobe overflow. In both cases, high levels of mass loss, either continuous or via short-lived outbursts associated with a luminous-blue-variable phase, can peel off the outer atmospheric layers and reveal successive phases of nuclear processed material. Specifically, the WN stars show anomalous atmospheric abundances reflecting the exposition of CNO-burning material, whereas the WC stars show the composition characteristic of helium-burned material, in which carbon and oxygen are produced from the previously formed helium. Moreover, the so-called WO stars, which have proportionately more oxygen and less helium and carbon, appear to exhibit the subsequent proceeds of α-capture products (see section 5.3). The quantitative abundance analyses of Wolf-Rayet stars have generally shown good agreement with the predicted chemical compositions of theoretical evolutionary models in the different WN-WC-WO phases. More detailed scenarios have been proposed.[17]

The Be and B[e] Stars

Among the group of anomalous early-type stars showing mass loss we must include the classical *Be stars*, which are nonsupergiant B-type stars whose spectra have, or had at some time, one or more Balmer lines in emission. These objects, which are the most rapidly rotating stars, very often display highly variable line profiles. There is now widespread agreement that matter is accumulating in the equatorial

---

[17]A. Maeder and P. S. Conti, *Annu. Rev. Astron. Astrophys.*, **32**, 227 (1994).

region of the Be stars, with the resultant circumstellar disk giving the emission seen in the hydrogen lines. Some of the Be stars also develop, from time to time, a network of deep and narrow absorption lines, and they are then called *shell stars*. Statistical studies have shown that these stars are normal Be stars viewed edge on: The difference in spectra is due to differences in the inclination of the rotation axes. As was noted in section 5.7, as a class the shell stars are the most rapidly rotating Be stars, with the $v_e \sin i$ values being as large as 500 km s^{-1}. This at once raises the following question: Do the Be and shell stars rotate at their break-up velocity, that is, the critical velocity at which centrifugal force balances gravity at the equator? The answer to that question is flatly no. Indeed, detailed numerical modelings of these stars have shown that there are no Be or shell stars with rotation anywhere near their theoretical break-up velocity. Like other O- and B-type stars, the ultraviolet spectral lines of the Be stars show evidence for substantial stellar winds, with mass-loss rates in the range 10^{-11}–$10^{-8} M_\odot$ yr^{-1} and terminal velocities of the order of 1,000 km s^{-1}.

Unexpectedly, in the early 1970s it was realized that another class of emission-line stars exists, which are now known as *B[e] stars*. These objects are B-type supergiants whose continua exhibit excesses of near-infrared radiation and forbidden emission lines. (Conti in 1976 suggested the designation B[e], following the standard notation for forbidden lines.) There are thus substantial differences between the Be and B[e] stars, in particular the missing dust infrared emission and the absence of pronounced forbidden lines in the classical Be stars. The Be and B[e] stars do share certain characteristics, however, like the appearance of Balmer lines and the intrinsic polarization observed in the optical continuum. This is commonly interpreted as evidence that the circumstellar material is substantially denser in the equatorial regions than it is near the poles. (In analogy with the Be stars, it is generally assumed that the B[e] stars as well as the Wolf-Rayet stars are rapid rotators.) Perhaps the most unusual characteristic of the B[e] stars is the presence of hot dust emission in their spectra. This is quite unusual because no other class of hot stars, other than some WC stars, show evidence of circumstellar dust in their winds. Their mass-loss rates are comparable to those observed in the Wolf-Rayet stars.

As explained in section 4.7, Struve in 1931 was the first to suggest that the emission lines in Be-star spectra arise from matter ejected from the equatorial belt of a rapidly rotating star. Then, in 1933 McLaughlin proposed another model that was a combination of Struve's rotating model and Beals' idea of an expanding atmosphere in the Wolf-Rayet stars. To account for the observed spectral variations, he assumed that temperature variations of the star would yield a variable radiation pressure force on the envelope material, thus producing expansion and contraction. Although this idea and a similar intermittently expanding-rotating model did not withstand the passage of time, on theoretical grounds they are important because they point to the idea that radiation pressure might be important for the Be stars; namely, that some other force apart from gravity and those arising from pressure gradient and centrifugal effects must be considered. Since the 1960s, several rotationally enhanced stellar wind models have been developed, first for the Be stars and then for the B[e] stars as well. Broadly speaking, these models can be separated into two classes: the 1D and the 2D models.

First, there are the 1D models, which postulate a nonspherically symmetric distribution of matter, with the circumstellar envelope strongly concentrated in the equatorial plane of the star. For these phenomenological models the flow is assumed to be primarily radial, and no flow occurs from one latitude zone to another. Observations can then be explained by the assumption of a two-component stellar wind, consisting of a radial radiation-driven wind (as observed in all hot high-luminosity stars) and an additional mass loss from the equatorial regions of the star's atmosphere. In this scenario the main difference between the Be and B[e] stars would be the radial component of the mass outflow, which is much larger in the B[e] supergiant stars than in the classical Be stars. The basic problem with these 1D models is to explain how rotation can lead to a large enhancement in the density of the radial streams of equatorial outflow.[18]

Second, there are the 2D models for which the enhancement in the equatorial density arises primarily from the flow of material down from higher latitude zones. A detailed hydrodynamical formulation of this problem was initiated in 1993 by Jon Eric BJORKMAN (b. 1956) and Joseph CASSINELLI (b. 1940) at the University of Wisconsin in Madison. They pointed out that a circumstellar disk is formed because a supersonic wind that leaves the stellar surface at high latitudes travels along trajectories that carry it toward the equatorial plane. If the rotation rate is high enough, this flow will attempt to cross the equator, where it collides with material from the opposite hemisphere. Since the flow velocities are assumed to be supersonic, this collision results in a pair of shocks above and below the equatorial plane. Between the shocks, the shock compression then produces a thin, dense, *wind-compressed disk*, as illustrated in figure 6.4. This is a most promising avenue of research because it has already predicted that for B2-type stars the onset of disk formation occurs at a rotational velocity of about 50 percent of the star's break-up velocity, whereas for O-type stars the disk can occur only for rotational velocities in excess of 90 percent of the break-up velocity. This is in agreement with the frequency distribution of the observed $v_e \sin i$ values for the Be stars, and indicates a maximum probability around spectral type B2. This model also explains why dust forms around B[e] stars and not around normal supergiants: The density of a spherically symmetric wind is too low for dust to form, but when the same mass loss is redistributed through a wind-compressed disk, as in the B[e] stars, the density becomes large enough for dust formation.[19]

---

[18]See, e.g., F. J. Zickgraf, B. Wolf, O. Stahl, C. Leitherer, and G. Klare, *Astron. Astrophys.*, **143**, 421 (1985). Detailed reviews of previous 1D models will be found in the following review papers: J. M. Marlborough, in *Be and Shell Stars*, edited by A. Slettebak, I.A.U. Symposium No. 70 (Dordrecht: Reidel, 1976), 355; J. M. Marlborough, in *Physics of Be Stars*, edited by A. Slettebak and T. P. Snow (Cambridge: Cambridge University Press, 1987), 316.

[19]J. E. Bjorkman and J. P. Cassinelli, *Astrophys. J.*, **409**, 429 (1993); S. P. Owocki, S. R. Cranmer, and J. M. Blondin, *ibid.*, **424**, 887 (1994). For a comprehensive review of other recent 1D and 2D models, see J. P. Cassinelli, in *B[e] stars*, edited by A. M. Hubert and C. Jaschek (Dordrecht: Kluwer, 1998), 177. See also H. Lamers and J. P. Cassinelli, *Introduction to Stellar Winds* (Cambridge: Cambridge University Press, 1999).

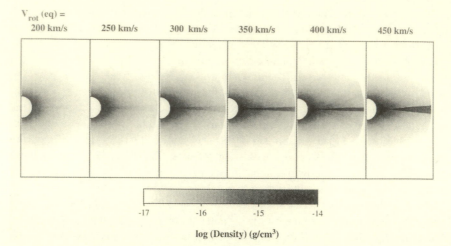

V_{rot} (eq) =

| 200 km/s | 250 km/s | 300 km/s | 350 km/s | 400 km/s | 450 km/s |

-17   -16   -15   -14

log (Density) (g/cm³)

Figure 6.4 Gray-scale plots of density for various equatorial velocities of a B2.5 main-sequence star. Note the clear increase in strength of the wind-compressed disk with increasing rotation. From S. P. Owocki, S. R. Cranmer, and J. M. Blondin, *Astrophys. J.*, **424**, 887 (1994).

The Chemically Peculiar Stars

We now turn to the *chemically peculiar stars* of the upper main sequence. These stars are identified by the presence of anomalously strong (or weak) absorption lines of certain elements in their spectra. They include the classical metallic-line (Am) and peculiar A-type (Ap) stars, as well as the HgMn stars and the He-rich/poor stars. Morgan in 1933 was the first to recognize that the Ap stars, ordered by predominant peculiarity (Mn, Si, Eu, Cr, Sr), correlate with temperature. Yet it was not until the late 1950s that the extent to which their peculiarities were anomalous was fully recognized. Several hypotheses have been advanced to account for the various chemically peculiar stars: transport to the surface of nuclearly processed material from the core, spallation of heavier nuclei in the stellar surface induced by high-energy particles accelerated in flares, surface contamination of a normal star by a nearby supernova, or selective magnetic accretion of interstellar matter. As more and more chemically peculiar stars were discovered, and their properties became better defined, the realization came that these explanations are statistically and physically untenable, at least in terms of providing a single physical mechanism for the bulk of these stars. One must for instance explain why *all* slowly rotating non-magnetic A-type stars are Am stars, as was noted by Helmut Abt and his associates. A major breakthrough came in 1970, when Georges MICHAUD (b. 1940), then at the Université de Montréal, published the idea he had developed at the California Institute of Technology that the observed abundance anomalies could be due to *diffusion processes* operating in the outer portion of stellar envelopes. As he pointed out, abundances anomalies appear, on the main sequence, in the atmosphere of stars most likely to have stable envelopes and atmospheres. These stars are slow rotators and so have less meridional circulation, they often have magnetic fields, and they have an effective temperature for which stellar envelope models give the weakest convection. In its simplest form, the diffusion model assumes that the region below the superficial convection layer of a chemically peculiar star is stable enough so that microscopic diffusion processes can separate the light elements from the heavy ones; that is to say, those chemical elements absorbing more of the outward going radiative flux per atom move to the surface, while those absorbing less sink to the interior. In subsequent years, Michaud and his associates developed detailed diffusion models for some of the more common classes of chemically peculiar stars. In the 1990s their models remained the leading contenders for the status of explanatory models for these stars.[20]

Although no serious discrepancy between theory and observation appears definitively established at the present level of uncertainties, chemical separation nevertheless presents a serious difficulty: For microscopic diffusion to be of importance, the various time scales for mixing (by circulation, convection, or turbulence) must be larger than the time scales for gravitational sorting to occur. In 1926 Eddington argued that meridional circulation would eliminate any chemical separation caused by diffusion. Unfortunately, his opinion carried enough weight for diffusion to be put aside as an important transport mechanism in stars for several decades. As was noted in section 5.3, Sweet and Mestel in the early 1950s were the first to calculate the circulation velocities of the meridional flow in a uniformly rotating star. Baker and Kippenhahn in 1959 performed a similar calculation for a differentially rotating star. All these formulations were incomplete, however, because their authors failed to prescribe that the meridional currents must stream along the star's surface. Their theoretical circulation velocities were therefore utterly inadequate to discuss the interaction between microscopic diffusion and large-scale meridional flow. No further progress was made until 1982, when the two of us—Jean-Louis Tassoul and Monique Tassoul (b. 1942)—pointed out that turbulent viscosity acting in the outmost surface layers of an early-type star is an essential ingredient of the problem. Thence, making allowance for a thin viscous boundary layer near the star's surface, we obtained a self-consistent solution for the thermally driven meridional flow, free of the unwanted mathematical singularities and other inconsistencies befalling previous calculations.[21]

Many characteristics of the chemically peculiar stars can be explained on the basis of microscopic diffusion in the presence of meridional circulation in their outer radiative envelopes. Indeed, when this large-scale flow is rapid enough to obliterate the settling of the diffusion of helium, no underabundance of this element is possible and the superficial He convection zone remains important, making the appearance of some of the abnormal abundances impossible. Comparing the meridional circulation velocities obtained by the Tassouls to diffusion velocities of helium below the He convection zone, Michaud originally showed that this zone disappears only in

---

[20]G. Michaud, *Astrophys. J.*, **160**, 641 (1970); see also J. Richer, G. Michaud, and S. Turcotte, *ibid.*, **529**, 338 (2000).

[21]J. L. Tassoul and M. Tassoul, *Astrophys. J. Suppl.*, **49**, 317 (1982); see also M. Tassoul and J. L. Tassoul, *Astrophys. J.*, **440**, 789 (1995).

stars with equatorial velocities smaller than about 90 km s^{-1}. This is in agreement with the cutoff velocity observed for the HgMn stars. Given this encouraging result, detailed bidimensional diffusion calculations have been carried out by Paul CHAR-BONNEAU (b. 1961), then a student with Michaud in Montréal, to determine with greater accuracy the maximum rotational velocity allowing the gravitational settling of helium. This upper limit was found to be 75 and 100 km s^{-1}, respectively, for the HgMn and FmAm stars. Now, because turbulent particle transport can also have drastic effects on chemical separation, Charbonneau and Michaud performed additional calculations that retain both meridional circulation and anisotropic turbulence. (Schatzman in 1969 had already stressed the importance of turbulent diffusion in stars.) Unfortunately, there is as yet no reliable theory that could provide firm analytical expressions or numerical values for coefficients of turbulent diffusivity. In fact, the practical evaluation of these coefficients is at least partly an art, not exactly a science: They are essentially free quantities that must be chosen, by trial and error, using the observed abundance anomalies to determine their values. Detailed bidimensional calculations have shown that small-scale turbulent motions cannot be ignored altogether, however, because they, too, can impede the gravitational settling of helium. At this writing, the relative importance of meridional circulation and anisotropic turbulence in reducing chemical separation remains uncertain.[22]

At this juncture it is perhaps appropriate to note that the accumulation of relatively recent observations has made it clear that, while we understand the fundamentals of stellar evolution, the so-called standard models are in error in a number of details. This is particularly true for early-type stars, and stellar rotation is currently the favorite candidate to explain the discrepancies. Indeed, a group of observers has found that all fast rotators among O-type stars show large surface helium abundances correlated with the rotation rate, which indicates that there is probably a link between rotation and turbulent mixing in these stars.[23] Unfortunately, because the choice of the diffusivity coefficients is far from being unique, we are not yet in a position to provide a fully quantitative explanation for the observational data.

6.3 THE SUN

It is generally accepted that the sun's energy is liberated by nuclear reactions in its central regions. Most of this energy goes into heat, which is transported to the solar surface and emitted as electromagnetic radiation. A small fraction of the energy is emitted in the form of *neutrinos*. These elementary particles, which have no charge and little or no mass, are created in the core of the sun in nuclear reactions such as those described in appendix G. Neutrinos have the unusual property of easily passing through matter so that they can travel directly from the sun's interior to the earth without significant absorption. They travel at the speed of light, just like

---

[22] G. Michaud, *Astrophys. J.*, **258**, 349 (1982); P. Charbonneau and G. Michaud, *ibid.*, **327**, 809 (1988); **370**, 693 (1991).
[23] A. Herrero, R. P. Kudritzki, J. M. Vilchez, D. Kunze, K. Butler, and S. Haser, *Astron. Astrophys.*, **261**, 209 (1992).

photons. Since the mid-1960s, efforts have been made to detect neutrinos coming from the sun. It now seems clear that these particles have been observed, directly confirming that nuclear fusion is actually taking place in the sun. As we shall see, however, for several decades the observed number of neutrinos was about two or three times smaller than expected on theoretical grounds. This is the so-called *solar neutrino problem*, which we shall briefly review in chronological order.

For a long time, the only attempt to detect solar neutrinos was the radiochemical experiment of Raymond DAVIS, Jr. (b. 1914) and his associates at the Brookhaven National Laboratory (Upton, N.Y.). Their detector consists of 600 tonnes of tetra-chlorethylene (C_2Cl_4), a common cleaning fluid, in a tank located 1,500 meters below the surface in a South Dakota gold mine, the ground above screening out unwanted cosmic ray particles. The central idea is that, on very rare occasions, a solar neutrino reacts with an atomic nucleus on earth, converting one of its neutrons into a proton and hence transmuting one element into another. One such case is

$$^{37}Cl + \text{neutrino from } ^{8}B \text{ beta decay} \Rightarrow {}^{37}Ar + \text{electron}, \qquad (6.1)$$

where the isotope of chlorine, ^{37}Cl (which constitutes 25 percent of all normal chlorine), is transmuted into the unstable isotope of argon, ^{37}Ar. The argon is extracted about once every two months and the amount (about one atom every two days) is measured by observing ^{37}Ar decays. In 1968, Davis and his associates announced the first results from their chlorine experiment: an upper bound on the solar neutrino flux of 3 SNU (where 1 SNU equals 10^{-36} interactions per target atom per second). In an accompanying paper, John BAHCALL (b. 1934) and coworkers found a rate of 7.5 ± 3.3 SNU for the standard solar model.[24] By the late 1990s the conflict between chlorine measurements and theoretical predictions still persisted, augmented by three decades of data from chlorine experiments and by new ones, this time with ^{71}Ga (instead of ^{37}Cl) as the target:

$$^{71}Ga + \text{neutrino from proton-proton fusion} \Rightarrow {}^{71}Ge + \text{electron}. \qquad (6.2)$$

So far, thus, several experiments have detected the proton-proton neutrinos as well as those obtained from ^{8}B beta decay in approximately the numbers (within a factor of two or three) predicted by the standard solar model.

Although these experiments establish empirically that the sun shines because of nuclear reactions, they all disagree quantitatively with the combined predictions of the currently accepted model for the sun and the standard model of particle physics, which assumes that nothing much happens to the neutrinos after they are created. As we shall see below, helioseismological measurements have strongly confirmed the predictions from solar-model calculations that include microscopic diffusion. By the late 1990s it was widely accepted, therefore, that the resolution of the solar neutrino problem would require some new particle physics that changes the neutrinos as they travel from the core of the sun to the earth. One very promising solution assumes that neutrinos do have a tiny mass, so that a resonance phenomenon in matter could efficiently convert many of the electron-type neutrinos created in the

---

[24]R. Davis, Jr., D. S. Harmer, and K. C. Hoffman, *Phys. Rev. Letters*, **20**, 1205 (1968); J. N. Bahcall, N. A. Bahcall, and G. Shaviv, *ibid.*, **20**, 1209 (1968).

sun's interior to muon-neutrinos and tau-neutrinos, which are much more difficult to detect. The first direct indication of a nonelectron flavor component in the solar neutrino flux was actually reported in June 2001 by an international team working at the new Sudbury Neutrino Observatory. (Its detector, which uses 1,000 tonnes of heavy water to intercept about 10 neutrinos per day, is located 2,000 meters belowground in a nickel mine near Sudbury, Ontario.) It would thus seem that the thirty-year-old mystery of the missing solar ^{8}B neutrinos has finally been resolved. Indeed, combining their results with those of a Japanese experiment, known as Super-Kamiokande, the Sudbury group has shown that the total flux of active ^{8}B neutrinos received from the sun agrees well (within 0.3 standard deviations) with the predictions of detailed solar models.[25] Admittedly, this will force modifications in the standard model of particle physics, which accommodates neither massive neutrinos nor the transformation of one type of neutrino into another.

A New Field: Helioseismology

A different line of enquiry was initiated by the discovery of the so-called *five-minute oscillations* in the solar photosphere. The first evidence for ubiquitous oscillatory motions was obtained at Mount Wilson during the summers of 1960 and 1961. These pioneering observations were made by Robert LEIGHTON (1919–1997), Robert NOYES (b. 1934), and George SIMON (b. 1934), who were then working at the California Institute of Technology. The discovery of the existence of these oscillations did not immediately result in the establishment of *helioseismology*. Indeed, for nearly a decade after their discovery, these motions were viewed as the response of the solar atmosphere to the turbulent motions emanating from the solar convection zone. In fact, it was not until 1968 that Edward FRAZIER (b. 1939), an American astronomer who was then working in Heidelberg, made the following remarks:[26]

> ... the well known 5 min oscillations are primarily standing resonant acoustic waves .... [They] are not formed directly from the "piston action" of a convective cell impinging on the stable photosphere, but rather are formed within the convection zone itself.

Two years later, Roger ULRICH (b. 1942) at the University of California in Los Angeles presented a detailed description of the phenomenon, showing that standing acoustic waves may be trapped in a layer beneath the solar photosphere. This model was independently proposed by his fellow countrymen John LEIBACHER (b. 1941) and Robert STEIN (b. 1935). In 1975, Franz-Ludwig DEUBNER (b. 1934), then at the Kiepenheuer-Institut für Sonnenphysik in Freiburg, obtained the first observational evidence for these trapped acoustic modes. Deubner's observations were quickly confirmed, in 1977, by Edward RHODES, Jr. (b. 1946), working in collaboration with Ulrich and Simon. In the first use of the five-minute oscillations as a tool for

---

[25] See their collective report in *Phys. Rev. Letters*, **87**, 071301 (2001). For a comprehensive review of the solar neutrino problem, see M. F. Altmann, R. L. Mössbauer, and L.J.N. Oberauer, *Rep. Prog. Phys.*, **64**, 97 (2001).

[26] *Zeit. Astrophys.*, **68**, 345 (1968).

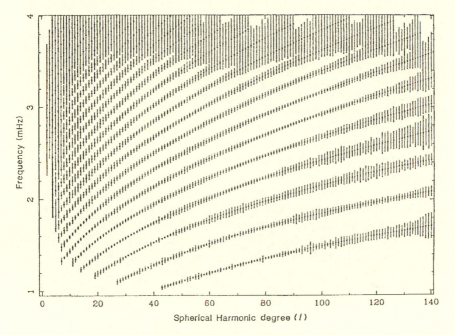

Figure 6.5 Schematic plot showing p-mode frequencies measured in 1986 at the Big Bear
            Solar Observatory in California. The usual 1σ error bars have been magnified by
            a factor of 1,000 to make them visible. Each ridge contains modes with a fixed
            number of radial nodes n, with the lowest-frequency ridge having $n = 1$. From
            K. G. Libbrecht and M. F. Woodard, *Nature*, **345**, 779 (1990).

helioseismology, they inferred that the solar convection zone is considerably deeper
than was generally accepted at the time.

Helioseismology, like terrestrial seismology, provides information about the interior of the body under study by using measurements of slight motions on the surface.
The sun is actually a very small-amplitude variable star. Its oscillations arise from
about 10 million discrete modes with periods ranging from a few minutes to several
hours (see figure 6.5).

The five-minute oscillations, which typically cause the photosphere to rise and
fall by less than 25 meters and to change temperature by just 0.005 K, have been
studied extensively since the mid-1970s.[27] They correspond to standing acoustic
waves (or p modes) that are trapped beneath the solar surface, with each mode
traveling within a well-defined shell bounded on the outside by the reflection due to
the density gradient near the solar surface and on the inside by refraction due to the
increasing sound speed. Since the properties of these modes are determined by the
stratification of the sun, accurate measurements of their frequencies thus provide a
new window on the hitherto invisible solar interior.

---

[27]The following pioneering papers may be noted: A. Claverie, G. R. Isaak, C. P. McLeod, H. B. van
der Raay, and T. Roca Cortez, *Nature*, **282**, 591 (1979); G. Grec, E. Fossat, M. Pomerantz, *ibid.*, **288**,
541 (1980).

In addition to providing a probe of overall characteristics of the sun's interior (such as the depth of its outer convection zone), helioseismology can also provide a probe of its internal rotation. Indeed, because axial rotation breaks the sun's spherical symmetry, it splits the degeneracy of the various oscillation modes with respect to the azimuthal angular dependence (see eq. [6.3] below). In 1979 Rhodes, Deubner, and Ulrich made use of this property to show that measurements of these frequency splittings could be used to provide a depth-dependent probe of the sun's internal angular velocity. This idea was originally exploited by Thomas DUVALL, Jr. (b. 1950), John HARVEY (b. 1940), and their associates, with observations obtained at the Kitt Peak National Observatory in Arizona.[28] Since the mid-1980s, several methods have been used to determine the solar internal rotation from the splittings of p modes that arise from the sun's differential rotation. Broadly speaking, these studies indicate that the rotation rate in the solar convection zone is similar to that at the surface, with the polar regions rotating more slowly than the equatorial belt. Below this zone is a thin layer, the so-called *tachocline*, of thickness no more than 5% of the solar radius, within which the latitude gradient disappears and the rotation becomes that of the deep interior. In contrast to the convective envelope, the radiative core appears to rotate nearly uniformly down to $r \approx 0.2R_\odot$, at a rate that is intermediate between the polar and equatorial rates of the photosphere. Within the central region $r \lesssim 0.2R_\odot$, the rotation rate is less precisely determined. Some studies suggest that the angular velocity increases with depth, implying rotation at a rate between two and four times that of the surface. According to other investigators, however, it is much more likely that this inner core rotates with approximately the same period as the outer parts of the radiative core. We shall not go into these disputes.

Another important step forward was made by Douglas GOUGH (b. 1941) in Cambridge, who developed an asymptotic inversion method of inferring the speed of sound at different depths in the sun.[29] The following year, 1985, this method was applied by the Danish astrophysicist Jørgen CHRISTENSEN-DALSGAARD (b. 1950) working in collaboration with Duvall, Gough, Harvey, and Rhodes to obtain the sound speed distribution from helioseismological measurements. By that time, however, the best solar models did not predict oscillation frequencies and a sound speed that match the observational data. These discrepancies led theoreticians to try many refinements—e.g., diffusion processes—to their models. By the mid-1990s current solar models were in sufficient agreement with the observational inferences, therefore suggesting that the disagreement between the predicted and observed fluxes of solar neutrinos was not caused by errors in the models. In particular, the sound

---

[28]T. L. Duvall, Jr., and J. W. Harvey, *Nature*, **310**, 19 (1984); T. L. Duvall, Jr., W. Dziembowski, P. R. Goode, D. O. Gough, J. W. Harvey, and J. W. Leibacher, *ibid.*, **310**, 22 (1984).

[29]The p modes observed at low or intermediate degrees generally have many nodes along the radius. If so, then, asymptotic expressions for their frequencies should be applicable. Paul Ledoux in Liège was the first to develop the asymptotic theory of high-order radial modes (*Bull. Acad. Roy. Belgique, Cl. Sci.*, **48**, 240 [1962]; **49**, 286 [1963]). A first-order approximation of the high-order nonradial p modes was originally published by Yuri Vasilevich VANDAKUROV (b. 1927) in Leningrad, now St. Petersburg (*Astron. Zh.*, **44**, 786, [1967] [*Soviet Astron.*, **11**, 630 (1968)]). However, better suited second-order approximations were not obtained until several years later (M. Tassoul, *Astrophys. J. Suppl.*, **43**, 469 [1980]; *Astrophys. J.*, **358**, 313 [1990]).

speed agreed to better than 1% throughout the sun, with a full percentage deviation only just beneath the outer convection zone, near $r = 0.7 R_\odot$.[30]

In the 1970s two distinct mechanisms were proposed for the excitation and damping of the solar acoustic modes.[31] One of the possibilities is that, in close analogy with the Cepheid variables, a surface layer in which hydrogen changes from neutral to ionized ($H \rightleftharpoons H^+$) might be the seat of a valve mechanism of the type discussed in section 5.6. Hiroyasu ANDO (b. 1946) and Yoji Osaki in Tokyo were the first to make a detailed linear stability analysis of the acoustic modes in the solar convection zone. Neglecting the effects of turbulent convection on these oscillatory motions, they found that a changing opacity in the hydrogen ionization zone could well be the dominant source of driving of the oscillations. Indeed, their analysis indicates that the most unstable modes have periods of about 300 seconds, which is in agreement with observations. Unfortunately, independent calculations have shown that the stability of the solar p modes depends sensitively on the treatment of turbulence as well as that of radiation in the optically thin layers. Since there is as yet no satisfactory theory of the interaction of turbulent convection with pulsation, we must therefore conclude that theoretical calculations cannot unequivocally resolve the question of the stability of these modes. This is why Peter GOLDREICH (b. 1939) from the California Institute of Technology and his associates have examined the possibility of stochastic excitation of the solar p modes by turbulent convection. At this writing, however, the calculations are still plagued by the lack of an adequate theory for the interaction of acoustic radiation with turbulence near the top of the solar convection zone. By the late 1990s neither observation nor theory has been able to establish which, if either, is the correct choice.

The Solar Differential Rotation

The problem presented by the observed solar differential rotation is one of long standing and many efforts have been made to formulate a plausible flow pattern that reproduces the large-scale motions in the solar atmosphere (see section 4.8). Following Lebedinsky's 1941 pioneering work in Leningrad (now St. Petersburg), many theories have been proposed to explain how the equatorial acceleration originated and is maintained in the solar convection zone. Broadly speaking, they can be divided into two classes, depending on the mechanism proposed to produce and maintain the equatorial acceleration: (i) the interaction of rotation with *global* turbulent convection in a rotating spherical shell, and (ii) the interaction of rotation with *local* turbulent convection. Until the late 1980s, however, the most detailed models invariably predicted rotation profiles that were constant on cylinders concentric with the rotation axis. Obviously, these solutions are at variance with the current observations, which indicate that the latitudinal angular velocity gradient

---

[30]Helioseismology is largely a cooperative endeavor, as illustrated by the whole range of collective review papers in *Science*, **272**, 1281–1309 (1996).

[31]H. Ando and Y. Osaki, *Publ. Astron. Soc. Japan*, **27**, 581 (1975); P. Goldreich and D. A. Keeley, *Astrophys. J.*, **212**, 243 (1977); P. Goldreich and P. Kumar, *ibid.*, **326**, 462 (1988); P. Kumar and P. Goldreich, *ibid.*, **342**, 558 (1989).

in the photosphere remains essentially constant with depth in the solar convection zone. Let us briefly review these models and explain what can be done to solve the disparities between the rotation profiles deduced from the helioseismological data and what has been predicted theoretically.

One class of models is based on the appealing assumption that the largest scales of convection are influenced by rotation, leading to a continuous redistribution of angular momentum, which we observe as a differential rotation. This approach was originally devised in the late 1960s by Bernard DURNEY (b. 1926) in Boulder and, independently, by Friedrich BUSSE (b. 1936) in Munich. Specifically, these *global-convection models* resolve numerically as many of the large scales as possible in a rotating spherical shell and parameterize, via eddy diffusivities, the transport of momentum and heat by all the smaller unresolved scales. Extensive numerical calculations were made, independently, by the American hydrodynamicists Peter GILMAN (b. 1941) and Gary GLATZMAIER (b. 1949) in the early 1980s. Although the numerical techniques employed in these models are quite different, the results obtained are qualitatively the same: Their simulated global convection in a rotating spherical shell tends to take the form of north-south (banana) rolls, the tilting of which yields stresses to drive the zonal flows that maintain differential rotation. Unfortunately, in these early models the simulated angular velocity in the convection zone is constant on cylinders coaxial with the rotation axis, which is not in agreement with the helioseismological data. In the 1990s work by Juri TOOMRE (b. 1940) and his associates at the University of Colorado showed that the numerical resolution of the Gilman-Glatzmaier models was actually insufficient to attain the fully turbulent regimes that are observed in the solar convection zone. Tridimensional numerical simulations of fully turbulent convection in a rotating spherical shell have already been produced. Although these extensions to fully turbulent regimes are quite promising, it is not yet clear to what extent the new global-convection models adequately describe the observed rotation profile in the solar convection zone.[32]

In the other class of models, the role of global convection is assumed to be unimportant. What is essential in these *mean-field models* is the interaction of rotation and local convective motions that are not greatly influenced by the sun's spherical shape. Lebedinsky in 1941 was the first to propose the concept of anisotropic turbulent viscosity to describe the dynamical consequences of turbulent motions in the solar convection zone. He was also the first to recognize the importance of this anisotropy in preventing the equalization of angular velocity in the solar photosphere. Unfortunately, no doubt due to a lack of communication with the Soviet Union during World War II, his results escaped notice in Western countries. Lebedinsky's concept of anisotropic turbulent viscosity, which was rediscovered in 1946 by Wasiutyński, was subsequently developed by Biermann and, later, by Kippenhahn and others. These

[32]B. R. Durney, *J. Atm. Sci.*, **25**, 771 (1968); F. H. Busse, *Astrophys. J.*, **159**, 629 (1970); B. R. Durney, *ibid.*, **161**, 1115 (1970); H. Yoshimura and S. Kato, *Publ. Astron. Soc. Japan*, **23**, 57 (1971). Detailed numerical results will be found in the following papers: P. A. Gilman and J. Miller, *Astrophys. J. Suppl.*, **61**, 585 (1986); G. A. Glatzmaier, in *The Internal Solar Angular Velocity*, edited by B. R. Durney and S. Sofia (Dordrecht: Reidel, 1987), 263. See also N. H. Brummell, N. E. Hurlburt, and J. Toomre, *Astrophys. J.*, **493**, 955 (1998). For a review of recent progress and unanswered scientific questions, see P. A. Gilman, *Solar Phys.*, **192**, 27 (2000).

authors appealed to the anisotropy of convection due to the preferred direction of gravity and showed that a complex flow, including differential rotation and a slow but inexorable meridional circulation, is always present in a region of efficient convection. Since the early 1970s, a variety of mean-field models have been calculated. However, some of them have meridional velocities at the surface that are too large, while others have pole-equator temperature differences that are too large. Then, in 1995, Leonid KITCHATINOV (b. 1955) in Irkutsk and Günther RÜDIGER (b. 1944) in Potsdam pointed out that the conflict between mean-field models and solar observations can be resolved by taking into account an anisotropic turbulent viscosity as well as an anisotropic turbulent heat transport. To the best of our knowledge, their mean-field model is the first one that satisfies almost all the observational constraints. Given this result, it thus seems highly probable that anisotropy plays a key role in the solar rotation problem, since calculations involving isotropic transport coefficients always yield angular velocities that are constant on cylinders in the models.[33]

Spin-Down of the Solar Interior

Now, in section 5.7 we pointed out that the sun and solar-type stars sustain a continuous loss of mass as a result of magnetized stellar winds and/or episodic mass ejections emanating from their outer convection zones. The central question is how the inexorable spin-down of the convective envelope that results from this mass loss will affect the rotational state of the radiative core. In the case of the sun, the absence of marked differential rotation in the bulk of the radiative interior implies that angular momentum redistribution within that region must be very efficient indeed. The problem of the solar spin-down was originally formulated in 1975 by the Japanese hydrodynamicist Takeo SAKURAI (b. 1931) from Kyoto University. By the 1980s, however, it was properly established that the thermally driven meridional currents in the sun's radiative core are so slow that, to a first approximation, the advection of angular momentum by these currents can be neglected. In fact, two categories of models have been proposed. In one of them, angular momentum redistribution is treated as a turbulent diffusion process, with advection by the meridional flow and magnetic fields being neglected altogether. The other group of models is based on the idea that this redistribution is dominated by magnetic stresses arising from the shearing of a preexisting poloidal magnetic field.[34]

---

[33]L. Biermann, *Zeit. Astrophys.*, **28**, 304 (1951); L. Biermann, in *Electromagnetic Phenomena in Cosmical Physics*, edited by B. Lehnert, I.A.U. Symposium No. 6 (Cambridge: Cambridge University Press, 1958), 248; R. Kippenhahn, *Astrophys. J.*, **137**, 664 (1963); H. Köhler, *Solar Phys.*, **13**, 3 (1970); B. R. Durney and I. W. Roxburgh, *ibid.*, **16**, 3 (1971). For a review of other mean-field models, see G. Rüdiger, *Differential Rotation and Stellar Convection* (New York: Gordon and Breach, 1989). Details about the Kitchatinov-Rüdiger model can be found in their paper: L. L. Kitchatinov and G. Rüdiger, *Astron. Astrophys.*, **299**, 446 (1995).

[34]T. Sakurai, *Mon. Not. Roy. Astron. Soc.*, **171**, 35 (1975). See also A. S. Endal and S. Sofia, *Astrophys. J.*, **243**, 625 (1981); M. H. Pinsonneault, S. D. Kawaler, S. Sofia, and P. Demarque, *ibid.*, **338**, 424 (1989); J. L. Tassoul and M. Tassoul, *Astron. Astrophys.*, **213**, 397 (1989). The effects of a large-scale magnetic field are discussed in the following papers: P. Charbonneau and K. B. MacGregor, *Astrophys. J.*, **417**, 762 (1993); G. Rüdiger and L. L. Kitchatinov, *ibid.*, **466**, 1078 (1996).

Detailed numerical calculations have shown that turbulent friction alone is most probably insufficient to explain a nearly constant rotation rate down to the sun's center. This is why Paul Charbonneau and Keith MACGREGOR (b. 1950) from the High Altitude Observatory in Boulder have investigated the spin-down of a solarlike model that includes both convection zone braking by a magnetized solar wind and internal angular momentum redistribution by turbulent friction and large-scale magnetic fields. For the sake of simplicity, however, they assumed the convective envelope to rotate as a solid body at all times. Their quantitative study is important because, for the first time, they demonstrated the existence of classes of internal fields that can accommodate rapid surface spin-down at early epochs while producing almost uniform rotation in the radiative core by the solar age.

In 1996 Rüdiger and Kitchatinov have also performed a large set of spin-down calculations, making allowance for differential rotation in the convective envelope. Their work thus combines differential rotation at the base of the solar convection zone, rotational braking due to a magnetically coupled solar wind, and an axially symmetric magnetic field in the radiative core. Nearly constant angular velocity in the core occurs only if the following two conditions are met: (1) the internal magnetic field does not penetrate into the outer convective envelope, where differential rotation prevails, and (2) viscosity is strongly enhanced compared to its microscopic value. The first constraint stems from Ferraro's 1937 finding that a poloidal magnetic field actually enforces a state of *isorotation* (that is, a rotation with constant angular velocity along each field line), although the constant angular velocity is in general different for each field line. If the internal poloidal field was anchored in the differentially rotating convective envelope, the latitudinal dependence of rotation in the outer layers would thus penetrate deep into the radiative interior, which is not observed. With the magnetic field fully embedded in the core, it is therefore possible, in principle, to reproduce the helioseismological data. The problem is then presented by the "dead zone" permeated by the field lines that never come close to the base of the convective envelope. This is why a sizable amount of turbulent viscosity is needed to link this region to the base of the solar convection zone. Specifically, the Rüdiger-Kitchatinov models require a turbulent viscosity that is four orders of magnitude larger than the microscopic viscosity. There is no contradiction at this point with the Charbonneau-MacGregor models, however, because the latter also require an amplification factor of the order of 10^4 in the viscosity coefficient to enforce almost uniform rotation. In both cases, the models are quite insensitive to the magnitude of the internal magnetic field, provided the poloidal field strength is larger than 10^{-3} G.[35]

---

[35] Hopefully, these robust quantitative results will put an end to the false rumor according to which there exists a weak poloidal magnetic field (perhaps as small as 10^{-6} G) that is symmetric about the rotation axis and that can quickly enforce solid-body rotation in a stellar radiative zone, with little or no turbulent friction being present. Historically, the existence of such a false rumor is quite interesting because it shows that scientific research is far from being free from argument of authority and wishful thinking.

Heating of the Solar Corona

As explained in section 5.8, another challenging problem arises from the observation that the widths of the iron emission lines in the solar corona indicate a temperature of about 2,000,000 K. For several decades, sound waves from the turbulent photosphere provided a widely accepted explanation for the heating of the solar corona. The low chromosphere does indeed seem to be heated by sound waves generated in the photospheric convection which propagate upward into the overlying atmosphere, where they dissipate and deposit their energy as heat in the surrounding gas. However, since the Skylab observations of ultraviolet spectral lines in the 1970s, it has become clear that these waves are dissipated and/or refracted before reaching the corona. Presumably, then, coronal magnetic fields are constantly displaced and disturbed in response to motions in the solar convection zone, thereby generating Alfvén waves that may reach the corona. Until the late 1990s, however, it was generally believed that these waves would propagate right through the corona without dissipating their energy into it. A major breakthrough occurred in 1998, when the Transition Region and Coronal Explorer (TRACE) spacecraft observed a powerful flare that was triggering waves in nearby coronal loops. Since these loops oscillated back and forth several times before settling down, it was therefore concluded that Alfvén waves could indeed deposit their energy into the corona. Ground-based observations during eclipses have further revealed localized oscillations along coronal loop structures. The periods are between 2 and 10 seconds. Elsewhere, however, the instruments detected no oscillations. These observational results strongly suggest, therefore, that Alfvén waves are likely to be present but not pervasive or strong enough to dominate coronal heating.[36]

Alternatively, it has been suggested by Parker (figure 6.6) and others in the 1980s that the steady heating of the corona can be regarded as the cumulative result of small-scale, intermittent explosions, which they referred to as *nanoflares*, each of them releasing magnetic energy in small, isolated magnetic loops. These sudden and low-level bursts of energy, which occur at random locations, are primarily a consequence of thin current sheets arising spontaneously in the surface bipolar magnetic fields by the continuous shuffling and intermixing of the footpoints of the field in the photospheric convection. They are assumed to be much more numerous than conventional flares, and to combine to generate the high-temperature corona. This hypothesis is supported by observations of localized impulsive heating events throughout the solar magnetic network. High-resolution ultraviolet observations from space have revealed the presence of intense, tiny, high-speed jets of matter directed upward. Intermittent spikes of hard X rays with short individual durations have also been observed. This surface activity was further confirmed by the detection of a nonstop succession of small explosions by the Solar and Heliospheric Observatory, or SOHO, in the late 1990s. These observations strongly suggest, therefore, that the sun's hot corona might well be understood as a swarm of nanoflares.[37]

---

[36] B. N. Dwivedi and K.J.H. Phillips, *Scientific American*, **284**, No. 6, 40 (2001).

[37] See especially E. N. Parker, *Astrophys. J.*, **330**, 474 (1988). For a brief outlline of some of the sun's more conspicuous challenges to physics, see E. N. Parker, *Physics Today*, **36**, No. 6, 26 (2000).

Figure 6.6 The American solar physicist Eugene Parker in 1968. He became internationally
known in the late 1950s for his pioneering theoretical work on the solar wind, and
has since remained at the forefront in theorizing about the sun and its surroundings.
Courtesy of Dr. E. Parker.

6.4 LATE-TYPE STARS

It has long been known that, on the main sequence, large mean rotational velocities
are common among the early-type stars and that the velocities decline steeply in the
F-star region, from 50–100 km s^{-1} to less than 20 km s^{-1} in the cooler stars (see
figure 5.22). By the 1960s, because the only observational technique available was
to determine line widths in stars from photographic spectra, rotational studies were
limited almost entirely to stars more massive than the sun ($M \gtrsim 1.5 M_\odot$). Higher-
resolution spectra were therefore required to measure rotational broadening in the
less massive main-sequence stars. In 1967 Robert Kraft pushed the photographic
technique to its limit to measure $v_e \sin i$ as low as 6 km s^{-1} in solar-type stars. Now,
as early as 1933, the British scientist John CARROLL (1899–1974) had suggested
the application of Fourier analysis to spectral-line profiles for rotational velocity
determinations. In 1973 the problem was reconsidered by David GRAY (b. 1938)
from the University of Western Ontario, who showed that high-resolution data make

it possible to distinguish between the Fourier-transform profile arising from rotation versus those arising from other broadening mechanisms. Since the mid-1970s, systematic studies of very slow rotators have been made by Gray, soon to be followed by his American colleagues Myron SMITH (b. 1944) and David SODERBLOM (b. 1948), and others. Current techniques limit the measurement accuracy of projected rotational velocities to 2 km s^{-1} in most stars.

Periodic variations in the light output due to dark or bright areas on some rotating stars have also been used to determine the rotation periods of these stars. Although the principle of rotational modulation was suggested as early as 1667 by Ismaël Boulliau, convincing detection of this effect was not made until 1947, when the American astronomer Gerald KRON (b. 1913) found evidence in the light curve of the eclipsing binary AR Lacertae for surface inhomogeneities in its G5 component. Kron's result was forgotten until 1966, when interest in the principle of rotational modulation was independently revived by Pavel CHUGAINOV (1933–1992) at the Crimean Astrophysical Observatory. A large body of literature has developed since the late 1960s. This technique has the advantage that a rotation period can be determined to much higher precision than $v_e \sin i$ and is free of the projection factor inherent to the spectrographic method. Moreover, very accurate rotation periods can be derived even for quite slowly rotating stars at rates that would be impossible to see as a Doppler broadening of their spectral lines.

Stellar Activity Cycles

A different line of inquiry was initiated in the mid-1960s by Olin Wilson (figure 6.7) at the Mount Wilson Observatory, who made systematic measurements of the variation in strength of the Ca II H and K emission lines which arise in chromospheric plages on late-type stars. From time-series observations spanning a decade of ninety-one young and old dwarf stars in a range of masses, Wilson in 1978 discovered that the solar activity cycle has measurable counterparts on other lower main-sequence stars. His original survey has been continued and expanded by Sallie BALIUNAS (b. 1953) and her associates (see figure 6.8). About 60% of their sample of lower main-sequence stars has periodic, or apparently periodic, records of Ca II H and K fluxes; about 25% of their records are variable, but with no obvious periodicity; and about 15% of all records are constant with time. Cycles occur in stars of all spectral types and all ages. The upper limit of periods is longer than thirty days (under the assumption that records with systematic trends eventually reverse); but in the older stars there is a cutoff for short periods: Older stars never show cycles shorter than seven years. The stars that vary erratically with time are generally young, rapidly rotating, and magnetically active.

By the mid-1980s they had also obtained precise rotation rates from the periodic modulation of chromospheric emission produced by localized active regions moving across the stellar disk. With this technique, rotation rates can be measured precisely for equatorial velocities as slow as 1 km s^{-1}, and independently of the aspect of the rotation axis. Rotation periods as large as 40–50 days have been measured among the K- and M-type dwarf stars. In several stars, changing

Figure 6.7 The American astronomer Olin Wilson, aged sixty-nine, at work examining a
stellar spectrum. Wilson, who spent his entire career at the Mount Wilson Observatory, is best known for his spectroscopic studies of stellar chromospheres and
stellar activity cycles in lower main-sequence stars. The total reach of his chromospheric investigations has come to be known as the solar-stellar connection.
Courtesy of the Observatories of the Carnegie Institution of Washington.

or dual periodicities observed in different seasons of chromospheric records have
been ascribed to surface differential rotation. If so, then the fractional differential rotation is at least 5–20% among the more rapidly rotating and chromospher-
ically active stars. During the 1980s, several independent observers have also
produced a number of Zeeman-broadening observations. These detections are interpreted to result from magnetic fields of up to several kilogauss covering up to
several tenths of the surface area of some stars, especially the active-chromosphere
areas.[38]

---

[38]For detailed reviews of stellar rotation and/or magnetic activity, see S. L. Baliunas, and A. H.
Vaughan, *Annu. Rev. Astron. Astrophys.*, **23**, 379 (1985); A. Slettebak, in *Calibration of Fundamental
Stellar Quantities*, edited by D. S. Hayes, L. E. Pasinetti, and A. G. Davis Philip, I.A.U. Symposium No.
111 (Dordrecht: Reidel, 1985), 163; J. R. Stauffer and L. W. Hartmann, *Publ. Astron. Soc. Pacific*, **98**,
1233 (1986); L. W. Hartmann and R. W. Noyes, *Annu. Rev. Astron. Astrophys.*, **25**, 271 (1987). See also
S. L. Baliunas, R. A. Donahue, W. Soon, and G. W. Henry, in *Cool Stars, Stellar Systems and the Sun*,

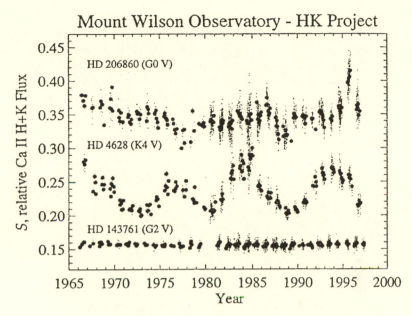

Figure 6.8 Three classes of Ca II flux variability: variable (*top*), cyclic (*middle*), and non-variable (*bottom*). The small points are nightly measurements; larger points are averaged every thirty days. From S. L. Baliunas, R. A. Donahue, W. Soon, and G. W. Henry, in *Cool Stars, Stellar Systems and the Sun*, edited by R. A. Donahue and J. A. Bookbinder, ASP Conference Series, vol. 154 (1998), 153. Courtesy of the Astronomical Society of the Pacific, San Francisco.

Observations of roughly periodic magnetic fields on the sun have led to the development of several dynamo models, in which magnetic activity results from the subsurface interaction of turbulent convection and differential rotation (see section 5.8). Until the 1960s, only the sun's magnetic activity and its periodicities could be intensively studied. The advent of sensitive electronic detectors and ultraviolet and X-ray satellite observatories made it possible to study various tracers of magnetic activity in late-type dwarf stars as well. It is now generally accepted that rotation plays a crucial role in the generation of stellar magnetic fields through a dynamo process.[39] This idea has been amply supported by observations that the average level of magnetic activity, which is always enhanced in more rapid rotators, is a function of mass and rotation (or, equivalently, age). It is also consistent with Schatzman's idea that magnetically controlled stellar winds and/or episodic mass ejections from stars with outer convection zones continuously decelerate these stars as they slowly evolve on the main sequence. However, there is now convincing observational

---

edited by R. A. Donahue and J. A. Bookbinder, ASP Conference Series, vol. 154 (San Francisco: Astron. Soc. Pacific, 1998), 153.

[39]The extreme complexity of dynamo phenomena is well illustrated by the many theoretical contributions in M. Núñez and A. Ferriz-Mas, eds., *Stellar Dynamos: Nonlinearity and Chaotic Flows*, ASP Conference Series, vol. 178 (San Francisco: Astron. Soc. Pacific, 1999).

evidence that the magnetic activity of a late-type star does not increase linearly with rotation at high angular velocities. This saturation was originally inferred in 1984 by the Finnish astronomer Osmi VILHU (b. 1944) from the observation that the chromospheric and coronal emission fluxes depend only weakly on rotation in the most rapid rotators. Moreover, observations made during the 1980s have substantially modified the old picture according to which both rotational velocities and Ca II emission always decline with advancing age as a $t^{-1/2}$ law (see section 5.7). Indeed, further complexity was added to the problem when the Dutch astronomers Floor VAN LEEUWEN (b. 1952) and Pieter ALPHENAAR (b. 1957) announced the discovery in 1982 of a number of rapidly rotating G and K dwarfs in the Pleiades, with equatorial velocities up to 150 km s^{-1}. This important result led to a flurry of interest in the rotational evolution of these low-mass stars, which spin down faster than predicted by Skumanich's 1972 law shortly upon arriving on the main sequence.[40]

Rotation of T Tauri and Cluster Stars

Another important discovery was made by the American astronomers Joanne AT-TRIDGE (b. 1966) and William HERBST (b. 1947) during their investigation of the rotational velocity properties of T Tauri stars, which are low-mass pre-main-sequence stars of age less than 10 million years. They found that the frequency distribution of rotation periods for the T Tauri stars in the Orion Nebula cluster is distinctly bimodal: There is a "rapid rotator" group with a mean period of 2.2 days and a "slow rotator" group with a mean period around 8.5 days. The following year, 1993, two independent teams found that the observed periods appear to be related to the presence or absence of a circumstellar accretion disk. Specifically, those stars that they infer to be surrounded by accretion disks (i.e., the *classical* T Tauri stars) are slow rotators with periods larger than 4 days, with a most probable period of 8.5 days, while those that lack accretion disk signatures (i.e., the *weak-line* T Tauri stars) cover a wide range of rotation periods, ranging from 1.5 to 16 days, including a significant number of objects with periods smaller than 4 days.[41]

At first sight, the presence of a circumstellar disk around some T Tauri stars poses a serious problem since they tend to accrete mass and hence angular momentum from the disk. Yet observations reveal that most of the T Tauri stars surrounded by accretion disks are rotating relatively slowly, with $v_e \sin i \lesssim 20$ km s^{-1}, which is one order of magnitude lower than their break-up velocity. Broadly speaking, two distinct angular momentum regulation mechanisms have been proposed, both of them relying on the interaction between the magnetosphere of a rotating star and

---

[40]F. van Leeuwen and P. Alphenaar, *The ESO Messenger*, No. 28, 15 (1982); O. Vilhu, *Astron. Astrophys.*, **133**, 117 (1984). For a short account of these and related problems, see T. Simon, in *Cool Stars, Stellar Systems and the Sun*, edited by R. J. García López, R. Rebolo, and M. R. Zapatero Osorio, ASP Conference Series, vol. 223 (San Francisco: Astron. Soc. Pacific, 2001), 235.

[41]J. M. Attridge and W. Herbst, *Astrophys. J. Letters*, **398**, L61 (1992); J. Bouvier, S. Cabrit, M. Fernández, E. L. Martín, and J. M. Matthews, *Astron. Astrophys.*, **272**, 176 (1993); S. Edwards, S. E. Strom, P. Hartigan, K. M. Strom, L. A. Hillenbrand, W. Herbst, J. Attridge, K. M. Merrill, R. Probst, and I. Gatley, *Astron. J.*, **106**, 372 (1993).

a circumstellar accretion disk. In 1991 Arieh Königl (b. 1951) from the University of Chicago suggested a model in which the material in the disk that spirals slowly inward moves along the closed field lines of a dipolar magnetic field, thus being channeled onto the star at high latitudes. That is to say, the dipolar field disrupts the inner parts of the disk and the central star becomes effectively coupled to the disk several radii out. This possibility was investigated by Königl, who found out that a kilogauss field could disrupt the disk at a distance of a few stellar radii from the center and that the spin-up torque applied by accreting material could be balanced exactly by the spin-down torque transmitted by the field lines that thread the disk beyond the corotation radius, where the Keplerian angular velocity of the disk matches the angular velocity of the star. Alternatively, Frank Shu and his associates have proposed a model in which shielding currents in the surface layers of the disk are invoked to prevent penetration of the stellar field lines everywhere except near the corotation radius. Exterior to this radius, matter diffuses onto field lines that bow outward, resulting in a magnetocentrifugally driven wind. Simultaneously, matter interior to the corotation radius diffuses onto field lines that bow inward and is funneled onto the star's surface. This flow actually results in a trailing-spiral configuration for the magnetic field that transfers angular momentum from the star to the disk. No commonly accepted model exists at the present time, however, since the fine details of the disk-star interaction are still to be modeled quantitatively.[42]

Nonetheless, ample evidence now exists that an accretion disk may play a fundamental role in regulating the rotation rate of a classical T Tauri star, holding its angular velocity almost fixed during its pre-main-sequence contraction. This locking results in a net transfer of angular momentum from the central star to the disk, so that the total angular momentum of the star steadily decreases in time until its regulating accretion disk is fully dissipated. If so, then the observed bimodal period distribution for T Tauri stars indicates that the fast rotators are stars that, for one reason or another, are not strongly locked to an accretion disk during their pre-main-sequence contraction. Hence, because they remain free to spin up in response to changes in moment of inertia as they contract, they also cover a wider range of rotation periods than their disk-locked counterparts.

Other clues to understanding the late pre-main-sequence/early main-sequence evolution of solar-type stars can be obtained from the rotational velocity properties of late-type stars in the α Persei cluster (age $\sim$ 50 million years), the Pleiades cluster (age $\sim$ 70 million years), and the Hyades cluster (age $\sim$ 600 million years). As was noted by Harvard astronomer John Stauffer (b. 1952) and others, the young cluster α Persei has a large number of very slowly rotating stars ($v_e \sin i < 10$ km s^{-1}) and a significant number of stars with $v_e \sin i$ greater than 100 km s^{-1}. Relatively rapid rotators ($v_e \sin i > 50$ km s^{-1}) are still present in the Pleiades among the K and M dwarfs but are nearly absent among the G dwarfs. This is in contrast to the much older Hyades cluster, in which all of the G and K dwarfs are slow rotators, although

[42]P. Ghosh and F. K. Lamb, *Astrophys. J.*, **232**, 259 (1979); **234**, 296 (1979); A. Königl, *Astrophys. J. Letters*, **370**, L39 (1991); A. C. Cameron and C. G. Campbell, *Astron. Astrophys.*, **274**, 309 (1993); F. Shu, J. Najita, E. Ostriker, and F. Wilkin, *Astrophys. J.*, **429**, 781 (1994).

there are still some late K and M dwarfs with moderate rotations ($v_e \sin i \approx 15\text{--}20$ km s^{-1}) in the cluster. The challenge for any theoretical modeling of the rotational deceleration of low-mass stars is to provide a convincing scenario that agrees with all of these observations. Since the early 1990s, much effort has been expended in trying to understand the rotational history of these stars, both before and during their main-sequence phase. Of particular interest is the pioneering work of MacGregor and his associates, who have constructed schematic evolutionary sequences while making use of a suitable parameterization for the angular momentum loss resulting from magnetized stellar winds. Perhaps the most important conclusion of their work is that a form of saturation has to be introduced to account for angular momentum loss in rapid rotators at the ages of α Persei and the Pleiades. This is in agreement with the observed saturation in chromospheric and coronal emission fluxes in the fastest rotators. Independent calculations have also shown that the saturation threshold is different for G, K, and M stars, with lower-mass stars saturating at lower angular velocities. Because lower-mass stars have deeper convective envelopes, this result seems to indicate that turbulent convection contributes significantly to the dynamo-generated magnetic fields of low-mass stars.[43]

Brown Dwarfs and Extrasolar Planets

Let us now turn to the very low mass stars, which populate the lower right-hand corner of the H-R diagram. To first approximation, these stars are among the easiest to model since they may be completely convective throughout.[44] As explained in section 3.2, they are thus well represented by a polytropic sphere of index $n = 3/2$. The existence of a lower mass limit along the main sequence was originally demonstrated in 1963 by the Indian-born astrophysicist Shiv Sharan KUMAR (b. 1939), who soon after became a faculty member of the University of Virginia in Charlottesville. Kumar pointed out that for sufficiently low mass the central temperature of a collapsing cloud of gas and dust will not become great enough to cause the hydrogen to burn at a rapid enough rate to stop the gravitational contraction. Specifically, he calculated that for $M \lesssim 0.08 M_\odot$ the contraction proceeds until it is stopped by electron degeneracy, and the configuration cools toward invisibility without ever consuming its hydrogen supply. These hypothesized degenerate configurations, which are roughly the size of Jupiter, were called black dwarfs or infrared dwarfs before the name "brown dwarf" was generally accepted by the astronomical community.

During the mid-1980s, there were a number of technological advances that enabled astronomers to begin an intensive search for these brown dwarfs. It was not until 1992, however, that a group of scientists working in Spain's Canary Islands proposed a method to help distinguish low-mass stars from brown dwarfs.[45] Their criterion exploits the fact that below a mass of about 60 Jupiter masses, a brown

---

[43] K. B. MacGregor and M. Brenner, *Astrophys. J.*, **376**, 204 (1991); D. R. Soderblom, J. R. Stauffer, K. B. MacGregor, and B. F. Jones, *ibid.*, **409**, 624 (1993); R. Keppens, K. B. MacGregor, and P. Charbonneau, *Astron. Astrophys.*, **294**, 469 (1995). See also S. Barnes and S. Sofia, *Astrophys. J.*, **462**, 746 (1996).

[44] See especially D. N. Limber, *Astrophys. J.*, **127**, 387 (1958); S. S. Kumar, *ibid.*, **137**, 1121 (1963).

[45] R. Rebolo, E. L. Martín, and A. Magazzù, *Astrophys. J. Letters*, **389**, L83 (1992).

dwarf never achieves the condition necessary to burn lithium in its core. Hence, the continued presence of lithium in an object is a sure sign that it has a substellar mass. For a number of years, this test was tried on several brown-dwarf candidates but none showed evidence of spectral lines produced by lithium. Then, in 1994, the discovery in the Pleiades of the exceptionally faint candidate PPL 15 by Stauffer and coworkers and the advent of the 10-meter Keck telescope on Mauna Kea in Hawaii provided the breakthrough: PPL 15—the 15th candidate in the Palomar Pleiades survey—was shown in February 1996 to exhibit a photospheric lithium line by a group of astronomers from the University of California at Berkeley.[46] Independently, a Canary Islands team reported the discovery in the Pleiades of the brown dwarfs Teide 1 and Calar 3, both named after Spanish observatories. The lithium detection was made in September 1996 with the newly built Keck telescope.[47] These three bodies have an inferred mass ranging from 50 to 70 Jupiter masses, with surface temperatures between 2,600 and 2,800 K. There seems little doubt that these objects are bona fide brown dwarfs, although their inferred mass is close to the upper end of the brown-dwarf range.

The first indisputable brown dwarf—Gliese 229B—was actually discovered by a group of astronomers from the California Institute of Technology and Johns Hopkins University, who had been looking for faint companions of nearby low-mass stars. Making use of the Palomar 1.5-meter telescope, they found a faint companion to the low-mass star Gliese 229. Since this object had the same proper motion as the star, they correctly inferred that it must be a companion of the star. Based on the photometry, they found that its intrinsic luminosity had to be well below that of the faintest possible star. From its infrared spectrum, which bears a remarkable resemblance to the spectrum of Jupiter, they also inferred the presence of methane in its atmosphere. Gliese 229's companion is clearly substellar since methane could not exist in the atmosphere of any star, only in much cooler atmospheres with temperatures of less than 1,200 K. At a meeting of the Cambridge Workshop on Cool Stars, Stellar Systems and the Sun in October 1995, the team therefore announced the discovery of Gliese 229B, the brown-dwarf companion to Gliese 229A.[48] By a strange coincidence, at the same meeting a Swiss team working at the Observatoire de Genève announced the discovery of the first extrasolar planet, a Jupiter-like body circling the star 51 Pegasi.[49] The following year, 1996, Jupiter-mass companions to two other solar-type stars—70 Virginis and 47 Ursae Majoris—were inferred from the observed periodic Doppler reflex motion of the primary.[50] By the end of the decade many more substellar bodies had been discovered. Within a relatively short

---

[46]J. R. Stauffer, D. Hamilton, and R. G. Probst, *Astron. J.*, **108**, 155 (1994); G. Basri, G. W. Marcy, and J. R. Graham, *Astrophys. J.*, **458**, 600 (1996).

[47]R. Rebolo, M. R. Zapatero Osorio, and E. L. Martín, *Nature*, **377**, 129 (1995); R. Rebolo, E. L. Martín, G. Basri, G. W. Marcy, and M. R. Zapatero Osorio, *Astrophys. J. Letters*, **469**, L53 (1996).

[48]T. Nakajima, B. R. Oppenheimer, S. R. Kulkarni, D. A. Golimowski, K. Matthews, and S. T. Durrance, *Nature*, **378**, 463 (1995); B. R. Oppenheimer, S. R. Kulkarni, K. Matthews, and T. Nakajima, *Science*, **270**, 1478 (1995).

[49]M. Mayor and D. Queloz, *Nature*, **378**, 355 (1995).

[50]R. P. Butler and G. W. Marcy, *Astrophys. J. Letters*, **464**, L153 (1996); G. W. Marcy and R. P. Butler, *ibid.*, **464**, L147 (1996).

span of time, the frantic search for brown dwarfs and extrasolar planets had come to a dramatic conclusion.[51]

6.5 THE PULSATION THEORY OF VARIABLE STARS (IV)

The variable stars hitherto discussed—the Cepheid variables, the RR Lyrae stars, and the long-period variables—are all of spectral type A or later (see section 2.7). Another interesting group of variable stars occurs near spectral type B2. They bear the name of "β Cephei stars" after the first known example, β Cephei, which was observed in 1906 by Edwin FROST (1866–1935) at the Yerkes Observatory.[52] Their most striking variations are those of radial velocity, but small changes of brightness are also present. Their periods are very small, from 3.5 to 6 hours, and the amplitudes of their light variation are almost always less than 0.1 magnitude. Moreover, in contrast to the periodic variations of δ Cephei, there is no appreciable phase lag between the mechanical pulsation of a β Cephei star and its brightness variations, in the sense that the maximum luminosity occurs when the star is most highly compressed (cf. figure 5.21). Another striking feature of many of the β Cephei stars is the coexistence of two or more periods, which are nearly alike and cause periodic amplitude fluctuations in the radial-velocity curve. As illustrated in figure 6.10, these stars form a narrow series in the observational H-R diagram.

Paul Ledoux (figure 6.9) in 1951 was the first to suggest that nonradial modes of oscillation may be responsible for the variability of β Canis Majoris. As is well known, the free oscillations of a spherically symmetric body correspond to the surface harmonics $Y_l^m(\theta, \varphi)$ with $|m| \leq l$, where θ is the colatitude and φ is the azimuthal angle. Because a spherical configuration has no preferred axis of symmetry, its deformations are independent of the azimuthal order m, so that to each value of the characteristic frequencies σ_l correspond $2l + 1$ displacements. In the presence of a slow rotational motion, however, this degeneracy is lifted and, to first order in the angular velocity Ω, the frequencies belonging to given values of l and m become

$$\sigma_{l,m} = \sigma_l \pm m\beta_l\Omega \qquad (m = 0, 1, 2, \ldots, l), \qquad (6.3)$$

where σ_l are the frequencies of the corresponding model in hydrostatic equilibrium and β_l are certain calculable constants which depend on the equilibrium structure. Ledoux's original theory postulates that the beat phenomenon of the β Cephei stars is the result of two different nonradial oscillations with $l = 2$ and $m = 2$ in a rotating star.

In 1962 a distinct mechanism was suggested by Chandrasekhar and Lebovitz, who pointed out that rotation couples two modes of oscillation that are purely radial and purely nonradial in a spherically symmetric configuration, that is, the lowest radial p

---

[51] For a detailed account of these and related matters, see the many review papers in R. Rebolo, E. L. Martín, and M. R. Zapatero Osorio, eds., *Brown Dwarfs and Extrasolar Planets*, ASP Conference Series, vol. 134 (San Francisco: Astron. Soc. Pacific, 1998). See also A. Boss, *Looking for Earths* (New York: John Wiley & Sons, 1998); G. Basri, *Scientific American*, **282**, No. 4, 76 (2000).

[52] The name β Canis Majoris has often been used to describe this group of variable stars. It is more correct historically to keep β Cephei as their prototype, however, since it was the first to be discovered.

Figure 6.9ㅤThe Belgian astrophysicist Paul Ledoux, a leading authority on stellar radial and
nonradial pulsations. From J. P. Swings, *Astrophys. Space Science*, **155**, 179
(1989).

mode and the lowest axisymmetric f mode. In the limit of hydrostatic equilibrium,
the corresponding characteristic frequencies are, to first approximation,

$$\sigma_0^2 = (3\gamma - 4)\frac{|W|}{I} \qquad \text{and} \qquad \sigma_2^2 = \frac{4}{5}\frac{|W|}{I}, \qquad (6.4)$$

where γ is the adiabatic exponent, W is the potential gravitational energy, and I is the
moment of inertia with respect to the star's center (see eqs. [4.19] and [3.13]). When
$\gamma = 1.6$, these two frequencies coincide and we have a case of degeneracy. We can
now picture the Chandrasekhar-Lebovitz mechanism as follows. It is first supposed
that the physical conditions prevalent in the β Cephei stars (in particular the adiabatic
exponent) are such that a degeneracy occurs between the frequencies σ_0 and σ_2.
According to the theory, rotation then removes this degeneracy and gives rise to two
axisymmetric modes of oscillation characterized by slightly different frequencies.

ㅤNo further progress was made until 1971, when Yoji Osaki calculated line-profile
variations for rotating stars undergoing nonradial oscillations and compared them
with the variations of observed line profiles in β Cephei stars. As he showed, the
use of two axisymmetric modes meets with serious difficulties in explaining the
observed variability in their line profiles. On the contrary, by identifying the two
oscillations with nearly equal periods as a wave traveling in the same direction as the

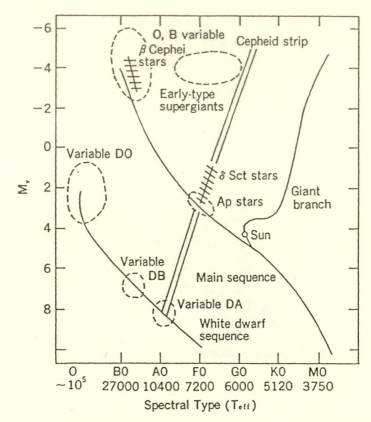

Figure 6.10 Location in the H-R diagram of stars for which nonradial oscillations are sus-
            pected to be involved and other related stars. From W. Unno, Y. Osaki, H. Ando,
            H. Saio, and H. Shibahashi, *Nonradial Oscillations of Stars*, 2nd edition (Tokyo:
            University of Tokyo Press, 1989).

rotation and a standing oscillation, Osaki found that the agreement between Ledoux's
theory and observation was fairly good. He rightfully concluded, therefore, that
nonradial oscillations may well explain most of the properties of the β Cephei stars,
in particular their multiperiod beating. If so, then what excitation mechanism is
responsible for preferentially selecting from all the modes a wave traveling around
the equator in the same direction as the rotation? This problem was discussed in 1974
by Osaki, who realized that the valve mechanism responsible for the radial pulsations
of the classical cepheids could not work in the β Cephei stars. Accordingly, he
proposed instead that the pulsation of these stars is excited by resonance between an
overstable g mode in the rapidly rotating convective core of the star and a nonradial
oscillatory mode in the radiative envelope. Subsequent work has shown that this
excitation mechanism may be considered also as an explanation of the short-period
light variations of many other massive B-type stars. As was pointed out by Myron
Smith, however, there is now ample evidence that the main pulsation mode of the

β Cephei stars is purely radial, therefore implying that both radial and nonradial modes must be involved in those β Cephei stars that exhibit the beat phenomenon.

Since the late 1960s, pulsations and oscillation-related phenomena have been observed in many stars that were hitherto regarded as nonpulsating stars.[53] In figure 6.10 we show the approximate locations on the H-R diagram of most of the relevant stars together with that of the Cepheid variables. (The RR Lyrae stars, which are not illustrated, have about $M_V \approx 0.6$ and occur in the spectral-type range A2–F6.) It is now generally believed that nonradial oscillations are responsible for the variability observed in most, if not all, of the newly discovered variable stars. They include early-type O and B variables, δ Scuti stars and rapidly oscillating Ap stars, and white-dwarf variables. The list is rapidly expanding.

The O and B variables are aligned along the main sequence between about $3 M_\odot$ and $15 M_\odot$. They can be further classified into three subclasses: the β Cephei stars, the 53 Persei variables, and the ζ Ophiuchi variables. The β Cephei stars show relatively large light- and radial-velocity variations with periods of a few hours. The other variables show mainly variations in their line profiles. The 53 Persei variables are slowly rotating, high-order g-mode pulsators with periods between 0.5 and 3 days. Their pulsations are recognized as g modes because the observed periods are larger than the expected period of the lowest radial mode. Apparent mode switching occurs in many of them within months or even days. The ζ Ophiuchi variables are characterized by rapid rotation ($v_e \sin i > 170$ km s^{-1}) and shorter oscillation periods ($\lesssim 1$ day); they include both emission-line B stars (type Be) and normal B stars. Like the 53 Persei variables, they lie within the β Cephei region on the H-R diagram. Their line-profile variations are generally interpreted in terms of sectorial nonradial modes, with spherical degree $l = |m|$ up to 16. A remarkable observational feature is that the traveling sectorial waves in 53 Persei stars run faster than rotation, whereas they run slower than rotation in ζ Ophiuchi stars.

The δ Scuti stars are located in the lower portion of the Cepheid instability strip. Their period range is limited to between 0.02 and 0.25 day, indicating low-order radial or nonradial p modes of low spherical degree l. Their pulsation amplitudes are usually small, with visual amplitudes near 0.01 magnitude, but can be as large as 0.8 magnitude. Only about one-third of them are observed to be photometrically variable, at least with amplitudes above the presently discernible limit of a few millimagnitudes. (There is also a recently discovered group of high-order g-mode pulsators—the γ Doradus stars—on the cool side of the δ Scuti instability strip.) Closer to the main sequence, at masses of about $2 M_\odot$, lies a small group of rapidly oscillating Ap-type (roAp) stars. These stars were first found in 1978 by Donald KURTZ (b. 1948) at the University of Cape Town in South Africa. Their light variation ranges from a few to roughly 50 millimagnitudes in the blue. Their multiple periods are confined to between 5 and 15 minutes. It is widely accepted that these

---

[53]For comprehensive reviews of these problems, see W. Unno, Y. Osaki, H. Ando, H. Saio, and H. Shibahashi, *Nonradial Oscillations of Stars*, 2nd edition (Tokyo: University of Tokyo Press, 1989); A. Gautschy and H. Saio, *Annu. Rev. Astron. Astrophys.*, **33**, 75 (1995); **34**, 551 (1996); S. A. Becker, in *A Half Century of Stellar Pulsation Interpretations: A Tribute to Arthur N. Cox*, edited by P. A. Bradley and J. A. Guzik, ASP Conference Series, vol. 135 (San Francisco: Astron. Soc. Pacific, 1998), 12.

short periods of oscillation are due to high-order p modes of low spherical degree l. One of the most conspicuous feature of the roAp stars is that the pulsation amplitudes are modulated in accordance with the phase of the magnetic-field strength, which itself varies with the rotation. As was shown by Kurtz in 1982, the basic pulsational behavior of these stars can be described by the so-called *oblique-pulsator model*. In this model, the pulsation is interpreted as nonradial, axisymmetric ($m = 0$) oscillations with low degree l, whose symmetry axis is coincident with the star's magnetic axis, which is itself oblique to the rotation axis. From a theoretical point of view, however, it is not yet clear how Kurtz's oblique pulsator is driven.

White-Dwarf Asteroseismology

Figure 6.10 depicts three distinct groups of white-dwarf variables. These are, in order of decreasing luminosity, the *DOV stars* (i.e., variable pre-white-dwarf stars), the *DBV stars* (i.e., variable white dwarfs with a helium envelope), and the *DAV stars* (i.e., variable white dwarfs with a hydrogen envelope, whose prototype is ZZ Ceti). The periods of their light variations, ranging from about 100 to 2,000 seconds, are consistent with the periods of nonradial g modes with low spherical degree l. The amplitudes of their luminosity variation are less than 0.3 magnitude. In most cases they show complex beat phenomena due to multiple modes being simultaneously excited. Historically, the DBV stars are of particular interest because, in contrast to the DAV and DOV stars which were found accidentally, the existence of the DBV stars was theoretically predicted by a group of North American scientists.[54] In fact, while studying the efficiency of partial hydrogen ionization ($H \rightleftharpoons H^+$) as a driving mechanism for the DAV stars, they pointed out that variable DB white dwarfs should exist in a narrow instability strip near $T \approx 19,000$ K because of the efficiency of partial helium ionization ($He^+ \rightleftharpoons He^{++}$) at these effective temperatures. After an extensive search, it was found that a DB white dwarf, GD 358, does indeed pulsate. This is one of the rare instances in which theory has preceded observation.[55]

A glance at figure 6.10 is sufficient to reveal that there are an impressive number of stars that are known to be multimode nonradial pulsators. Unfortunately, the number of pulsation modes detected in these stars is by many orders of magnitude smaller than that in the sun (cf. figure 6.5). Moreover, there is no hope, at least in the foreseeable future, of obtaining the analogue of the highly spatially resolved data of solar oscillations. The lack of spatially resolved data and the small amount of observed modes has the consequence that mode identification is difficult. This is why the asymptotic analysis of high-order modes has proven so useful in de-

---

[54] D. E. Winget, H. M. Van Horn, M. Tassoul, C. J. Hansen, G. Fontaine, and B. W. Carroll, *Astrophys. J. Letters*, **252**, L65 (1982).

[55] It might have been found for the wrong reason, however, because there is now growing evidence that it is convective driving—rather than partial ionization—that is responsible for most of the excitation of g modes in the DAV and DBV stars! See A. J. Brickhill, *Mon. Not. Roy. Astron. Soc.*, **251**, 673 (1991); A. Gautschy, H.-G. Ludwig, and B. Freytag, *Astron. Astrophys.*, **311**, 493 (1996); P. Goldreich and Y. Wu, *Astrophys. J.*, **511**, 904 (1999); Y. Wu, in *The Impact of Large-Scale Surveys on Pulsating Star Research*, edited by L. Szabados and D. W. Kurtz, ASP Conference Series, vol. 203 (San Francisco: Astron. Soc. Pacific, 2000), 508.

riving global informations about the internal structure of nonradial pulsators. It also explains why two degenerate dwarfs with over 100 excited modes each—PG 1159–035 and GD 358—are on top of the observer's list of multimode pulsating stars.

The asymptotic theory of high-order g modes in a spherically symmetric star containing both convectively stable and unstable regions was originally developed by one of us, Monique Tassoul, at a time when this kind of work was not much considered outside academic circles.[56] For high-order g modes belonging to spherical harmonics of degree l, she found that the periods are given, to first approximation, by

$$P_{l,n} = \Pi_l(n + \epsilon), \tag{6.5}$$

when n is the number of nodes in the radial direction and ϵ is a small number. Here we have let

$$\Pi_l = \frac{2\pi^2}{[l(l+1)]^{1/2}} \left(\int \frac{N(r)}{r} \, dr \right)^{-1}, \tag{6.6}$$

where $N(r)$ is the buoyancy frequency and the integral is over the region of g-mode propagation. One thus has

$$P_{l,n_1} - P_{l,n_2} = \Pi_l(n_1 - n_2), \tag{6.7}$$

which implies that the high-order g modes are approximately uniformly spaced in period, with a spacing that decreases with increasing l as $[l(l+1)]^{-1/2}$. Equations (6.5)–(6.7) are particularly useful because they allow quantitative analysis of the pulsation properties of stellar models based only on their structural properties.

Since its discovery in 1979 by a group of American astronomers, the pulsating DOV pre-white-dwarf star PG 1159–035 has undergone intense scrutiny by observers in a variety of wavelengths and has been the subject of numerous theoretical studies. In particular, observations made in Texas and South Africa from 1979 to 1984 revealed that this object pulsates in at least eight modes with periods ranging from 390 to 830 seconds.[57] Their identification with g modes was not made until 1988, however, when Steven KAWALER (b. 1958) from Yale University demonstrated that PG 1159–035 shows very strong evidence for period spacings that are integer multiples of some uniform period interval. Making use of eq. (6.7), Kawaler found that its periods show spacings that correspond very closely to theoretical models for g modes with $l = 1$ and/or $l = 3$. Since the integral in eq. (6.6) is sensitive to total stellar mass alone, he was also able to show that this property of the period spectrum of PG 1159–035 strictly constrains its mass to be about $0.6 M_\odot$.[58]

[56]M. Tassoul, *Astrophys. J. Suppl.*, **43**, 469 (1980).

[57]J. T. McGraw, S. G. Starrfield, J. Liebert, and R. F. Green, in *White Dwarfs and Variable Degenerate Stars*, edited by H. M. Van Horn and V. Weidemann, I.A.U. Colloquium No. 53 (Rochester, N.Y.: University of Rochester Press, 1979), 377; D. E. Winget, S. O. Kepler, E. L. Robinson, R. E. Nather, and D. O'Donoghue, *Astrophys. J.*, **292**, 606 (1985).

[58]S. D. Kawaler, in *Advances in Helio- and Asteroseismology*, edited by J. Christensen-Dalsgaard and S. Frandsen, I.A.U. Symposium No. 123 (Dordrecht: Reidel, 1988), 329.

The next important progress in white-dwarf asteroseismology was made in March 1989, when an international team of astronomers thoroughly monitored the rapid luminosity variations of PG 1159–035, using this time a network of photometric telescopes located at different longitudes to minimize or eliminate gaps in the observational coverage. This network, which is generally known as the Whole Earth Telescope (WET), provided what is perhaps the most revealing power spectrum ever obtained for a star. It was fully resolved into 125 individual frequencies, of which 101 have been identified with high-order g modes belonging to $l = 1$ and $l = 2$. The total stellar mass derived from the period spacings was found to be $0.586M_\odot$ and, from eq. (6.3), its rotation period 1.38 days. However, regular departures from uniform period spacing were observed, which result from resonant mode trapping induced by a discontinuity in the chemical stratification. A slow increase of the 516-second period was also found, in perfect agreement with the theoretical predictions originally made in 1983 by Donald WINGET (b. 1955), Carl HANSEN (b. 1933), and Hugh VAN HORN (b. 1938).[59] Observations of the pulsation spectrum of the DB white dwarf GD 358 with the Whole Earth Telescope have further provided rich detail on the normal mode spectrum of nonradial g modes in this star. Although the potential of asteroseismology has barely been tapped, much progress has already been made during the 1990s. There is no doubt, therefore, that this frontier science will soon become a major contributor to our understanding of the internal structure of these and other pulsating stars.

6.6 FINAL STAGES OF STELLAR EVOLUTION

As explained in section 5.4, by the early 1970s most astronomers believed that stars initially less massive than about $4M_\odot$ will lose mass to such an extent that the remnant core will become a white dwarf, whereas the ignition of carbon in the electron-degenerate core of single stars in the initial mass range 4–$8M_\odot$ will trigger a supernova event. This carbon detonation model of type I supernovae soon ran into difficulty, however, because several phenomenological studies have shown that mass loss during stellar evolution is so strong as to stop core growth before carbon burning begins. By the late 1990s it was therefore widely accepted that white dwarfs represent the ultimate fate of all stars with initial masses less than about $8M_\odot$. By then it was also known that stars having initial masses greater than about $8M_\odot$ do not develop an electron-degenerate carbon core, but evolve instead to nondegenerate carbon ignition. As we shall see below, current indications are that a dramatic implosion-explosion does indeed ensue in most of these stars, which end their lives in spectacular fashion as supernovae, leaving behind a neutron star or a black hole.[60]

---

[59] D. E. Winget, C. J. Hansen, and H. M. Van Horn, *Nature*, **303**, 781 (1983); D. E. Winget, R. E. Nather, and many other participants to the WET program, *Astrophys. J.*, **378**, 326 (1991); see also S. D. Kawaler and P. A. Bradley, *ibid.*, **427**, 415 (1994).

[60] For comprehensive reviews of these and related topics, see I. Iben, Jr., *Astrophys. J. Suppl.*, **76**, 55 (1991); S. D. Kawaler, I. Novikov, and G. Srinivasan, *Stellar Remnants*, Saas-Fee Advanced Course 25 (Berlin: Springer-Verlag, 1997).

White Dwarfs

The physical mechanisms of stellar evolution that lead to the formation of a white dwarf are fairly well understood. Following exhaustion of hydrogen in the core of main-sequence stars with masses between $0.08M_\odot$ and about $8M_\odot$, hydrogen fusion proceeds in a thin expanding shell above a rapidly contracting and heating core composed essentially of pure helium. The star swells in size and cools, becoming a red giant. Eventually, conditions in the center become so extreme that fusion reactions begin in the core helium, converting this element into carbon and oxygen. For stars with masses smaller than about $2M_\odot$, a helium runaway is initiated in the core—the so-called *helium flash*—and this runaway continues until electron degeneracy is lifted. Thereafter, these stars experience a period of quiescent core burning in a manner similar to that in stars in the intermediate mass range 2–$8M_\odot$, with the helium burning core surrounded by the hydrogen burning shell. Following exhaustion of helium in the core, a helium burning shell then begins to expand outward in the path of the hydrogen burning shell. The star swells even more, becoming a red supergiant (see figure 2.12). As this star evolves, its degenerate carbon-oxygen core contains between $0.5M_\odot$ and $1.4M_\odot$ of material in a region about the size of the planet earth! The remainder of the stellar mass lies in an increasingly tenuous envelope, reaching a radius of several hundred solar radii. At this time the gravitational binding energy of the envelope is so weak that the latter begins to evaporate into space, creating a planetary nebula around the remainder of the star. From theoretical evolution studies and empirical space density arguments, there is thus overwhelming evidence that the core of a supergiant star is a white-dwarf precursor which is slowly shedding the surrounding cocoon of the stellar envelope. These advanced stages of stellar evolution actually feature large and quasiperiodic *helium shell flashes* so that the principal energy source oscillates between helium and hydrogen shell burning.[61] Accordingly, depending on the phase of these oscillations at which the star leaves the supergiant region in the H-R diagram, it could be left with a hydrogen- or helium-rich envelope.

Parenthetically, note that the mean mass for single white dwarfs in the solar neighborhood is about $0.6M_\odot$, with a dispersion of about $0.1M_\odot$. For some poorly understood reason, the process of mass loss in white-dwarf precursors is therefore regulated by mechanisms that are tuned finely enough to leave remnants with about the same masses. Concomitant angular momentum loss during these phases is probably responsible for the fact that most white dwarfs do not rotate appreciably. Note also that large magnetic fields have been detected in some of these stars.[62] By the late 1990s the lower limit for detectable fields was about 10^4 G. The magnetic white dwarfs comprise about 4% of all known white dwarfs and are distributed more or less uniformly throughout the magnetic range 10^6–10^9 G. From kinematical and

---

[61] Helium shell flashes were first discovered in supergiant models of low-mass population II stars by Martin Schwarzschild and Richard Härm in 1965, and in supergiant models of intermediate-mass population I stars by their German colleague Alfred WEIGERT (1927–1992) in 1966.

[62] The first observational evidence of a magnetic field as large as 10^7 G was obtained by measuring circularly polarized light from the white dwarf Grw $+ 70°8247$; see J. C. Kemp, J. B. Swedlund, J. D. Landstreet, and J.R.P. Angel, *Astrophys. J. Letters*, **161**, L77 (1970). For a review of these and related matters, see G. Chanmugam, *Annu. Rev. Astron. Astrophys.*, **30**, 143 (1992).

space-density arguments, it is most likely that the magnetic white dwarfs are the descendants of the magnetic main-sequence Ap stars.

At this juncture it is worth noting that there is as yet no unanimous agreement about the origin of magnetic fields in any of the major classes of magnetic stars. Broadly speaking, two major theories have been invoked to explain the observed fields in these stars. According to the "fossil field" theory, the magnetic field existing in a star is the slowly decaying relic of the field present in the gas from which the star formed. The dynamo theory supposes that the motion of material across the magnetic lines of force of a small initial field generates electric currents which maintain this "seed" field. For the sun and late-type stars, it is widely believed that their field is dynamo generated, although the details of the mechanism are not yet fully understood. However, there is no consensus about the relative importance of fossil magnetism and dynamo maintenance in the Ap stars. These and other theories have been invoked also to explain the strong fields of white dwarfs and pulsars.[63]

Now, it has been known for several decades that white-dwarf surface abundances are peculiar by ordinary standards. Typically, their spectra are dominated by a single element, either hydrogen or helium. They thus divide into two dominant composition sequences, those with hydrogen-rich atmospheres (denoted DA) and those with helium-rich atmospheres (denoted DB and DO). DA stars have spectra that are dominated by the Balmer lines of hydrogen. DB stars show strong lines of neutral helium, whereas the DO stars show lines of singly ionized helium. The DA stars comprise about 80% of all white dwarfs hotter than 10,000 K, with the remaining 20% being referred to as helium-rich white dwarfs. The breakdown of white dwarfs by surface composition also shows remarkable changes with effective temperature. At the very hot end of the white-dwarf sequence, nearly all of them appear to be DO or DOV stars (see figure 6.10). The hottest DA stars have temperatures of about 80,000 K. As one proceeds to lower temperatures, these stars increase in number compared to the helium-rich white dwarfs until, at about 45,000 K, the DO stars disappear altogether. Below 30,000 K, however, the helium-rich DB stars emerge, to reach about 25% of all white dwarfs. While the spectral features are related to temperature changes, there remains a peculiar distribution of surface abundances as a function of effective temperature, indicating that forces are at work that cause changes in the surface composition of all white dwarfs as they cool.

The observation that white dwarfs have very pure surfaces of either hydrogen or helium was originally explained by Schatzman (figure 6.11) in 1945.[64] He demon-

---

[63]For a clear introductory report on these complex matters, see D. Moss, in *The Astronomy and Astrophysics Encyclopaedia*, edited by S. P. Maran (New York: Van Nostrand–Reinhold, 1991), 757.

[64]This work has had a long gestation, which is well worth recalling here. Evry Schatzman was born in Neuilly in 1920. He was a young physics student in Paris when Hitler's forces invaded France. An armistice was signed in June 1940, which provided for the German occupation of the northern part of France (including Paris) and the entire Atlantic coast, the rest being nominally sovereign. As early as October 1940, Jews were given especially harsh treatment. Facing the danger of being arrested and sent to a concentration camp in Germany, Schatzman fled in January 1942 to Lyon, in unoccupied France, where he completed his undergraduate studies. In July 1943 he joined the Observatoire de Haute-Provence, where he worked as a night assistant. According to his own testimony, this was an incredibly safe place to hide from the German troops, which were by then occupying the whole of France. There he started his long career in astrophysics and found that, due to the high gravitational field in white dwarfs, hydrogen

Figure 6.11 Evry Schatzman, a versatile French theoretician with a strong leaning toward phenomenology, who has contributed to many branches of astrophysics. The picture was taken in 1969. Courtesy of Dr. E. Schatzman.

strated that in stars with such high surface gravities, heavier elements will sink while lighter ones will float to the top on time scales that are relatively short compared to the stellar ages. The purity of white-dwarf atmospheres is not exactly perfect, however, since pollution may come from above, in the case of accretion from the interstellar medium, or from below in the form of diffusive processes operating in the outer layers. Detailed calculations have been performed by a number of independent groups for specialized problems, which show that several physical processes—such as gravitational settling, radiative levitation, convective mixing, and accretion—are indeed at work in the outer layers of white dwarfs.

In order to understand the relative number of hydrogen-rich and helium-rich white dwarfs at different effective temperatures, consideration of the abundances in the supergiant phase is critical. Prima facie, one might assume that helium-rich white dwarfs are formed by stars that leave the supergiant phase during a helium shell flash. These stars would shed their outer mass, begin with helium-rich surfaces, and evolve to very low temperatures with permanent helium abundances. If so, then, hydrogen-rich white dwarfs are formed when their precursors leave the supergiant

---

should float to their surface. This result was originally published in French (*Annales d'astrophysique*, **8**, 143 [1945]). Schatzman obtained his doctorate in 1946 and, a year later, went to Denmark to pursue his research on white-dwarf atmospheres. This expanded work, which was also written in French, was belatedly published in 1950 by the Copenhagen Observatory; it remained largely unknown until 1958, when it was incorporated into his influential monograph entitled *White Dwarfs*. See E. Schatzman, *Annu. Rev. Astron. Astrophys.*, **34**, 1 (1996).

phase between helium shell flashes, that is, during times of hydrogen shell burning. Although this scenario explains some of the observational features, it faces the major difficulty of explaining why we observe no helium-rich white dwarfs between 45,000 and 30,000 K. Failure to explain this gap has led to the idea that spectral evolution takes place among white dwarfs so that hydrogen-rich stars become helium-rich objects, and vice versa, as they cool. Of great interest is the general scenario suggested by Gilles FONTAINE (b. 1948) and François WESEMAEL (b. 1954) at the Université de Montréal.[65] It involves a white dwarf starting off with a predominantly helium surface, but with trace amounts of hydrogen, upon leaving the supergiant phase during a helium shell flash. This star will initially appear as a DO object. Then, as it cools, the hydrogen moves to the surface while heavier elements sink, eventually producing a hot DA star when the surface hydrogen-rich layer becomes deeper than a few times $10^{-16} M_\odot$. Eventually, subsurface convection in the helium-rich interior mixes this hydrogen back down, and the white dwarf becomes a DB star once again. In this scenario, the temperature gap in the helium-rich stars can be explained as the effective temperatures at which the hydrogen has reached the surface but has not yet been mixed down. Unfortunately, the fine details of such a model depend on convection and on the final states of supergiants—both of which are difficult to formulate with any certainty.

Another area of active interest during the 1980s and 1990s has focused on the final cooling of white dwarfs. Since these stars represent the ultimate fate of all stars with initial masses less than about $8M_\odot$, they thus hold a valuable record of the past history of the galaxy. One property of particular interest in this context is the fact that the cooling time of a white dwarf increases as its effective temperature decreases. (This was originally shown by Mestel in 1952, and has been confirmed by more realistic cooling calculations.) Accordingly, if we assume a constant rate of white-dwarf formation, many more faint white dwarfs than bright ones should be present in a given volume of space. This is indeed what the distribution of observed white dwarfs generally shows. As was found by James LIEBERT (b. 1946) from the University of Arizona and his associates, however, there is a very real deficit of low-luminosity white dwarfs: The observed luminosity function turns downward at luminosities below about $10^{-4} L_\odot$. It is commonly accepted that this deficit is caused by the finite age of the galaxy so that its oldest white dwarfs have not yet had the time to cool to invisibility, beyond the reach of our telescopes. By comparing the location of the low-luminosity cutoff in the white-dwarf distribution with theoretical cooling rates, it is therefore possible to infer the age of the oldest white dwarfs in the galaxy and, hence, a lower limit for the age of the universe. This method was originally suggested by an international team of scientists, who derived an age for the galactic disk comprised between 7 and 11 billion years.[66]

---

[65] G. Fontaine and F. Wesemael, in *The Second Conference on Faint Blue Stars*, edited by A.G.D. Philip, D. S. Hayes, and J. Liebert, I.A.U. Colloquium No. 95 (Schenectady, N.Y.: L. Davis Press, 1987), 319.

[66] D. E. Winget, C. J. Hansen, J. Liebert, H. M. Van Horn, G. Fontaine, R. E. Nather, S. O. Kepler, and D. Q. Lamb, *Astrophys. J. Letters*, **315**, L77 (1987).

Supernovae

The final decades of the twentieth century have also seen great advances in our understanding of supernovae.[67] As was noted in section 5.4, their true nature as exploding stars was first recognized in the early 1940s. For several decades, the empirical classification scheme for supernovae was based upon their optical spectra near maximum light. The first broad categories were enumerated as type I, for those that displayed no evidence of hydrogen in their spectra, and type II, for those that did exhibit hydrogen. During the 1980s, however, important subdivisions of these basic categories have been elucidated. They will be described briefly in turn.

The majority of type I supernova events have a very characteristic spectral development. Specifically, about 75% of all type I supernovae display a spectrum near maximum light having a 15,000 K black-body-type continuum with absorption lines of Si II, Ca II, S II, Mg II, and O I. Their early-time spectra are always characterized by the presence of a strong blueshifted absorption Si II line near 6150 Å, whereas their late-time spectra are dominated by Fe II emission lines. This sort of type I event has been classified as type Ia. There exist other hydrogen-deficient events which do not display the characteristic silicon feature of type Ia supernovae near peak brightness. These particular outbursts can be further differentiated by the presence or absence of strong helium lines in their spectra. They have been classified as type Ib or type Ic, depending on whether they exhibit strong helium lines or show only weak evidence for helium. Another important difference between these three types can be found in the late-time spectra: Broad emission lines of neutral oxygen and magnesium and once ionized calcium appear in type Ib and Ic, whereas iron features dominate in type Ia spectra.

Type Ia supernova events have been observed to occur in both elliptical and spiral galaxies. Like type II supernovae, the type Ib and Ic events are found in regions of spiral galaxies characterized by ongoing star formation. These evidences suggest that type Ia supernovae come from stars that live long lives before exploding, whereas the other types arise in stars of at least moderate mass. Models of light curves for type Ib and Ic events suggest that the ejected mass is about $5M_\odot$ or $6M_\odot$. This is too big to represent the conditions that lead to thermonuclear explosion, and has therefore led to the commonly accepted idea that they arise from hydrogen-denuded cores of massive stars. This is in contrast to the type II supernovae, which are stellar explosions involving stars that have not lost their hydrogen envelope. Type II supernovae actually comprise an inhomogeneous group with at least two distinct light curves (see figure 6.12). Those with an extended plateau in their postmaximum decline, producing a nearly constant luminosity for two or three months, are called type II-P; the others exhibit little or no plateau, and are called type II-L after the linear nature of their decline in logarithmic coordinates. Type II-P supernovae are thought to arise from explosions in stars with

---

[67] Supernova research is by far the most complex and intricate field of modern astrophysics. For succinct introductory reviews, see the many papers devoted to supernovae in S. P. Maran, ed., *The Astronomy and Astrophysics Encyclopedia* (New York: Van Nostrand–Reinhold, 1991), 883–904. More elaborate reviews can be found in R. McCray and Z. Wang, eds., *Supernovae and Supernova Remnants*, I.A.U. Colloquium No. 145 (Cambridge: Cambridge University Press, 1996).

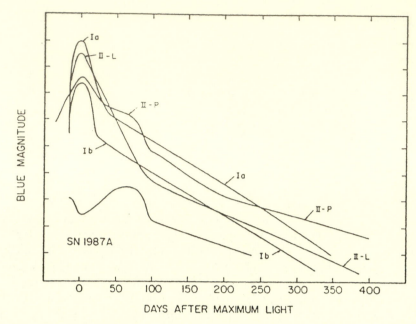

Figure 6.12 Schematic light curves for supernovae of types Ia, Ib, II-L, and II-P and supernova SN 1987A. From J. C. Wheeler and R. P. Harkness, *Rep. Prog. Phys.*, **53**,
1467 (1990).

extended hydrogen-rich red supergiant envelopes. Detailed calculations suggest
that the supergiant progenitors of type II-L supernovae are less massive that those
of type II-P.

Fred Hoyle and William Fowler in the 1960s were the first to suggest two basic
processes by which stars explode. One involves a thermonuclear runaway resulting
from the ignition of nuclear fuel in an electron-degenerate core. This process was
predicted to completely disrupt the star. The other involves the implosion of nondegenerate core material to form a neutron star. Some of the gravitational energy
liberated during this collapse was presumed to generate a shock wave which blows
off the star's outer layers. Throughout the 1970s, there was a tendency to associate
the former mechanism with type I supernovae, and the latter with those of type
II. Recent observational and theoretical advances have cast serious doubt on these
simple categorizations.

As was explained, isolated white dwarfs are simply cooling stars that eventually
end their lives as cold inert bodies. A white dwarf in a close binary system evolves
differently, however, because its companion star expands and transfers material to
the white dwarf at some point in their evolution. Thermonuclear explosions of
accreting white dwarfs have been considered to be the most promising models for
type Ia supernovae. Unfortunately, the exact binary evolution that leads to such an
event has not yet been firmly identified. The outcome of an accreting white dwarf
depends on the accretion rate $\dot{M}$, the composition of materials transferred from the
companion star, and the initial mass of the white dwarf. When a certain amount

of hydrogen is accumulated on the white-dwarf surface, hydrogen shell burning is ignited. For slow accretion rates ($\dot{M} \lesssim 10^{-8} M_{\odot}$ yr^{-1}), this burning is unstable and tends to produce flashes, which lead to the ejection of most of the accreted matter from the white-dwarf surface; the strongest flash then grows into a *nova* explosion (see section 5.5). For accretion rates in the intermediate range 10^{-8}–$10^{-6} M_{\odot}$ yr^{-1}, these flashes are of moderate strength, thereby increasing the white dwarf mass toward the $1.4 M_{\odot}$ critical mass. At this point, central densities reach about 3×10^9 g cm^{-3}, and explosive carbon burning starts in the central regions. In one particularly successful model, the flame front propagates at a subsonic speed as a deflagration wave due to heat transport across the front. In the inner regions of the star, nuclear reactions are rapid enough to incinerate the material into iron-peak elements, mostly ^{56}Ni, while in its outer layers the peak temperature is too low to complete silicon burning and thus only Ca, Ar, S, and Si are produced from oxygen burning. In the intermediate layers, explosive burning of carbon and neon synthesizes S, Si, and Mg. In the outermost layers, the deflagration wave dies so that carbon and oxygen remain unburned. Calculated light curves and synthetic spectra for this model account well for the observed light curves and spectra at both early and late times of type Ia supernovae.[68]

It is widely accepted that the progenitors of type Ib and Ic supernovae are stars more massive than about $12 M_{\odot}$ that have lost their hydrogen-rich envelope through strong winds as in Wolf-Rayet stars, or through Roche-lobe overflow in close binary systems. Since these stars evolve in the same manner as helium cores in massive stars, one thus expects that they experience core collapse to a neutron star, thereby initiating a spectacular explosion as in type II supernovae. Specifically, as the central density of a collapsing core reaches and then exceeds the density of the atomic nucleus, 2.7×10^{14} g cm^{-3}, the outer parts of the core rebound, causing a shock wave to move outward. Since energy is required to propagate this wave out of a star, another mechanism is therefore needed to transfer the energy of the compressed core to the outer layers so as to blow them off. As was originally suggested by Stirling Colgate (figure 6.13) and Richard White in 1966, such a mechanism is found in the very-high-temperature neutrinos that are emitted from the high-temperature core shock wave (see section 5.4). However, it was not until the 1980s that the viability of this mechanism was first demonstrated by James WILSON (b. 1922) from the Lawrence Livermore National Laboratory, Hans Bethe from Cornell University, and others: Although the neutrinos streaming from the hot inner core lose only a small fraction of their energy (at most a few tenths of one percent) in the outer layers, this energy deposition may be sufficient to heat matter and eject all the material exterior to the collapsed core with high velocity. Independent supernova simulations have shown that convection inside and around the forming neutron star is the crucial ingredient to neutrino-driven supernova explosions.[69] A strong shock wave can therefore find its way into the heavy-element layers surrounding the core. As this shock transits the shells of silicon and oxygen

---

[68] K. Nomoto, F. K. Thielemann, and K. Yokoi, *Astrophys. J.*, **286**, 644 (1984).

[69] For a more detailed history of supernova neutrino theory, see A. S. Burrows, in *Supernovae*, edited by A. G. Petschek (Berlin: Springer-Verlag, 1990), 144–152.

242

CHAPTER 6

Figure 6.13   The American astrophysicist Stirling Colgate at the focus of the automated supernovae search telescope on Magdalena Ridge, Langmuir Observatory, New
Mexico Institute of Mining and Technology (1970). He is best known for his
seminal role in predicting the generation of neutrinos in core collapse and elu-
cidating the importance of the neutrinos for the dynamics of type II supernova
explosions. Courtesy of Dr. S. Colgate.

just outside the core, the high temperatures it produces lead to the formation of
heavy elements in explosive nucleosyntheses. One of the most abundant of these
elements is the radioactive nucleus ^{56}Ni. This nucleus decays with a half-life of
6.1 days to ^{56}Co, which in turn decays to ^{56}Fe with a half-life of 77.2 days. Analysis of the supernova SN 1987A in the Large Magellanic Cloud has shown that
$0.075 M_\odot$ of ^{56}Ni was produced in that explosion. Detection of neutrinos from this
object has given further support to our current understanding of type II supernova
events.

Neutron Stars and Pulsars

Following Gold's 1968 original suggestion, it is now commonly accepted that pulsars are highly magnetized, rapidly rotating neutron stars which gradually convert
their stored rotational energy into electromagnetic radiation, some of which we
observe over a broad band of radio frequencies (see section 5.4).[70] The obvious as-

[70]For a comprehensive discussion of pulsar electrodynamics, see L. Mestel, *Stellar Magnetism* (Oxford: Clarendon Press, 1999), 532.

sociation between the Crab pulsar and the remains of the supernova explosion of A.D. 1054 leads naturally to the suggestion that all neutron stars originate in supernova explosions. If so, then, we might expect to find them in close proximity of known supernova remnants. Actually, only a very few of the youngest pulsars—notably the Crab and the Vela pulsars—are so associated. Direct measurements of pulsar proper motions have shown, however, that pulsars are high-velocity objects, with velocities between 30 and 300 km s^{-1}. For some reason not clearly understood, neutron stars are thus asymmetrically expelled from their birthplace fast enough to move away from the supernova remnant during its lifetime of some 30,000 years or so. The observations are consistent with neutron stars being formed during supernova explosions.

In contrast to the frequent occurrence of binary systems among main-sequence stars, the first hundred pulsars to be discovered were solitary objects. The first binary pulsar, PSR 1913+16, was discovered in 1975 by Russell HULSE (b. 1950) and Joseph TAYLOR (b. 1941) during a systematic search for new pulsars being carried out at the Arecibo Observatory in Puerto Rico.[71] It has a pulsation period of 59 milliseconds and is a member of a binary system with an orbital period of 0.32 day. Since no eclipses were observed, it was therefore concluded that the unseen companion is a compact object, most likely another neutron star. Thus for the first time it was possible to observe the gravitational interaction of a pulsar and another massive object. The Hulse-Taylor pulsar was soon to become the most important and also the most successful laboratory for verifying various predictions of Einstein's 1915 theory of gravitation (see figure 6.14). The most spectacular result of timing this radio pulsar was, of course, the clear demonstration that it is losing orbital energy (presumably by emitting gravitational radiation) at the rate predicted by the general theory of relativity, thereby providing the most direct evidence yet that gravitational radiation exists.

Another important step forward was made in 1982, when an isolated pulsar, PSR 1937+21, was discovered at the Arecibo Observatory with a pulsation period of 1.6 millisecond.[72] This is a very peculiar object indeed, since the very slow rate of decay of its period indicates that it has a characteristic age of 4×10^8 years, two orders of magnitude greater than the typical age of normal pulsars. By the end of the decade over twenty other millisecond pulsars had been discovered. Almost half of them are in low-mass binary systems, whereas binary pulsars are a minority compared to the population of solitary pulsars. The presumption is, therefore, that the binary and millisecond pulsars are in a different category from the general population. One scenario assumes that millisecond pulsars descend from low-mass X-ray binaries. Mass and angular momentum are therefore transferred from a rapidly evolving companion in the late stages of stellar evolution to an old neutron star. Here the mass transfer via an accretion disk provides the thermal energy for the intense X rays. During this phase the old neutron star, which is initially rotating slowly, will be spun up to an equilibrium state in which the corotation speed at the magnetospheric boundary equals the Keplerian speed at that distance. Once the companion star has

---

[71] R. A. Hulse and J. H. Taylor, *Astrophys. J. Letters*, **195**, L51 (1975).

[72] D. C. Backer, S. R. Kulkarni, C. Heiles, M. M. Davis, and W. M. Goss, *Nature*, **300**, 615 (1982).

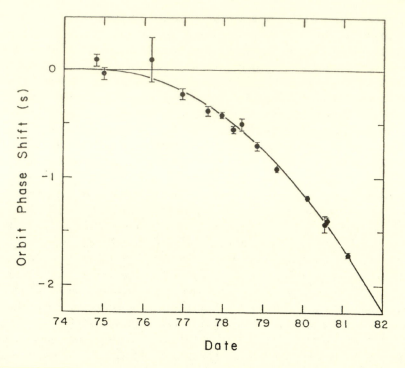

Figure 6.14 A pictorial representation of the observed change of orbital period for the Hulse-Taylor binary pulsar. If the orbital period remained constant, the points would be expected to lie on a straight line. The curvature of the parabola drawn through the points corresponds to the general relativistic prediction for loss of energy to gravitational radiation. From J. H. Taylor and J. M. Weisberg, *Astrophys. J.*, **253**, 908 (1982).

lost its envelope, accretion of matter ceases, and the now rapidly rotating neutron star begins its second lease of life as a millisecond pulsar.[73]

Observations made during the late 1980s have shown that low-mass X-ray sources and millisecond pulsars are both common in globular clusters, giving strong support to the idea that they are associated in an evolutionary sequence. If so, then how can one explain the origin of solitary millisecond pulsars? The discovery in 1988 of an *eclipsing*, binary millisecond pulsar, PSR 1957+20, at the Arecibo Observatory suggests a possible theory.[74] The period of this pulsar is 1.6 millisecond, and its orbital period is 0.38 days. The mass of the companion, presumably a white dwarf, is only about $0.01 M_\odot$, but the duration of the occultation shows that the occulting

---

[73]The first compelling evidence for this scenario did not come until the late 1990s, however, when X-ray bursts with millisecond pulsations were observed from several low-mass X-ray binaries. Since these observations strongly suggest the presence of rapidly rotating neutron stars, there is little doubt, therefore, that some of them may become millisecond radio pulsars once accretion of matter shuts off. See, e.g., L. Bildsten and T. Strohmayer, "New Views of Neutron Stars," *Physics Today*, **52**, No. 2, 40 (1999).

[74]A. S. Fruchter, D. R. Stinebring, and J. H. Taylor, *Nature*, **333**, 237 (1988).

disk is larger than the sun. This can be explained by a wind that is being ablated from the surface of the pulsar's companion, probably through bombardment by energetic charged particles from the pulsar. It seems possible, therefore, that its low-mass companion will disappear in about 10^8 years or so, leaving behind an isolated millisecond pulsar. Thus it may be that pulsars spun up to very short periods in close binary systems will eventually blow away their companions which rejuvenated them.

Black Holes and X-Ray Sources

For the sake of completeness, let us now briefly review the vexing problem of black holes in binary star systems.[75] As explained in section 5.4, the most significant hint that a hidden body in an X-ray source is a black hole is its mass. Black-hole physics tells us that there is no upper limit on the mass of such an object. This is not the case for white dwarfs and neutron stars, which have an upper mass limit of $1.4M_\odot$ and $3M_\odot$, respectively. Massive stars may lose excess material, either smoothly and continuously, to leave behind a white dwarf, or violently and explosively, to leave in some cases a neutron star. In other cases, however, it may be that the formation of a collapsing neutron star is followed by a failed explosion, so that accretion continues until the whole mass of the star falls into a black hole. The object thus formed has a mass that exceeds the largest mass that neutron stars can have. Observation suggests that both black-hole formation and explosive ejection of matter occur in nature. By the end of the twentieth century, however, a deep understanding of the causes and origins of these two distinct modes of stellar demise was still eluding us.

Since the discovery in 1972 of the first black-hole candidate, Cygnus X-1, only a handful of possible candidates with masses well above $3M_\odot$ have been found. One might therefore conclude that black holes in binary star systems are exceedingly rare objects. This is in contrast to the many hundreds of neutron stars that have been identified in binaries. A very plausible explanation for the relative paucity of black-hole binaries was originally suggested by the Dutch astronomer Edward VAN DEN HEUVEL (b. 1940) in 1983. He estimates that the evolutionary stages when an accreting black hole continuously radiates X rays may last only 10^4 years. Since we can observe such an object during this relatively short period only, he therefore concludes that the population of black-hole binaries could be much larger than we can presently detect.

In closing this section we would like to stress the fact that, observationally, all we know about these black-hole candidates is that there is an ultradense body involved whose physical properties are not known, except for bounds on its mass and radius. Although these bounds might suggest the existence of a body that has the properties of a black hole, it is important to recall that the physics of ultradense matter beyond the neutron-star stage is still poorly understood. We must also keep in mind that we do not know whether the basic equations of Einstein's general theory of relativity

---

[75]For a detailed review of this and related problems, see W.H.G. Lewin, J. van Paradijs, and E.P.J. van den Heuvel, eds., *X-Ray Binaries* (Cambridge: Cambridge University Press, 1995).

still adequately describe natural events at these extreme densities. This is perhaps best expressed in FitzGerald's translation of Omar Khayyam's celebrated quatrain:

> There was the Door to which I found no Key;
> There was the Veil through which I might not see
>   Some little talk awhile of ME and THEE
> There was—and then no more of THEE and ME.

The Persian poet Omar Khayyam (c. A.D. 1047–1122) was also a renowned astronomer and mathematician in his day.

Epilogue

It is undeniable that our scientific knowledge of the sun and the stars has progressed continuously since the early seventeenth century, when Galileo and his contemporaries first directed telescopes at the heavens. Yet it was not until the middle of the nineteenth century that pioneering scientists developed instruments and theories capable of providing detailed information about the physical and chemical nature of these objects. In this book we have attempted to present the history of solar and stellar physics from the viewpoint of the theoretician, which is why we had to quote differential equations at some places in chapters 3–6. Thus, throughout these chapters, we have presented theories as they were successively developed over the years, from about the mid-1840s to the late 1990s. Despite the fact that most of the early theoretical work had to be discarded, it is our firm belief that these failed theories are still worth considering because they shed light on the processes by which a theory is formed, then undergoes successive revisions to accommodate new facts, and is finally discarded if it can neither fit the most recent observational data nor satisfy consistency tests. Since the mid-1840s, the fields of solar and stellar physics have passed through four distinct periods, each one of them providing an increasingly detailed and robust picture of the sun, the stars, and their surroundings.

During the first period—for the most part the second half of the nineteenth century—theoretical progress was slow and depended mostly on the new science of thermodynamics. In this period, a half-dozen scientists from Germany, Scotland, and the United States came to envision the sun and stars as polytropic gas spheres in convective equilibrium, and they hypothesized that a slow but inexorable contraction of these objects was the source of heat and radiation. As we know, this simple model did not withstand the passage of time. August Ritter's 1880 idea that variable stars are pulsating in the lowest radial mode did survive, however.

The transition between the nineteenth and twentieth centuries was marked by Karl Schwarzschild's 1906 paper concerning the radiative equilibrium of the solar atmosphere. During the second period—essentially the years between the two world wars—theoreticians revised their opinion and came to the conclusion that the sun and stars are gaseous bodies with central energy sources, probably of subatomic origin, and that the energy liberated is transported radially outward by radiation. In the same period, the new science of quantum physics paved the way for a quantitative analysis of stellar spectra. No doubt due to the scientific reputation of Arthur Eddington and his closest competitors, Cambridge University in England was then the mecca of theoretical stellar studies. Recall, however, that important contributions were also made during this period in Germany, in the Scandinavian countries, and in the United States. In this respect, the effects of World War I should not be underestimated, since it was the lack of communication between enemy countries during the war that forced Vilhelm Bjerknes and his associates in Norway to develop new weather forecasting methods. This led to important progresses in hydrodynamics which, in turn, had a definite influence on the nascent sciences of solar and stellar physics.

The eve of World War II marks the beginning of the third period, which lasted for about three decades. During the 1940s, the Cambridge leadership rapidly passed on to a few select universities in the United States, where the development of fast electronic computers made it possible to carry out extensive numerical calculations. Advances in nuclear physics by Hans Bethe and others enabled some dozen or so theoreticians, several of them refugees from Hitler's Germany and German-occupied territories, to investigate the structure and evolution of stars as they shine by fusing light chemical elements into heavier ones in their central regions. Martin Schwarzschild and coworkers at Princeton University were at the forefront in theorizing about these matters. Solar physics also progressed at a very rapid pace during the postwar era, culminating in the 1950s with Eugene Parker's epoch-making work on the solar wind. During this period, despite the devastating effects of World War II, Germany and Great Britain remained at the forefront of theoretical stellar studies, while Japan and the Soviet Union rapidly developed a solid tradition in these research areas. By the early 1970s, then, solar and stellar physics had become two mature sciences.

In all human ventures, whether it be in the arts or in the sciences, there occur from time to time epochs that can fitly be called "golden ages." These epochs never last for long, however, and the three glorious decades 1940–1970 are no exception: by the late 1960s it was already apparent that enquiries into stellar structure and evolution had reached the point of diminishing returns. As we were entering the 1970s, several theoreticians turned their attention to galaxies and cosmology—a field that had been recently made fashionable again—while others embarked on the study of intricate problems that had been hitherto neglected. Fortunately, the open-end period from 1970 onward has seen great technical advances, which in turn have led to an impressive accumulation of new information about the sun and stars. International collaboration has become the norm, therefore, witnessed by the creation of the Global Oscillation Network Group (GONG), the Whole Earth Telescope (WET), the Solar Heliospheric Observatory (SOHO), and many other collaborative endeavors. As was expected, such a large collection of new results led to frequent collaborations between observers and theoreticians, so that the latter can now include a growing number of increasingly precise parameters in their descriptions of the sun and stars. The fact remains, however, that progress in our understanding of fluid motions in these objects has lagged well behind the progress generated by the gathering of new and often unexpected observational results. Until such time as an adequate hydrodynamical theory of the phenomena of turbulence and convection has been formulated, there is thus no hope of describing the sun and stars from first principles alone.

Not too surprisingly, wealthy North America has retained its supremacy during the past three decades but, as can be seen in chapter 6, the geographical distribution of the most active centers of solar and stellar studies has changed considerably. The fact is that there has been a dramatic increase in the number of astronomy graduates since the 1960s. A shortage of permanent academic positions was therefore inevitable, which resulted in a rapid turnover of young researchers in the field. Publication practices have also changed over the years, since multiauthor publications have rapidly become the rule rather than the exception. In the limit, one can thus imagine that this will lead, in the not too distant future, to a situation not too different from that prevailing in some branches of physics; that is, a situation in which research has become a collective and largely anonymous activity, with only the names of a few highly visible spokesmen being remembered by the scientific community. This has not yet happened, has it?[1]

---

[1]To those who like to be kept informed of recent developments in a most informal manner, we recommend reading the annual chronicles by Virginia Trimble and Markus Aschwanden (*Publ. Astron. Soc. Pacific*, **111**, 385 [1999]; **112**, 434 [2000]; **113**, 1025 [2001]).

Appendix A

Lane's Fully Convective Gas Spheres

Lane's formulation of the problem is surprisingly simple.[1] Making use of eqs. (3.3) and (3.6) and integrating eq. (3.5) over the sphere of radius r, he obtained

$$\frac{\gamma}{\gamma - 1} \mathcal{R}_g T_c \left[1 - \left(\frac{\rho}{\rho_c} \right)^{\gamma - 1} \right] = G \int_0^r \frac{m(r)}{r^2} \, dr, \tag{A.1}$$

where the subscript "c" denotes the value at the center $r = 0$. Letting

$$r = ax \qquad \text{with} \qquad a = \left(\frac{\gamma}{\gamma - 1} \frac{\mathcal{R}_g T_c}{4\pi G \rho_c} \right)^{1/2}, \tag{A.2}$$

he next rewrote eq. (3.4) in the form

$$m(ax) = 4\pi \rho_c a^3 \mu(x), \tag{A.3}$$

where

$$\mu(x) = \int_0^x \frac{\rho}{\rho_c} x^2 \, dx, \tag{A.4}$$

so that eq. (A.1) became

$$1 - \left(\frac{\rho}{\rho_c} \right)^{\gamma - 1} = \int_0^x \frac{\mu(x)}{x^2} \, dx. \tag{A.5}$$

Lane integrated eqs. (A.4) and (A.5) by successive approximations. The relation between the central density ρ_c and the mean density $\bar{\rho}$ follows at once from eqs. (A.2) and (A.3). Indeed, when evaluated at $r = R$ and $m = M$, these equations imply that

$$\rho_c = \frac{M}{4\pi R^3} \frac{x_0^3}{\mu_0} = \frac{1}{3} \bar{\rho} \frac{x_0^3}{\mu_0}, \tag{A.6}$$

where x_0 and μ_0 are the values of x and μ corresponding to $\rho = 0$, or the outer boundary of the supposed atmosphere.

---

[1] *Amer. J. Sci. Arts*, 2nd ser., **50**, 57 (1870).

Appendix B

Ritter's Polytropic Gas Spheres

Ritter's method of resolution of eqs. (3.3)–(3.5) and (3.8) can be described as follows.[1] Combining eqs. (3.3) and (3.8), one readily sees that

$$\rho = \rho_c \left(\frac{T}{T_c}\right)^n \qquad \text{and} \qquad p = \mathcal{R}_g \rho_c T_c \left(\frac{T}{T_c}\right)^{n+1}, \tag{B.1}$$

where the subscript "c" denotes a central value. Making use of these two relations, we can rewrite eq. (3.5) and the derivative of eq. (3.4) in the forms

$$\frac{Gm}{r^2} = -(n+1)\mathcal{R}_g T_c \frac{d}{dr}\left(\frac{T}{T_c}\right) \tag{B.2}$$

and

$$\frac{dm}{dr} = 4\pi\rho_c r^2 \left(\frac{T}{T_c}\right)^n. \tag{B.3}$$

Next, letting

$$y = \frac{T}{T_c} \qquad \text{and} \qquad x = \frac{r}{a}, \tag{B.4}$$

where

$$a = \left[\frac{(n+1)\mathcal{R}_g T_c}{4\pi G\rho_c}\right]^{1/2}, \tag{B.5}$$

and eliminating the variable m from eqs. (B.2) and (B.3), we eventually obtain the equation

$$\frac{d^2 y}{dx^2} + \frac{2}{x}\frac{dy}{dx} + y^n = 0, \tag{B.6}$$

from which the temperature distribution in a polytropic gas sphere can be obtained. The physical boundary conditions on its solutions are $y = 1$ and $dy/dx = 0$ at $x = 0$. One can also show that

$$\rho_c = \frac{1}{3}\bar{\rho}\left(\frac{x}{|dy/dx|}\right)_{x=x_0}, \tag{B.7}$$

which is strictly equivalent to Lane's eq. (A.6).

---

[1] *Ann. Phys. Chemie*, **11**, 332 (1880).

Appendix C

Ritter's Theory of Pulsating Stars

Following Ritter's analysis closely,[1] we shall describe the radial oscillations of a spherically symmetric star by means of the displacement ξ, which is the change that a mass element experiences relative to a mean state of equilibrium. The relative displacement η is, therefore,

$$\eta = \frac{\xi}{r_0} = \frac{r - r_0}{r_0},\qquad\text{(C.1)}$$

where the subscript "0" indicates an undisturbed value. By virtue of Newton's second law, eq. (3.1), the fundamental equation governing the radial oscillations of a star is

$$\frac{d^2\xi}{dt^2} = -\frac{Gm}{r^2} - 4\pi r^2 \frac{dp}{dm},\qquad\text{(C.2)}$$

where m is the mass inside the sphere of radius r (see eq. [3.4]). In accordance with eq. (3.5), the mean state of equilibrium satisfies the condition

$$\frac{Gm}{r_0^2} + 4\pi r_0^2 \frac{dp_0}{dm} = 0\qquad\text{(C.3)}$$

since $dm = 4\pi r^2 \rho\, dr = 4\pi r_0^2 \rho_0 dr_0$.

Assuming next that the relative displacement does not depend on the radial coordinate, one has

$$r = r_0[1 + \eta(t)],\qquad\text{(C.4)}$$

which corresponds to homologous expansions and contractions (i.e., $\xi \propto r_0$). Making use of eq. (C.4), we can thus write

$$\frac{Gm}{r^2} = \frac{Gm}{r_0^2}(1 + \eta)^{-2}.\qquad\text{(C.5)}$$

Since a volume element necessarily varies as $(1+\eta)^3$, conservation of mass further implies that

$$\rho = \rho_0(1 + \eta)^{-3};\qquad\text{(C.6)}$$

hence, by virtue of eq. (3.6), one has

$$p = p_0(1 + \eta)^{-3\gamma} \qquad\text{and}\qquad dp = dp_0(1 + \eta)^{-3\gamma}.\qquad\text{(C.7)}$$

---

[1] *Ann. Phys. Chemie*, **8**, 157 (1880).

Combining eqs. (C.2)–(C.5) and (C.7), we obtain

$$\frac{d^2\xi}{dt^2} = -\frac{Gm}{r_0^2}\left[(1+\eta)^{-2} - (1+\eta)^{2-3\gamma}\right]. \tag{C.8}$$

For small-amplitude oscillations ($\eta \ll 1$), this equation can be rewritten in the form

$$\frac{d^2\xi}{dt^2} + (3\gamma - 4)\frac{Gm}{r_0^3}\xi = 0. \tag{C.9}$$

If we further assume that the density ρ_0 is uniform throughout the model, one readily sees from eq. (3.4) that $m \propto r_0^3$. If so, then eq. (C.9) implies that

$$\frac{d^2\eta}{dt^2} + (3\gamma - 4)\frac{GM}{R^3}\eta = 0, \tag{C.10}$$

where M is the total mass and R is the outer radius. This equation, which is exact, was obtained for the first time by Ritter in 1880. When $\gamma > 4/3$, it describes periodic oscillations with frequency σ, where

$$\sigma^2 = (3\gamma - 4)\frac{GM}{R^3}. \tag{C.11}$$

This is also a good albeit approximate solution for centrally condensed stars.

Appendix D

Radial and Nonradial Stellar Pulsations

In appendix C we considered the homologous expansions and contractions of a spherically symmetric star (see eq. [C.4]). Following Eddington's original discussion,[1] here we shall relax Ritter's assumption and investigate the properties of a more general class of radial oscillations. In agreement with appendix C, the relative displacement has the form

$$\eta = \frac{\xi}{r_0} = \frac{r - r_0}{r_0}. \tag{D.1}$$

We shall also let

$$p_1 = \frac{\delta p}{p_0} = \frac{p - p_0}{p_0} \quad \text{and} \quad \rho_1 = \frac{\delta \rho}{\rho_0} = \frac{\rho - \rho_0}{\rho_0}, \tag{D.2}$$

where p_0 and ρ_0 denote the undisturbed values of the pressure and density, respectively. We shall consider small oscillations and neglect the square of their amplitude. If the period of pulsation is $2\pi/\sigma$, then the quantities p_1, ρ_1, ξ, and η will contain the factor $\cos \sigma t$.

From the condition of mass conservation, $r^2 \rho \, dr = r_0^2 \rho_0 \, dr_0$, we obtain

$$\frac{\delta \rho}{\rho_0} + 2 \frac{\xi}{r_0} + \frac{d\xi}{dr_0} = 0, \tag{D.3}$$

so that, by virtue of eqs. (D.1) and (D.2),

$$\rho_1 = -2\eta - \frac{d}{dr_0} (r_0 \eta) = -3\eta - r_0 \frac{d\eta}{dr_0}. \tag{D.4}$$

For adiabatic changes, eq. (3.6) implies that

$$p_1 = \gamma \rho_1. \tag{D.5}$$

Combining eqs. (D.4) and (D.5), we can thus write

$$p_1 = -\gamma \left(3\eta + r_0 \frac{d\eta}{dr_0} \right), \tag{D.6}$$

where γ is the adiabatic exponent.

The basic equation of motion, eq. (C.2), has the form

$$4\pi \frac{dp}{dm} = -\frac{Gm}{r^4} + \frac{\sigma^2}{r^2} \xi. \tag{D.7}$$

---

[1] *Mon. Not. Roy. Astron. Soc.*, **79**, 2 (1918).

For small-amplitude motions, this equation becomes

$$4\pi \frac{d}{dm}(p_0 + \delta p) = -\frac{Gm}{r_0^4}\left(1 - \frac{4\xi}{r_0}\right) + \frac{\sigma^2}{r_0^2}\xi. \tag{D.8}$$

Making use of eqs. (C.3) and (D.2), we thus have

$$4\pi \frac{d}{dm}(p_0 p_1) = \left(\frac{4Gm}{r_0^4} + \frac{\sigma^2}{r_0}\right)\eta, \tag{D.9}$$

since $\xi = r_0\eta$. This equation is equivalent to

$$p_0\frac{dp_1}{dr_0} - p_0 g_0 p_1 = p_0\left(4g_0 + \sigma^2 r_0\right)\eta, \tag{D.10}$$

where $g_0 = Gm/r_0^2$.

Next, eliminating p_1 from eqs. (D.6) and (D.10), we obtain

$$\frac{d^2\eta}{dr^2} + \frac{4 - \mu}{r}\frac{d\eta}{dr} + \left[\frac{\sigma^2\rho}{\gamma p} - \left(3 - \frac{4}{\gamma}\right)\frac{\mu}{r^2}\right]\eta = 0, \tag{D.11}$$

where $\mu = \rho g r / p$, which has the dimensions of a pure number. (Note that, without confusion, we have deleted the subscript "0.") This equation must be solved with the following boundary conditions: $\xi = 0$ at the center $r = 0$, and $\delta p = 0$ at the free surface $r = R$. This is an eigenvalue problem which admits a solution (eigenfunction) only for certain values (eigenvalues) of the parameter σ^2. When $\gamma > 4/3$, there is an infinite discrete set of positive eigenvalues: $0 < \sigma_0^2 < \sigma_1^2 < \sigma_2^2 < \cdots$; to these corresponds a complete set of eigenfunctions, with the number of nodes along the radius varying in proportion to the order of the mode.

For example, if we assume that the density is uniform throughout the equilibrium model, the eigenvalues are

$$\sigma_j^2 = \frac{4}{3}[(3\gamma - 4) + j(2j + 5)\gamma]\pi G\rho, \tag{D.12}$$

with $j = 0, 1, 2, \ldots$. This exact solution was originally obtained by Theodore Eugene STERNE (1907–1970) at Harvard College Observatory.[2] As was expected, the lowest eigenvalue is strictly equivalent to Ritter's solution, eq. (C.11), and corresponds to the eigenfunction $\eta = 1$ (i.e., $\xi \propto r \cos \sigma_0 t$).

At this juncture it is worth noting also the solution obtained by Pekeris for the nonradial oscillations of a homogeneous compressible fluid sphere. He found[3]

$$\sigma_{l,j}^2 = \frac{4}{3}\left\{D_j \pm \left[D_j^2 + l(l + 1)\right]^{1/2}\right\}\pi G\rho, \tag{D.13}$$

where

$$2D_j = (3\gamma - 4) + [j(2j + 5 + 2l) + 2l]\gamma \tag{D.14}$$

and $j = 0, 1, 2, \ldots$. Thus, for any given spherical harmonic of order l, we have two discrete spectra—one of positive eigenvalues tending toward infinity as j increases,

[2] *Mon. Not. Roy. Astron. Soc.*, **97**, 582 (1937).
[3] *Astrophys. J.*, **88**, 189 (1938).

and another of negative eigenvalues tending toward zero as j increases. (In the limiting case $l = 0$, one recovers of course eq. [D.12].)

As was shown by Cowling,[4] the positive values obtained by Pekeris correspond to stable oscillations of very short period, with motions which are chiefly radial. They are usually called *acoustic modes* (or p modes). The origin of the negative values can be traced back to the fact that in a compressible model with constant density the temperature lapse rate is superadiabatic (see eq. [4.15]), so that convective motions always prevail in such a system. In the absence of convection, however, these modes would have been stable, with largely horizontal motions which are due to the action of gravity in attempting to smooth out the density differences on the equipotential surfaces. They are usually called *gravity modes* (or g modes). Actually, Ledoux and Smeyers in 1966 showed that these modes split into two discrete sets of modes when the system has both convectively stable and unstable regions, a stable one and an unstable one.[5] Cowling in 1941 also showed that there exists a third set of oscillations, the *Kelvin modes* (or f modes), which mimic the volume-preserving oscillations of an incompressible fluid sphere (see eq. [3.13]).

Later, in 1968, these three classes of modes were complemented by the so-called *toroidal modes* as a result of a group-theoretical study by Jean PERDANG (b. 1940) at the Institut d'Astrophysique de Liège.[6] In a spherically symmetric star these additional modes correspond to trivial displacements with vanishing frequencies ($\sigma = 0$). As a rule, however, these trivial eigenvalues give rise to nonzero frequencies under the influence of rotation, tides, and magnetic fields. For example, in the presence of a slow rotational motion, one obtains nonradial oscillations with periods of the order of the rotation period. Four classes of modes are therefore essential for the description of motions in a rotating star.

---

[4] *Mon. Not. Roy. Astron. Soc.*, **101**, 367 (1941).

[5] *Comptes-rendus Acad. Sci. Paris*, **262**, 841 (1966).

[6] *Astrophys. Space Science*, **1**, 355 (1968).

Appendix E

Bohr's Model of the Atom

In its simplest form, Bohr's theory[1] embodies the following principles:

1. In a system consisting of a positively charged nucleus of very small dimensions and a single electron describing closed orbits around it, the electron is in a state of relative circular motion under the action of their mutual electrical attraction.

2. The system may remain for extended periods of time in a given state without radiating electromagnetic energy, provided the electron moves in an orbit for which its orbital angular momentum is an integral multiple of Planck's constant h divided by 2π.

3. Radiation is emitted whenever an electron moves discontinuously from one of the allowed states of energy E_i to another of lower energy E_f. When radiation is emitted, its frequency is determined by Einstein's condition $h\nu = E_i - E_f$.

For the sake of simplicity, we shall initially assume that the mass of the electron, m, is negligible compared to the mass of the nucleus, M, so that the nucleus remains fixed in space.

Consider a system consisting of an electron of charge $-e$ in circular orbit about a nucleus of charge $+Ze$. Following Bohr's first postulate, the condition of mechanical equilibrium of the electron is

$$\frac{Ze^2}{r^2} = m\frac{v^2}{r}, \tag{E.1}$$

where v is the velocity of the electron on its orbit, and r is the radius of the orbit. (Compare with eq. [2.8].) This equation merely expresses the fact that the centripetal force maintaining the circular orbit, mv^2/r, is provided by the electrical attraction between the electron and the nucleus.[2] Bohr's second postulate implies the additional constraint

$$mvr = n\hbar \qquad (n = 1, 2, 3, \ldots) \tag{E.2}$$

---

[1] *Phil. Mag.*, 6th ser., **26**, 1 (1913). This and related papers are reprinted with an introduction by L. Rosenfeld in Bohr's posthumous book *On the Constitution of Atoms and Molecules* (Copenhagen: Munksgaard, 1963).

[2] The first effective verification of this electrical attractive force was made by the French engineer Charles Augustin DE COULOMB (1736–1806) in 1785.

where $\hbar = h/2\pi$. Combining these two equations, we thus have

$$r = n^2 \frac{\hbar^2}{me^2 Z} \qquad (n = 1, 2, 3, \ldots). \qquad (E.3)$$

The application of Bohr's second postulate has therefore restricted the permitted orbits to those of radii given by eq. (E.3). Note that these radii increase as n^2, with the smallest one occurring when the quantum number n equals 1. Inserting the known values of h, m, and e, one finds that the first Bohr's orbit has the radius $r = 5.3 \times 10^{-9}$ cm for a hydrogen atom ($Z = 1$). This is in good agreement with the estimate that the order of magnitude of an atomic radius is 10^{-8} cm.

Next we calculate the total energy of an electron moving in one of its allowed orbits. In direct analogy to the case of universal gravitation, we may write

$$E = \frac{1}{2}mv^2 - \frac{Ze^2}{r}. \qquad (E.4)$$

Making use of eqs. (E.1) and (E.3), we obtain

$$E = -\frac{1}{n^2}\frac{me^4 Z^2}{2\hbar^2} \qquad (n = 1, 2, 3, \ldots). \qquad (E.5)$$

Again inserting the known values of h, m, and e, one finds that $E = -13.6(Z^2/n^2)$ in electron volts. (One eV equals about 1.6×10^{-12} erg.) Note that the orbit corresponding to $n = 1$ has the lowest possible energy, and this is called the ground state. For hydrogen, the ground state has an energy of -13.6 eV on the scale in which zero energy corresponds to escape velocity for the electron.

Making use of eq. (E.5), we can also calculate the frequency of the radiation corresponding to the transition from orbit n_i to n_f. Following Bohr's third postulate, we have

$$\nu = \frac{E_i - E_f}{h} = \frac{me^4}{4\pi \hbar^3} Z^2 \left(\frac{1}{n_f^2} - \frac{1}{n_i^2} \right). \qquad (E.6)$$

In terms of the wave number $k = 1/\lambda = \nu/c$, this becomes

$$\frac{1}{\lambda} = \frac{me^4}{4\pi c \hbar^3} Z^2 \left(\frac{1}{n_f^2} - \frac{1}{n_i^2} \right), \qquad (E.7)$$

where c is the speed of light.

As we recall, in deriving eq. (E.7) we have explicitly assumed the mass of the electron to be infinitely small compared to the mass of the nucleus (i.e., $m \ll M$). For the case of a finite mass ratio m/M, one can show that all the equations are identical with those derived above, except that the electron mass m is replaced by the reduced electron mass $m/(1 + m/M)$. In particular, the formula for the wave numbers of the spectral lines becomes

$$\frac{1}{\lambda} = R_M Z^2 \left(\frac{1}{n_f^2} - \frac{1}{n_i^2} \right), \qquad (E.8)$$

where

$$R_M = \frac{mM}{m + M} \frac{e^4}{4\pi c\hbar^3} \tag{E.9}$$

is the Rydberg constant for a nucleus of mass M.

Equation (E.8) is the principal measure of the success of Bohr's theory of the one-electron atom: it has exactly the form of eq. (4.1) deduced by Balmer for the spectral lines of hydrogen that fall in the visible part of the spectrum. (These lines correspond to $n = 2$ and $m \geq 3$ in eq. [4.1].) The success of Bohr's theory was particularly impressive in the 1910s because it not only predicted the general form of other series of lines but gave a quantitatively accurate value for Rydberg's constant, which was previously merely an empirical constant. If we evaluate that constant for hydrogen from eq. (E.9), we obtain $R_M = 109,681$ cm^{-1}, which agrees with the spectroscopic data to within three parts in 100,000. Note that Bohr's theory works equally well when applied to one-electron atoms with $Z = 2$, i.e., singly ionized helium atoms. In fact, the spectrum of ionized helium is very similar to that of neutral hydrogen ($Z = 1$), except that all lines are reduced in wavelength by almost exactly a factor of 4. This is explained very easily in terms of the Bohr atom, by setting $Z^2 = 4$ in eq. (E.8).

Appendix F

Einstein's Mass-Energy Relation

Newton's second law of motion, which we have already expressed by eq. (3.1), has the form

$$\mathbf{f} = \frac{d\mathbf{p}}{dt}, \tag{F.1}$$

where $\mathbf{f}$ is the applied force and

$$\mathbf{p} = m\mathbf{v} \tag{F.2}$$

is the momentum of the particle. Specifically, this vectorial equation is valid only if the frame of reference defined by the coordinates x, y, z, t (say) is an *inertial frame*, that is, a frame of reference in which a body not under the influence of forces, and initially at rest, will remain at rest. Once an inertial frame has been found, one can show that eq. (F.1) holds in any frame of reference moving uniformly with respect to the first. Consider a (primed) frame moving in the x direction with a constant speed u relative to our inertial (unprimed) frame. We thus have

$$x' = x - ut, \quad y' = y, \quad z' = z, \quad \text{and} \quad t' = t, \tag{F.3}$$

which is known as a *Galilean transformation*. Substituting this transformation into eqs. (F.1) and (F.2), one readily sees that Newton's second law transforms to the same law in the primed coordinates. That is to say, the basic law of classical mechanics remains invariant in form with respect to uniform translations, so that it is impossible to tell whether an inertial frame of reference is moving or not. This is the Newtonian principle of relativity.

At the end of the nineteenth century, the theory of physics was based upon two sets of equations: (a) Newton's laws of motion, which govern mechanical phenomena, and (b) Maxwell's equations of the electromagnetic field, which describe electricity, magnetism, and light in one uniform system. However, whereas all inertial frames were equivalent as far as Newtonian mechanics was concerned, in regard to electromagnetic phenomena they were not equivalent. Actually, it was known at that time that Maxwell's equations were invariant in form with respect to the following transformation of coordinates:

$$x' = \frac{x - ut}{\sqrt{1 - u^2/c^2}}, \quad y' = y, \quad z' = z, \tag{F.4}$$

and

$$t' = \frac{t - ux/c^2}{\sqrt{1 - u^2/c^2}}, \tag{F.5}$$

where c is the speed of light, which is about 300,000 km s^{-1}. These equations are known as the *Lorentz transformation* equations.

The foregoing results led Poincaré to suggest that Maxwell's equations as well as Newton's equations should be the same in all inertial frames of reference, despite the fact that these frames may be in uniform translation with respect to each other. Since Maxwell's equations did not seem to obey the Newtonian principle of relativity, Einstein then proposed that all the physical laws should be of such kind that they remain invariant in form with respect to a Lorentz transformation. In other words, he suggested that we should change, not the laws of electromagnetism (including the fact that the speed of light must have the same value c in either frame of reference), but the laws of classical mechanics. This can be achieved by retaining the form of eq. (F.1), provided one replaces the constant mass in eq. (F.2) by the expression

$$m = \frac{m_0}{\sqrt{1 - v^2/c^2}}, \tag{F.6}$$

where the rest mass m_0 represents the mass of a body that is not moving, and v is the magnitude of the velocity $\mathbf{v}$. This is Einstein's modification of Newton's second law of motion.[1] It is immediately apparent that, as long as v is small compared to c, the relativistic momentum is nearly the same as in Newtonian mechanics. But when v approaches c, the square-root expression in eq. (F.6) approaches zero so that the relativistic mass and hence the momentum both go to infinity.

Now, in prerelativity days scientists had established two separate conservation principles: conservation of mass and conservation of energy. The new theory of special relativity, as it became known, suggested the possibility of converting mass into energy, and conversely. This was properly demonstrated by Einstein in 1907.[2]

Consider a particle, initially at rest, which is accelerated up to a velocity v_f by a force of magnitude f in the x direction. The rate of change of energy with time equals the force times the velocity. Making use of eq. (F.1), we can write

$$\frac{dT}{dt} = fv = \frac{dp}{dt}v = \frac{d}{dt}(pv) - p\frac{dv}{dt}. \tag{F.7}$$

The total work done on the particle is, therefore,

$$T = [pv]_{v=0}^{v=v_f} - \int_0^{v_f} p\,dv. \tag{F.8}$$

Substituting next for the momentum from eqs. (F.2) and (F.6), one can easily perform the integration in eq. (F.8). This leads to the following result:

$$T = \frac{m_0 c^2}{\sqrt{1 - v^2/c^2}} - m_0 c^2, \tag{F.9}$$

---

[1] *Annalen der Physik*, **17**, 891 (1905), trans. in *The Collected Papers of Albert Einstein*, vol. 2 (Princeton, N.J.: Princeton University Press, 1989), 140.

[2] *Annalen der Physik*, **23**, 371 (1907), trans. in *The Collected Papers of Albert Einstein*, vol. 2 (Princeton, N.J.: Princeton University Press, 1989), 238.

where, without confusion, we have omitted the subscript "f." For small values of the ratio v/c, this is equivalent to

$$T = m_0 c^2 \left(1 + \frac{1}{2} \frac{v^2}{c^2} + \cdots - 1 \right) \approx \frac{1}{2} m_0 v^2, \tag{F.10}$$

which agrees with the classical expression for the kinetic energy.

Identifying T with the *relativistic kinetic energy* of the particle and using eq. (F.6), we can rewrite eq. (F.9) in the compact form

$$mc^2 = T + m_0 c^2. \tag{F.11}$$

Obviously, mc^2 is some energy associated with the particle when its speed is v, whereas $m_0 c^2$ is some energy associated with the particle when it is at rest. The energy mc^2 is called the *total relativistic energy*. Einstein interpreted the other term, $m_0 c^2$, to be part of the total energy of the particle, an intrinsic energy known as the *rest mass energy*. This is not an abstract result, however, since it was translated into practice in the form of an atomic bomb in 1945. Nuclear reactors also generate energy from the masses of the reacting particles. And it is the same principle that governs the generation of energy inside the stars for extended periods of time.

Appendix G

Three Important Nuclear Reactions

In the temperature range of 10^6 to 10^9 K, the following three nuclear reactions provide the main source of stellar energy: (i) the proton-proton (pp) chain, (ii) the carbon-nitrogen (CN) cycle, and (iii) the helium (3α) reactions. Below 10^7 K the pp chain dominates, while above approximately 2×10^7 K, the CN dominates. In the middle range, both of them compete with each other. The 3α reaction comes into operation near about 10^8 K.

The pp chain takes place through the following three alternative schemes[1]:

$$
\begin{aligned}
{}^1\mathrm{H} + {}^1\mathrm{H} &\rightarrow {}^2\mathrm{D} + e^+ + \text{neutrino}, \\
{}^2\mathrm{D} + {}^1\mathrm{H} &\rightarrow {}^3\mathrm{He} + \gamma, \\
{}^3\mathrm{He} + {}^3\mathrm{He} &\rightarrow {}^4\mathrm{He} + {}^1\mathrm{H} + {}^1\mathrm{H}
\end{aligned}
$$

or

$$
\begin{aligned}
{}^1\mathrm{H} + {}^1\mathrm{H} &\rightarrow {}^2\mathrm{D} + e^+ + \text{neutrino}, \\
{}^2\mathrm{D} + {}^1\mathrm{H} &\rightarrow {}^3\mathrm{He} + \gamma, \\
{}^3\mathrm{He} + {}^4\mathrm{He} &\rightarrow {}^7\mathrm{Be} + \gamma, \\
{}^7\mathrm{Be} + e^- &\rightarrow {}^7\mathrm{Li} + \text{neutrino}, \\
{}^7\mathrm{Li} + {}^1\mathrm{H} &\rightarrow {}^4\mathrm{He} + {}^4\mathrm{He}
\end{aligned}
$$

or

$$
\begin{aligned}
{}^1\mathrm{H} + {}^1\mathrm{H} &\rightarrow {}^2\mathrm{D} + e^+ + \text{neutrino}, \\
{}^2\mathrm{D} + {}^1\mathrm{H} &\rightarrow {}^3\mathrm{He} + \gamma, \\
{}^3\mathrm{He} + {}^4\mathrm{He} &\rightarrow {}^7\mathrm{Be} + \gamma, \\
{}^7\mathrm{Be} + {}^1\mathrm{H} &\rightarrow {}^8\mathrm{B} + \gamma, \\
{}^8\mathrm{B} &\rightarrow {}^8\mathrm{Be}^* + e^+ + \text{neutrino}, \\
{}^8\mathrm{Be}^* &\rightarrow {}^4\mathrm{He} + {}^4\mathrm{He}.
\end{aligned}
$$

The notation is standard, with the atomic weight as a superscript; γ is a gamma ray, e^+ is a positron, e^- is an electron, and $^2\mathrm{D}$ stands for the nucleus of deuterium (i.e., heavy hydrogen). An asterisk denotes an excited state. Note that the nucleus of $^8\mathrm{Be}^*$, which is unstable, breaks up into two alpha particles. The *total* amount of energy available in making one helium atom is 4.2×10^{-5} erg. The neutrino losses relative to this figure for the three alternative chains are 2%, 4%, and 29%, respectively.

---

[1] H. A. Bethe and C. L. Critchfield, *Phys. Rev.*, **54**, 248 (1938); H. A. Bethe, *ibid.*, **55**, 434 (1939); W. A. Fowler and C. C. Lauritsen, *ibid.*, **76**, 314 (1949); E. Schatzman, *Comptes-rendus Acad. Sci. Paris*, **232**, 1740 (1951); H. D. Holmgren and R. L. Johnson, *Bull. Amer. Phys. Soc.*, 2nd ser., **3**, 26 (1958).

The basic CN cycle is as follows[2]:

$$
\begin{aligned}
{}^{12}\text{C} + {}^{1}\text{H} &\rightarrow {}^{13}\text{N} + \gamma, \\
{}^{13}\text{N} &\rightarrow {}^{13}\text{C} + e^{+} + \text{neutrino}, \\
{}^{13}\text{C} + {}^{1}\text{H} &\rightarrow {}^{14}\text{N} + \gamma, \\
{}^{14}\text{N} + {}^{1}\text{H} &\rightarrow {}^{15}\text{O} + \gamma, \\
{}^{15}\text{O} &\rightarrow {}^{15}\text{N} + e^{+} + \text{neutrino}, \\
{}^{15}\text{N} + {}^{1}\text{H} &\rightarrow {}^{12}\text{C} + {}^{4}\text{He}.
\end{aligned}
$$

The net effect of this sequence is to change four hydrogen atoms into one atom of helium, with carbon acting as a catalyst. This cycle liberates 4×10^{-5} erg per helium atom formed, an estimate which includes the 6% energy loss through the escape of neutrinos.

The 3α reaction can be written as[3]:

$$
\begin{aligned}
{}^{4}\text{He} + {}^{4}\text{He} &\rightarrow {}^{8}\text{Be} + \gamma, \\
{}^{8}\text{Be} + {}^{4}\text{He} &\rightarrow {}^{12}\text{C} + \gamma.
\end{aligned}
$$

This reaction converts three alpha particles into one carbon atom, with the resultant energy production of about 1.2×10^{-5} erg per carbon atom formed. One of the most dramatic features of the 3α reaction is the very strong temperature dependence. Near $T = 10^{8}$ K, the energy-production rate varies approximately as T^{30} and is proportional to $\rho^{2} Y^{3}$, where ρ is the density and Y is the helium abundance per weight.

The 3α reaction is perhaps the first example of what is now called a "two-stage nuclear process." As was first noted by Hoyle, however, there must be an excited nuclear state of carbon possible, that had not been previously predicted, in order for the overall reaction ${}^{8}\text{Be} + {}^{4}\text{He} \rightarrow {}^{12}\text{C}$ to form carbon nuclei. Knowing the masses of both ${}^{8}\text{Be}$ and ${}^{4}\text{He}$, he was able to predict that this excited state had to lie at a resonance energy of 7.6 MeV in the carbon nucleus. The work of William Fowler and coworkers at the California Institute of Technology subsequently demonstrated the existence of such a resonant state, which is now known as the *Hoyle resonance*.

---

[2]C. F. von Weizsäcker, *Phys. Zeit.*, **39**, 633 (1938); H. A. Bethe, *Phys. Rev.*, **55**, 434 (1939).

[3]E. J. Öpik, *Proc. Roy. Irish Acad.*, **54A**, 49 (1951); E. E. Salpeter, *Astrophys. J.*, **115**, 326 (1952); F. Hoyle, *Astrophys. J. Suppl.*, **1**, 121 (1954).

General Bibliography

Detailed biographies of astronomers and physicists will be found in the following:

American Men and Women of Science, 8 vols., 20th edition. New York: R. R. Bowker, 1998–1999.

The Biographical Dictionary of Scientists: Astronomers, Abbott, D., general editor. London: Blond Educational, 1984.

Dictionary of Scientific Biography, 18 vols., Gillispie, C. C., and Holmes, F. L., editors-in-chief. New York: Charles Scribner's Sons, 1970–1990.

The New Encyclopaedia Britannica, 15th edition. Chicago: Encyclopaedia Britannica, 1991.

Who's Who in Science in Europe, 9th edition. London: Cartermill International, 1995.

The following books contain reprints and English translations of many fundamental papers:

Lang, K. R., and Gingerich, O., eds., *A Source Book in Astronomy and Astrophysics 1900–1975*. Cambridge, Mass.: Harvard University Press, 1979.

Meadows, A. J., *Early Solar Physics*. London: Pergamon Press, 1970.

Munitz, M. K., ed., *Theories of the Universe: From Babylonian Myth to Modern Science*. New York: Free Press, 1957.

Sambursky, S., ed., *Physical Thought from the Presocratics to the Quantum Physicists: An Anthology*. New York: Pica Press, 1975.

Schove, D. J., ed., *Sunspot Cycles*. Stroudsburg, Pa.: Hutchinson Ross Publishing Company, 1983.

Shapley, H., ed., *Source Book in Astronomy 1900–1950*. Cambridge, Mass.: Harvard University Press, 1960.

Among the books which deal with topics treated in the text, our own preference goes to the following:

Abetti, G., *The Sun*. London: Faber and Faber, 1957.

Baker, R. H., and Fredrick, L. W., *An Introduction to Astronomy*, 7th edition. New York: Van Nostrand–Reinhold Company, 1967.

Berry, A., *A Short History of Astronomy*. London: John Murray, 1898; New York: Dover Publications, 1961.

Brandt, J. C., *The Sun and Stars*. New York: McGraw-Hill Book Company, 1966.

Cajori, F., *A History of Physics*. New York: Macmillan, 1929; New York: Dover Publications, 1962.

Clerke, A. M., *Problems in Astrophysics*. London: Adams and Charles Black, 1903.

————, *A Popular History of Astronomy during the Nineteenth Century*, 4th edition. London: Adams and Charles Black, 1908.

de Vaucouleurs, G., *Discovery of the Universe*. London: Faber and Faber, 1957.

Gamow, G., *A Star Called the Sun*. New York: Viking Press, 1964.

Hearnshaw, J. B., *The Analysis of Starlight: One Hundred and Fifty Years of Astronomical Spectroscopy*. Cambridge: Cambridge University Press, 1986.

————, *The Measurement of Starlight: Two Centuries of Astronomical Photometry*. Cambridge: Cambridge University Press, 1996.

Herrmann, D. B., *The History of Astronomy from Herschel to Hertzsprung*. Cambridge: Cambridge University Press, 1984.

Hoskin, M., ed., *The Cambridge Illustrated History of Astronomy*. Cambridge: Cambridge University Press, 1997.

Hoyle, F., and Narlikar, J., *The Physics-Astronomy Frontier*. San Francisco: W. H. Freeman and Company, 1980.

Hufbauer, K., *Exploring the Sun*. Baltimore: Johns Hopkins University Press, 1991.

Kippenhahn, R., *100 Billion Suns*. New York: Basic Books, 1983.

Kopal, Z., *Of Stars and Men: Reminiscences of an Astronomer*. Bristol: Adam Hilger, 1986.

Kuhn, T. S., *The Essential Tension*. Chicago: University of Chicago Press, 1977.

Lang, K. R., *Sun, Earth, and Sky*. Berlin: Springer-Verlag, 1995.

————, *The Sun from Space*. Berlin: Springer-Verlag, 2000.

Leverington, D., *A History of Astronomy from 1890 to the Present*. Berlin: Springer-Verlag, 1995.

Motz, L., and Weaver, J. H., *The Story of Physics*. New York: Plenum Press, 1989.

Pannekoek, A., *A History of Astronomy*. London: Allen and Unwin, 1961; New York: Dover Publications, 1989.

Segrè, E., *From X-Rays to Quarks*. San Francisco: W. H. Freeman and Company, 1980.

Singer, C., *A Short History of Scientific Ideas to 1900*. Oxford: Oxford University Press, 1959.

Struve, O., and Zebergs, V., *Astronomy of the 20th Century*. New York: Macmillan, 1962.

Index of Names

Page numbers in *italics* refer to material in the figure captions, notes, and bibliography. Page numbers in **boldface** indicate a major entry.

Index of Subjects